Jacopo Iannacci

Practical Guide to RF-MEMS

Related Titles

Hierold, C. (ed.)

Carbon Nanotube Devices

Properties, Modeling, Integration and Applications

2008

Print ISBN: 978-3-527-31720-2, also available in electronic formats.

Bechtold, T., Schrag, G., Feng, L. (eds.)

System-level Modeling of MEMS

2013

Print ISBN: 978-3-527-31903-9, also available in electronic formats.

Korvink, J.G., Smith, P.J., Shin, D. (eds.)

Inkjet-based Micromanufacturing

2012

Print ISBN: 978-3-527-31904-6, also available in electronic formats.

Ramm, P., Lu, J.J., Taklo, M.M. (eds.)

Handbook of Wafer Bonding

2012

Print ISBN: 978-3-527-32646-4, also available in electronic formats.

Garrou, P., Bower, C., Ramm, P. (eds.)

Handbook of 3D Integration

Volumes 1 and 2: Technology and Applications of 3D Integrated Circuits

2012

Print ISBN: 978-3-527-33265-6

Garrou, P., Koyanagi, M., Ramm, P. (eds.)

Handbook of 3D Integration

Volume 3: 3D Process Technology

2014

Print ISBN: 978-3-527-33466-7, also available in electronic formats.

Baltes, H., Brand, O., Fedder, G.K., Hierold, C., Korvink, J.G., Tabata, O. (eds.)

Enabling Technologies for MEMS and Nanodevices

Advanced Micro and Nanosystems

2013

Print ISBN: 978-3-527-33498-8

Baltes, H., Brand, O., Fedder, G.K., Hierold, C., Korvink, J.G., Tabata, O. (eds.)

CMOS-MEMS

2013

Print ISBN: 978-3-527-33499-5

Kockmann, N. (ed.)

Micro Process Engineering

Fundamentals, Devices, Fabrication, and Applications

2013

Print ISBN: 978-3-527-33500-8

Tabata, O., Tsuchiya, T. (eds.)

Reliability of MEMS

Testing of Materials and Devices

2013

Print ISBN: 978-3-527-33501-5

Jacopo Iannacci

Practical Guide to RF-MEMS

Verlag GmbH & Co. KGaA

The Author

Dr. Jacopo Iannacci
Fondazione Bruno Kessler
Ctr. for Materials & Microsystems
via Sommarive 18
38123 Trento
Italy

All books published by **Wiley-VCH** are carefully produced. Nevertheless, authors, editors, and publisher do not warrant the information contained in these books, including this book, to be free of errors. Readers are advised to keep in mind that statements, data, illustrations, procedural details or other items may inadvertently be inaccurate.

Library of Congress Card No.:
applied for

British Library Cataloguing-in-Publication Data:
A catalogue record for this book is available from the British Library.

Bibliographic information published by the Deutsche Nationalbibliothek
The Deutsche Nationalbibliothek lists this publication in the Deutsche Nationalbibliografie; detailed bibliographic data are available on the Internet at http://dnb.d-nb.de.

© 2013 WILEY-VCH Verlag GmbH & Co. KGaA, Boschstr. 12, 69469 Weinheim, Germany

All rights reserved (including those of translation into other languages). No part of this book may be reproduced in any form – by photoprinting, microfilm, or any other means – nor transmitted or translated into a machine language without written permission from the publishers. Registered names, trademarks, etc. used in this book, even when not specifically marked as such, are not to be considered unprotected by law.

Print ISBN 978-3-527-33564-0
ePDF ISBN 978-3-527-67392-6
ePub ISBN 978-3-527-67394-0
mobi ISBN 978-3-527-67393-0

Cover Design Grafik-Design Schulz, Fußgönnheim
Typesetting le-tex publishing services GmbH, Leipzig
Printing and Binding Markono Print Media Pte Ltd, Singapore

Printed in Singapore
Printed on acid-free paper

Intellectual honesty is like a breeze for a sailor. It is invisible and intangible, but it makes a difference.

Intelligence is not all you need in life. You should be intelligent enough to practice it too.

J. Iannacci, June 2012.

Happiness only real when shared.

C.J. McCandless, August 1992.

To my parents, and to their unconditional presence.

Contents

Foreword XI

Preface XV

1	**RF-MEMS Applications and the State of the Art**	*1*
1.1	Introduction *1*	
1.2	A Brief History of MEMS and RF-MEMS from the Perspective of Technology *2*	
1.3	RF-MEMS Lumped Components *3*	
1.3.1	Variable Capacitors *4*	
1.3.2	Inductors *10*	
1.3.3	Ohmic and Capacitive Switches *12*	
1.4	RF-MEMS Complex Networks *20*	
1.4.1	Reconfigurable Impedance-Matching Networks *20*	
1.4.2	Reconfigurable RF Power Attenuators *23*	
1.4.3	Reconfigurable Phase Shifters and Delay Lines *26*	
1.4.4	Reconfigurable Switching Matrices *26*	
1.5	Modeling and Simulation of RF-MEMS Devices *28*	
1.5.1	The Finite Element Method Approach *28*	
1.5.2	Compact Modeling of RF-MEMS *28*	
1.5.3	Mixed-Domain Electromechanical Simulation Environment *30*	
1.6	Packaging of RF-MEMS *31*	
1.7	Brief Overview of Exploitation of RF-MEMS in RF Systems *33*	
1.8	Conclusions *38*	
2	**The Book in Brief** *41*	
2.1	Introduction *41*	
2.2	A Brief Introduction to the FBK RF-MEMS Technology *42*	
2.3	An RF-MEMS Series Ohmic Switch (Dev A) *44*	
2.4	RF-MEMS Capacitive Switches/Varactors *49*	
2.4.1	Design 1 (Dev B1) *49*	
2.4.2	Design 2 (Dev B2) *50*	
2.4.3	RF-MEMS Ohmic Switch with Microheaters (Dev C) *50*	
2.4.4	MEMS-Based Reconfigurable RF Power Attenuator (Dev D) *52*	

| 2.4.5 | MEMS-Based Reconfigurable Impedance-Matching Network (Dev E) 55
| 2.5 | Conclusions 55

3 Design 57
3.1 Introduction 57
3.2 Design Rules of the Fondazione Bruno Kessler RF-MEMS Technology 58
3.3 Design of an RF-MEMS Series Ohmic Switch (Dev A) 60
3.4 Generation of 3D Models Starting from the 2D Layout 77
3.5 Conclusions 83

4 Simulation Techniques (Commercial Tools) 85
4.1 Introduction 85
4.2 Static Coupled Electromechanical Simulation of the RF-MEMS Ohmic Switch (Dev A) in ANSYS Multiphysics™ 86
4.2.1 Block 1: Definition of the Geometry and Properties of the Material 88
4.2.2 Block 2: Meshing of the Structure 92
4.2.3 Block 3: Generation of the Elements for the Electromechanical Coupling 93
4.2.4 Block 4: Definition of the Mechanical Boundary Conditions 95
4.2.5 Block 5: Definition of the Simulation 97
4.2.6 Block 6: Simulation Execution 99
4.2.7 Block 7: Postprocessing and Visualization of Results 99
4.3 Modal Analysis of the RF-MEMS Capacitive Switch (Dev B2) in ANSYS Multiphysics 101
4.4 Coupled Thermoelectromechanical Simulation of the RF-MEMS Ohmic Switch with Microheaters (Dev C) in ANSYS Multiphysics 104
4.5 RF Simulation (S-parameters) of the RF-MEMS Variable Capacitor (Dev B1) in ANSYS HFSS™ 121
4.6 Conclusions 130

5 On-Purpose Simulation Tools 133
5.1 Introduction 133
5.2 MEMS Compact Model Library 134
5.2.1 Suspended Rigid Plate Electromechanical Transducer 134
5.2.2 Flexible Beam 140
5.2.3 Simulation Validation of a MEMS Toggle Switch 144
5.3 A Hybrid RF-MEMS/CMOS VCO 149
5.4 Excerpts of Verilog-A Code Implemented for MEMS Models 151
5.4.1 Anchor Point 152
5.4.2 Force Source 154
5.4.3 Flexible Beam 157
5.5 Conclusions 165

6 Packaging and Integration 167
6.1 Introduction 167
6.2 A WLP Solution for RF-MEMS Devices and Networks 168

6.2.1	Package Fabrication Process *169*
6.2.2	Wafer-to-Wafer Bonding Solutions *173*
6.3	Encapsulation of RF-MEMS Devices *177*
6.3.1	The Issue of Wafer-to-Wafer Alignment *178*
6.3.2	Hybrid Packaging Solutions for RF-MEMS Devices *180*
6.4	Fabrication Run of Packaged Test Structures *181*
6.5	Electromagnetic Characterization of the Package *185*
6.5.1	Validation of the S-parameter Simulations of Packaged Test Structures *186*
6.5.2	Parameterized S-parameter Simulation of Packaged Test Structures *187*
6.6	Influence of Uncompressed ACA on the RF Performance of Capped MEMS Devices *191*
6.7	Conclusions *194*
7	**Postfabrication Modeling and Simulations** *195*
7.1	Introduction *195*
7.2	Electromechanical Simulation of an RF-MEMS Varactor (Dev B2) with Compact Models *196*
7.3	RF Modeling of an RF-MEMS Varactor (Dev B2) with a Lumped Element Network *202*
7.4	Electromechanical Modeling of an RF-MEMS Series Ohmic Switch (Dev A) with Compact Models *213*
7.5	Electromagnetic Modeling and Simulation of an RF-MEMS Impedance-Matching Network (Dev E) for a GSM CMOS Power Amplifier *218*
7.5.1	Introduction *219*
7.5.2	Electromagnetic Design and Optimization of the RF-MEMS Impedance-Matching Network *219*
7.5.3	Deign of a Reconfigurable Class E PA *224*
7.5.4	Experimental Results for the Hybrid RF-MEMS/CMOS PA *227*
7.6	Electromagnetic Simulation of an RF-MEMS Capacitive Switch (Dev B1) in ANSYS HFSS™ *229*
7.7	Electromagnetic Simulation of a MEMS-Based Reconfigurable RF Power Attenuator (Dev D) in ANSYS HFSS *234*
7.8	Conclusions *238*
Appendix A	**Rigid Plate Electromechanical Transducer (Complete Model)** *241*
A.1	Introduction *241*
A.2	Mechanical Model of the Rigid Plate with Four DOFs *242*
A.3	Extension of the Mechanical Model of the Rigid Plate to Six DOFs *249*
A.3.1	Placement of Nodes along the Edges of a Rigid Plate *253*
A.4	Contact Model for Rigid Plates with Four and Six DOFs *255*
A.5	Electrostatic Model of the Rigid Plate *258*
A.5.1	The Four DOFs Condition *258*
A.5.2	Extension to the Case of Six DOFs *262*
A.5.3	Curved Electric Field Lines Model *266*
A.6	Electrostatic Model of the Plate with Holes *268*

A.7	Electrostatic Model of the Fringing Effect *272*
A.7.1	Fringing Effect on the Vertical Faces of the Plate *278*
A.8	Viscous Damping Model *281*
A.8.1	Squeeze-Film Viscous Damping of the Rigid Plate with Holes *281*
A.8.2	Viscous Damping Model for Lateral Movements *284*
A.8.3	Effect of the Mean Free Path of Gas Molecules *285*
A.9	Conclusions *286*

Appendix B Flexible Straight Beam (Complete Model) *287*

B.1	Mechanical Model of the Flexible Beam with Two Degrees of Freedom *287*
B.1.1	The Stiffness Matrix *288*
B.1.2	The Mass Matrix *290*
B.2	Mechanical Model of the Flexible Beam with 12 DOFs *292*
B.2.1	The Stiffness Matrix for 12 DOFs *293*
B.2.2	The Mass Matrix for 12 DOFs *297*
B.3	Complete Mechanical Model of the Euler Beam with 12 DOFs *302*
B.4	Electrostatic Model of the Euler Beam with 12 DOFs *304*
B.4.1	Fringing Effect Model *310*
B.5	Viscous Damping Model *312*
B.6	Conclusions *315*

References *317*

Index *331*

Foreword

The world, as we all know it, has three dimensions. Some of us enjoyed playing with construction bricks as a child, and some still enjoy building stuff, as a hobby or for work. Through the years different branches of engineering have studied how to build complex, efficient, reliable, 3D structures, either movable or still. Indeed, the progress made in the fields of mechanical and civil engineering in recent decades is impressive, making possible the construction of amazing bridges and buildings that literally fight against the force of gravity. In parallel, the achievements obtained by electronic engineering have been continuously changing our life styles and deserve a quick recap. Since the invention of the planar fabrication process at the end of the 1950s by Jean Hoerni, and the successive demonstration of the first CMOS circuit by Frank Wanlass at the beginning of the 1960s, the evolution of microelectronic fabrication technologies has literally exploded, pushing to a few nanometers the lithographic resolution achievable nowadays, and making it possible to integrate tens of millions of transistors in a few square millimeters. However, as already mentioned, we live in a 3D world, so why not exploit the advantages offered by the planar fabrication process to build 3D, even movable structures with dimensions well below the millimeter scale? It is likely that Harvey Nathanson had the same question, and in 1964 with his team at Westinghouse he answered it by producing the first batch-fabricated electromechanical device. The device, a resonant transistor, joined for the first time a mechanical component and electronic circuitry. In other words, he gave birth to the first microelectromechanical system (MEMS). The term "microelectromechanical system" (MEMS) originated in the United States and was followed by "microsystem technology" in Europe and "micromachining" in Japan. Despite the different terminology, the essence is the demonstration that it is possible to co-integrate electrical and 3D mechanical components with dimensions (well) below 1 mm using a fabrication process to produce microelectronic devices. A couple of decades since the first MEMS demonstration, the field has experienced a huge expansion. Nowadays, extremely complex MEMS, such as high-resolution accelerometers and gyroscopes, are commonly implemented in many portable guidance and entertainment systems, and in state-of-the-art automotive and avionic applications. This is very exciting from an engineering point of view and, from a business perspective, we are speaking about multi-billion-dollar markets. In parallel to those and other kinds of successful micromachined sensors/actuators, there has been great interest in the last two decades in developing 3D devices for radio frequency (RF) applications. Good examples are transmission lines, suspended inductors, ohmic/capacitive switches, varactors, resonators, and so on. Given the superior performance demonstrated by MEMS compared with traditional solid-state/fully mechanical technologies, such as higher linearity and frequency bandwidth, smaller volume occupancy and weight, ultralow power consumption, and lower batch production cost and fabrication process complexity, it has been straightforward to exploit such kinds of devices in next-generation portable and regular telecommunications systems. As said before, these are exciting and remunerative markets! Given the great interest in the field, many books have appeared on the design of MEMS. Many of them were meant to give the reader a deep theoretical understanding of the working principles of devices. Fewer works preferred to investigate the field of reliability, showing how it is easy to damage, or even destroy, these wonderful, little, 3D movable structures if they are not properly protected from potentially critical electrical, mechanical, or environmental overstress.

The monograph of Iannacci is different from what has already been presented in the literature, filling the gap between pure theoretical books and experimental ones. Indeed, this book takes the reader by the hand, showing how to fabricate, design, simulate, and characterize electrostatic-based MEMS for RF applications. The starting point is the introduction of the Fondazione Bruno Kessler MEMS process flow and a design software tool (L-Edit) in the first chapters, with an easy to follow and practical approach. But the design and optimization of any new device requires spending some time on structure simulation. Any RF design engineer knows very well the importance of a good design and the considerable amount of time it can take to predict the response of the device. Playing with 3D movable devices, everything is complicated by the need to couple the electromechanical physics with electromagnetism. Iannacci did not forget this often cumbersome problem, and has devoted ample space to introduce the reader to many finite element method and RF simulation tools, always with a "from the scratch" approach, and always comparing simulation results with experimental data. In the end, it is worth remembering that you will sell actual devices, not just simulation data. And the differences with previous literature continue. Instead of simply introducing the basic blocks of MEMS, Iannacci provides a full description of two MEMS-based, complex systems: a GSM MEMS-based power amplifier and an RF-MEMS-based reconfigurable power attenuator. These examples can be definitely beneficial to experienced "solid-state" engineers, who can easily compare the pros and cons offered by a 3D technology, as well as to students, who can understand the possible problems in integrating different technologies, including differences and problems encountered in MEMS packaging. Despite the practical approach of the book, theory has not been forgotten, and in the appendices Iannacci describes the theoretical basis behind the electromechanical actuator, the heart of each electrostatic-based MEMS, as well as the flexible beam, that is, the suspension of the MEMS movable structure. In conclusion, the field of RF-MEMS is so ably described by Iannacci that both professionals and students, experienced or newbies, will benefit from his presentation and experimental point of view. As a last, personal, comment, I really enjoyed reading the book of my friend Jacopo. Actually, I just realized I have never called him by his surname so often as in this foreword!

Pittsburg, USA *Augusto Tazzoli*
May 2012

Preface

Microelectromechanical systems (MEMS) technology is characterized by great flexibility, making it suitable for sensor and actuator applications spanning from the biomedical to the automotive sector. A rather recent exploitation of MEMS technology which emerged in the research community about 15 years ago is in the field of radio frequency (RF) passive components. RF-MEMS devices – RF-MEMS is an acronym identifying microsystem-based RF passive components – range from lumped components, such as microrelays, variable capacitors (i.e., varactors), and inductors, to complex elements based on a combination of the previously listed basic components, such as tunable filters, phase shifters, and impedance-matching tuners. All the RF-MEMS components mentioned are characterized by very high performance, for example, low loss, high quality factor (Q factor), and good isolation, as well as by wide reconfigurability. Such characteristics offer the potential to extend the operability and to boost the performance of RF systems, such as transceivers, radar systems, cell phones, and smartphones, and this is the main reason for which significant effort is now being expended (at the research and industrial level) in order to solve the issues still impeding the full integration of MEMS technology in standard RF circuits, they being mainly reliability, packaging, and integration.

Given the wide interdisciplinary behavior of MEMS and RF-MEMS devices, the aspects to be faced as well as the knowledge required to handle their development are multiple, regardless of the specific phase – design, simulation, fabrication, testing – one is dealing with. For example, the proper design of an innovative RF-MEMS component implies propaedeutic knowledge of the physics of semiconductors, structural mechanics, dynamics, electrostatics, and electromagnetism. Consequently, MEMS/RF-MEMS being a novel discipline demanding a straightforward definition of its pertinence, several remarkable handbooks and textbooks were written in the last decade, covering both the comprehensive discussion of the field as a whole and the close examination of specific device development phases, such as modeling, simulation, microfabrication, and packaging.

Since it would have made no sense to publish another text containing the theory of RF-MEMS that, in the best case, could have been as good as already existing texts, a different approach was chosen and pursued in the preparation of this book. The philosophy behind it can be summarized as follows: if relevant books dealing with the theory of RF-MEMS and hands-on texts already exist, why not aim at something in between? And this is it.

This book covers some of the most critical phases that have to be faced in order to develop novel RF-MEMS device concepts according to a very practical approach, it being the one followed by the author in the research activities he pursued in the last decade. A very limited set of RF-MEMS devices – including both lumped components and complex networks – is chosen at the beginning of the book as reference examples, and they are then discussed from different perspectives while progressing from the design to the simulation, packaging, testing, and postfabrication modeling. Theoretical bases are introduced when necessary, while several practical hints are reported concerning all the development steps discussed, providing the book with an engineering flavor.

In conclusion, given its practical approach, this book is meant to give a helping hand to a wide audience, ranging from scientists and researchers directly involved in the RF-MEMS field, to those who want to gain insight into what working in the field of microsystems for RF applications means.

Trento, Italy *Jacopo Iannacci*
June 2012

1
RF-MEMS Applications and the State of the Art

Abstract: This introductory chapter provides a comprehensive overview of the state of the art in radio frequency (RF) microelectromechanical systems (MEMS) technology and its applications. The exploitation of MEMS technology in the field of RF circuits and systems (i.e., RF-MEMS) represents a rather recent exploitation of microsystems if compared with the field of sensors and actuators, and can be framed in the last 10–15 years. Firstly, the chapter discusses some of the history of MEMS technology, focusing on the development of suitable and appropriate technological steps for the manufacturing of microsystems. Subsequently, the focus is moved to RF-MEMS technology, and a comprehensive scenario of the most relevant devices manufactured with such a technology is provided. The discussion of microsystem-based RF passive components is arranged according to an increasing complexity fashion, with the basic passive lumped components, namely, switches, variable capacitors, and inductors, being presented first. Then, the potential of RF-MEMS technology is framed by showing how such basic elements can be combined in order to realize complex functional RF subblocks, such as phase shifters and filters, characterized by large reconfigurability and very high performance. To conclude this introductory chapter, information about the state of the art in modeling and packaging of RF-MEMS devices is provided, while other relevant aspects, such as testing and reliability, are not described in detail as they will not be treated in this work. A brief overview of the exploitation of RF-MEMS devices in RF systems is also given.

1.1
Introduction

What is the relationship, if it truly exists, between the progress of technology and the progress of human kind? This is one of those complicated questions that do not admit a unique answer; in fact, it is a question with no answer, according to the point of view of an engineer. Nonetheless, trying for a very short while not to be a scientist, but just a human being, which is the basis of a scientist, as well as of a lawyer, a worker, a secretary, and so on, one can maybe address, even though partially, the initial question. The intricacy between the progress of human kind and that of technology is clearly bidirectional. The advancements in technology, and their influence on the daily life of people, definitely improved, and are seamlessly improving, our conditions of life. Let us think about how radio, television, cell phones, and other equipment have brought about a revolution in our lifestyle as well as in our habits, both of which concern ourselves and our social life. Nevertheless, the same advancements have generated a series of negative consequences for human kind, outlining a fundamental paradox. What is the point if the improvement of one aspect of life causes the degradation of other traits? The point is that positive and negative consequences of each change, including the no-change option, are unavoidable. Consequently, real progress is not represented by the sole boost given by technology, but rather by the conscious evaluation of all the aspects and consequences that the employment of a new solution will cause. And this delicate aspect is not solved by the ultimate step forward taken by technology, as it is a responsibility resting on the shoulders of human beings. This is the perspective within which the progress of technology can really become progress also for human kind. These considerations apply to all the changes we have faced, are facing, and will face in the future, in technology as well as in society (concerning laws, regulations, health, etc.), and, of course, also apply to radio frequency (RF) microelectromechanical systems (MEMS) technology, which is the real topic of this book. In other words, it is time for a human being to remember to be an engineer after all.

1.2
A Brief History of MEMS and RF-MEMS from the Perspective of Technology

MEMS are, by definition, submillimeter systems (i.e., microsystems) fabricated by means of the same technological steps used in the manufacturing of integrated circuits (ICs). If we focus on the most important electronic component, namely, the transistor, its manufacture is the result of a sequence of steps in which different doses of dopant are selectively implanted or diffused within a substrate (typically silicon) in order to locally obtain certain electrical properties. The same implantation/diffusion, or alternatively the digging of deep trenches, can be used in order to enhance the isolation and reduce the cross talk between adjacent devices. Moreover, conductive and insulating layers are selectively deposited/grown, or deposited/grown everywhere and then selectively removed, in order to redistribute the electrical signals from the intrinsic devices to the external world.

The most important steps in the manufacturing of ICs are ion implantation, diffusion, epitaxial growth, chemical vapor deposition/physical vapor deposition and their variations, wet and dry etching, sputtering, evaporation, and electrodeposition of metals. The selection of the areas that have to undergo one or more of the previously listed steps is always performed by means of lithography [1].

MEMS devices and components are manufactured using the same steps as just discussed, despite the fact that their number and sequence are different. The transistor is a device built into the silicon substrate, and the insulating and metal layers processed on top of the silicon surface are needed mainly for the routing of electrical signals to and from the intrinsic transistor. Differently, a MEMS device is typically defined by a certain sequence of layers patterned on top of a silicon substrate.

Because of these considerations, it is unclear when MEMS technology was born. From the point of view of the fabrication process, the steps developed together with the diffusion of semiconductor technologies started in the 1950s. However, the exploitation of such techniques aimed at the fabrication of microsystems started in earnest in the early 1970s.

The development of semiconductor technology based on silicon motivated the scientific community to investigate in detail also the mechanical properties of the materials involved, beside the critical aspects related to microelectronics. Significant contributions in this direction are, for instance, the works of several authors concerning the mechanical properties of both bulk materials [2] and deposited layers [3–5] which date back to the period from the mid 1950s to the mid 1960s.

However, the exploitation of such techniques aimed at the manufacturing of microdevices with movable parts and membranes started later, in the period from the second half of the 1970s to the beginning of the 1980s. Examples on how to use anisotropic etching in order to obtain a variety of 3D suspended structures starting from a silicon substrate are given in [6]. Such techniques, together with the ones typically exploited for the fabrication of transistors and ICs, brought about the realization of miniaturized pressure sensors [7], accelerometers [8, 9], switches [10, 11], and other devices for various applications, such as in the optical and biomedical fields. A remarkable article summarizing the state of the art of microsystems technology and providing a comprehensive snapshot of various applications was written by Petersen [12] at the beginning of the 1980s.

Nevertheless, it was with the further maturation of the surface micromachining fabrication technique [13] that the development of microsystems started to receive a significant boost, leading to the concept of MEMS sensors and actuators as we know it today. A significant contribution is represented by the work of Howe and Muller [14] in 1983, in which microcantilever and double supported beams were realized in polysilicon, and suspended above the silicon substrate using silicon oxide as a sacrificial layer. Since then, a wide variety of MEMS-based sensors, actuators, and various mechanisms (e.g., gears and micromotors) have been developed, tested, and presented [15–17].

1.3
RF-MEMS Lumped Components

The progress and maturation of the technology processes mentioned above enabled the realization of a large variety of MEMS-based devices and components that can be successfully exploited for various sensor and actuator applications. The same consideration applies to RF-MEMS devices, as the technology platforms for the fabrication of microsystems empowered, in the past decade, the manufacturing of

MEMS components for RF applications. Among the wide variety of RF-MEMS, two main categories can be identified, namely, lumped components, such as tunable reactive elements (capacitors/inductors) and microswitches, and complex networks, such as reconfigurable filters and phase shifters, which are based on a suitable combination of the basic RF-MEMS components belonging to the first category. This section will discuss RF-MEMS lumped components in more details, while the next one will focus on the implementation of complex networks and functional RF subblocks based on RF-MEMS devices.

1.3.1
Variable Capacitors

Capacitors are passive components playing an important role in the realization of bandpass and bandstop filters [18], matching networks, and more generally, the large part of telecommunication systems [19]. The most important characteristics that lumped capacitors exhibit are the tuning range and the quality factor (Q factor), which both should be as large as possible. A wide tunability range for the capacitance enables a correspondingly large reconfigurability of the functional block that employs it [18]. On the other hand, a large Q factor ensures a high selectivity concerning passive filters, and more generally, better performance in terms of low losses [20]. Variable capacitors (i.e., varactors) are widely realized in standard semiconductor technology by reverse biasing of diodes [21], with some limitations in terms of performance. A significant alternative solution to obtain varactors with better performance and characteristics is to fabricate them with MEMS technology, and this possibility has been widely investigated at the research level in the past decade [22]. The concept of a varactor realized with MEMS technology is based on having (at least) one of the two capacitor plates movable with respect to the other. The displacement of one electrode modifies the distance between the two plates and, in turn, tunes the capacitance of the entire passive element. The displacement of the movable plate can occur in multiple ways. Electrostatic, piezoelectric, thermal, and magnetic actuation are the commonest ones [23]. A typical implementation of a variable capacitor based on MEMS technology is based on two parallel plates, the lower one being fixed and the one on top being movable [24]. The latter is kept suspended by means of flexible parts (e.g., slender beams or folded deformable structures), and when a DC bias is applied between the two plates, the one on top starts to move toward the fixed (bottom one), because of the electrostatic attraction force, increasing, in turn, the capacitance. The schematic in Figure 1.1 shows a typical geometry for an RF-MEMS electrostatically controlled tunable component (or capacitive switch).

Looking at the cross section of the schematic reported in Figure 1.2, we can better understand the working principle of an electrostatically controlled RF-MEMS component. When no bias is applied between the movable membrane and the underlying electrode, the MEMS structure is in the rest position, as Figure 1.2a shows. In this case, the distance between the two electrodes is the maximum possible, and the capacitance, in turn, assumes the minimum value. On the other hand, when a

1.3 RF-MEMS Lumped Components

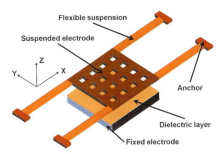

Figure 1.1 An electrostatically controlled radio frequency (RF) microelectromechanical system (MEMS)-based variable capacitor based on an electrode kept suspended by deformable slender beams over a fixed electrode.

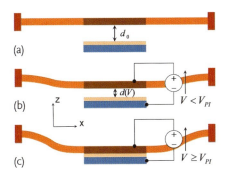

Figure 1.2 Cross section of the variable capacitor in Figure 1.1 when (a) the switch is in the rest position, (b) the applied bias is smaller than the pull-in voltage ($V < V_{PI}$), and (c) the applied bias is larger than the pull-in voltage ($V \geq V_{PI}$).

bias (i.e., a DC voltage) is applied between the two electrodes, the electrostatic attraction force causes the suspended plate to move toward the underlying electrode, as it is anchored to flexible suspensions.

A scanning electron microscopy (SEM) photograph [25] of an anchoring point for a slender beam flexible suspension of an RF-MEMS device based on the concept reported in the schematic in Figure 1.1 is shown in Figure 1.3.

The smaller the distance between the suspended and the fixed electrode, the larger is the capacitance realized by the RF-MEMS variable capacitor. By increasing the applied voltage, one can further increase the capacitance. However, the whole distance d_0 cannot be used to tune seamlessly the capacitance. Indeed, when the downward displacement of the suspended plate reaches $d_0/3$ (i.e., one-third of the initial air gap), the balance between the attractive electrostatic force and the restoring mechanical force, induced by the deformed suspensions, reaches a condition of instability, and the plate collapses onto the underlying surface. This phenomenon is known as pull-in, and the bias level causing such an abrupt snap down of the movable membrane is referred to as pull-in voltage (V_{PI}) [26]. Figure 1.2b shows the movable plate configuration when the applied bias is smaller than the pull-in

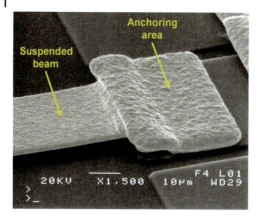

Figure 1.3 Scanning electron microscopy (SEM) photograph of an anchoring point to which a flexible and suspended straight beam is anchored in a physical RF-MEMS device based on the concept in Figures 1.1 and 1.2.

voltage ($V < V_{PI}$), while Figure 1.2c reports what happens when the applied bias is larger than the pull-in voltage ($V \geq V_{PI}$). In the latter case, the insulating layer deposited above the fixed electrode prevents an electrical short between the two conductive layers, and the capacitance jumps to the maximum value.

Typically, the difference between the values the capacitance can assume when it is tuned before the pull-in and after it occurs is rather large (up to two to three orders of magnitude). This means that an RF-MEMS device, like the one depicted in Figures 1.1 and 1.2, can be exploited as a variable capacitor with a small and continuous tuning range [27]. Additionally, it can also be exploited as a two-state capacitor, with a large difference between the two capacitance values, that is, rest position and pulled-in position [28]. In the latter case, the RF-MEMS device could also be exploited as a capacitive switch for RF signals. Indeed, if the variable capacitor is inserted in shunt-to-ground configuration on the RF line, when the MEMS is pulled in, its large capacitance to ground realizes a low-impedance path for the RF signal, which is shorted to ground, and consequently does not reach the output termination (i.e., open switch) [29]. After the pull-in has been reached, if the applied bias starts to decrease, the device approaches the pull-out level, which is the critical voltage causing the release of the MEMS switch [23]. After the RF-MEMS release, if the applied bias is further decreased, the vertical position of the movable membrane will smoothly approach the initial rest position (corresponding to no bias applied). The measured vertical position of the movable suspended RF-MEMS device, similar to the one discussed in the previous figures, in response to an applied varying controlling voltage is reported in Figure 1.4. As is visible in the plot at the top of this figure, the applied controlling signal is a zero mean value triangular symmetric pulse with a period of 50 ms (i.e., frequency of 20 Hz).

The voltage ranges between ±20 V (Figure 1.4, top plot), and the plot at the bottom of Figure 1.4 shows the response of the RF-MEMS device concerning the vertical position of the movable membrane. The initial applied bias, that is, −20 V, is

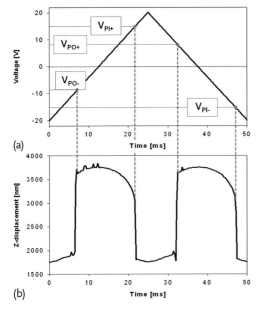

Figure 1.4 Triangular symmetric zero mean value controlling bias applied to an RF-MEMS variable capacitor/switch (a). The frequency of the signal is 20 Hz and the voltage varies in the ±20 V range. Response of the RF-MEMS device (i.e., vertical displacement of the movable membrane) to the applied triangular bias (b). Pull-in and pull-out transitions are visible in both the positive and the negative range of the applied voltage, and they are $V_{PI+} \simeq 15\,V$, $V_{PO+} \simeq 8\,V$, $V_{PI-} \simeq -15\,V$, and $V_{PO-} \simeq -8\,V$, respectively.

larger than the RF-MEMS pull-in voltage, and the movable membrane collapses onto the underlying electrode. Afterward, the applied voltage decreases toward zero, and the RF-MEMS releases as the pull-out voltage is reached ($V_{PO-} \simeq -8\,V$). When the bias crosses the zero axis, the movable plate reaches its rest position, and then starts to move down as the bias increases (positive applied voltage). The RF-MEMS device then collapses onto the underlying surface when the pull-in is reached ($V_{PI+} \simeq 15\,V$). The second half of the measurement registers another two transitions of the MEMS device, namely, the pull-out in the positive range of the applied voltage and the pull-in in the negative range of the applied voltage, being $V_{PO+} \simeq 8\,V$ and $V_{PI-} \simeq -15\,V$, respectively. This measurement is obtained by means of a white-light 3D profiling system based on optical interferometry [30]. The vertical displacement just discussed can be more easily reported versus the applied bias, highlighting the pull-in/pull-out characteristic of a certain MEMS device, as shown in Figure 1.5. The range of continuous tunability, corresponding to a vertical displacement at most equal to $d_0/3$ (see Figure 1.2), is highlighted in the plot.

On the basis of the working principle of the RF-MEMS variable capacitors and switches discussed in the previous pages, several geometries of the suspended MEMS membrane are possible, depending on the requirements, the application, and the expected performance and specifications. For instance, Figure 1.6 reports

1 RF-MEMS Applications and the State of the Art

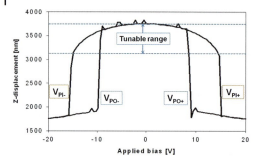

Figure 1.5 Pull-in/pull-out characteristic of a MEMS switch. The range of continuous tunability is highlighted in the plot.

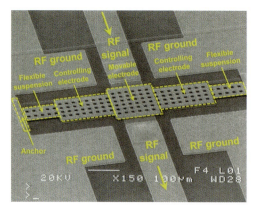

Figure 1.6 SEM photograph of an RF-MEMS variable capacitor (or capacitive switch). The suspended capacitor plate is connected to two controlling electrodes and to flexible suspensions.

a SEM photograph of a capacitive switch (shunt capacitance loading the RF line). The suspended plate of the capacitor is the large one in the central part of the device. It is connected to other two rectangular plates that are meant to control the vertical position of the central membrane. Underneath these plates there are two fixed electrodes that, when biased, cause a decrease of the gap between the capacitor plates. Finally, two narrow membranes connect the capacitor and the actuation plates to the surrounding fixed metal, realizing the flexible suspensions for the whole structure. The anchoring part is visible and highlighted on the left of the suspended membrane. The RF-MEMS variable capacitor (or capacitive switch) is framed within a coplanar waveguide (CPW) structure, with two lateral RF ground planes, and a central RF signal line, compatible with on-wafer microprobe measurements. The variable capacitor realized by the suspended MEMS structure loads the RF line in shunt-to-ground configuration. Figure 1.7 shows a close-up of the RF-MEMS variable capacitor concerning a region of the central suspended plate and the left-hand-side actuation electrode. The underlying RF line constituting the fixed capacitor plate is visible, as are the square openings on the suspended MEMS

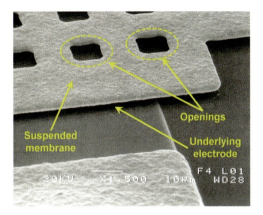

Figure 1.7 Close-up of a portion of the central suspended metal membrane. The underlying RF line and the openings on the suspended metal layer are visible.

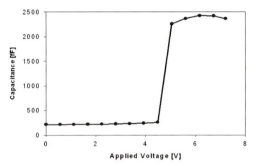

Figure 1.8 Experimental C–V characteristic of an RF-MEMS variable capacitor.

membrane, which are necessary for the complete removal of the sacrificial layer during the release of the air gaps.

A typical experimental C–V characteristic (i.e., capacitance versus applied voltage) of an RF-MEMS variable capacitor is reported in Figure 1.8. A sudden large increase in the capacitance with increasing pull-in (actuation) of the microdevice is visible. The capacitance, that is, in the range of about 200 fF when the RF-MEMS device is not actuated, shifts to about 2.5 pF when the suspended membrane collapses onto the underlying electrode.

Other examples concerning the surface micromachining of RF-MEMS variable capacitors are discussed in the literature, such as in the work by Park *et al.* [31], and Goldsmith *et al.* [32]. Beside standard implementations of micromachined RF-MEMS variable capacitors, the attention of research has also been focusing on the improvement of performance and characteristics. For instance, Liang *et al.* [33] discussed a solution to extend the tuning range of RF-MEMS varactors. Recently, Mahameed *et al.* [34] proposed a new design of a zipper RF-MEMS varactor with multiple sensing and controlling interdigitated electrodes in order to improve device robustness versus charge accumulation within the oxide, on one hand, and to better control the capacitance tuning, on the other. Among the different topologies of RF-MEMS variable capacitors (and capacitive switches), a rather interesting solution to increase the tuning range is constituted by the so-called toggle-switch geometry. In this case, a push–pull mechanism, operated by means of multiple biasing electrodes, enables one to control the displacement of the suspended capacitor plate both toward the underlying electrode and upward [35]. When the central capacitor plate moves upward, the tuning range is extended to capacitances lower than the capacitance corresponding to the RF-MEMS rest position.

1.3.2
Inductors

MEMS technology also proved to be suitable for the fabrication of high-performance inductors. Spiral coils can be easily electrodeposited above a substrate. For this purpose, the work reported in [36] shows the realization of a metal spiral inductor framed within a CPW structure compatible with the experimental characterization on a probe station. The center of the spiral coil is connected to the output by means of an overpass, that is, a suspended metallization, that makes it possible to cross all the windings and to bring the RF signal to the output. The Q factor of a MEMS inductor can be significantly increased by choosing a low-loss substrate [37] as well as by reducing the coupling of the inductor windings with the substrate by depositing a good insulating layer in between [38]. However, MEMS technology also enables other solutions at manufacturing level that significantly reduce losses of inductors and, in turn, enhance the Q factor. Such solutions consist in having the metal inductor coil suspended above an air layer rather than a silicon one [39]. Cross-sectional schematic views of suspended MEMS inductors are reported in Figure 1.9. In particular, Figure 1.9a shows the cross section of an RF-MEMS inductor kept suspended by means of a sacrificial layer, which is then removed to release the floating structure [40]. This is a typical approach based on surface micromachining fabrication steps, and is similarly used for the release of the suspended RF-MEMS variable capacitors discussed in the previous section. Differently, the other two cross sections show how it is possible to suspend MEMS inductors through a bulk micromachining approach, that is, through the removal of the substrate material (e.g., silicon). In Figure 1.9b, the air cavity underneath the inductor coil is obtained by removing the substrate material from the top side of the wafer [41]. On the other hand, in Figure 1.9c the removal is done from the bottom side of the substrate until a thin membrane is left nonetched in order to mechanically sustain the inductor metallization [42]. An example of a MEMS inductor based on gold and suspended by means of a sacrificial layer (i.e., surface micromachining) is reported in Figure 1.10. The RF signal is brought from the central part of the coil to the output by means of an underpass. This means that a vertical transition brings the electrical signal from the gold level to a conductive buried layer. RF-MEMS-based inductors can also be easily integrated within more complex networks, for example, also comprising microswitches, variable capacitors, and so on. The microphotograph reported in Figure 1.11 shows a close-up of a complex

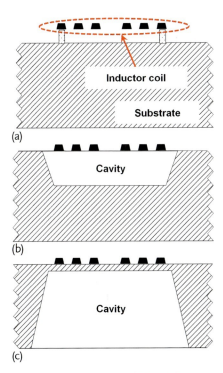

Figure 1.9 Cross sections of RF-MEMS inductors kept suspended according to different technology solutions: (a) sacrificial layer (surface micromachining); (b) removal of the substrate material from the front side of the wafer (bulk micromachining); (c) removal of the substrate material from the back side of the wafer (bulk micromachining).

RF-MEMS network for the conditioning of RF signals that highlights the presence of suspended MEMS inductors. The reconfigurability typical of MEMS technology has also been investigated concerning lumped inductors, and there are various techniques enabling such a tuning capability. One of the most straightforward techniques employs the use of RF-MEMS switches to vary the length of the metal line realizing the inductor [43, 44]. Moreover, self-assembly techniques are also used in order to realize high-Q-factor inductors that can be tuned by thermally stressing the device [45]. Another solution consists in deploying a suspended movable metal plate on top of a spiral planar inductor. The metal plate is electrostatically actuated, and when it approaches the underlying coil, the interaction of the plate with its magnetic field changes the inductance [46].

More exotic approaches to reconfigure the inductance of RF-MEMS metal coils are also discussed in the literature. For instance, in [47] the inductance is modified by using a micropump that injects a fluid between the spirals, shortening the length of the electrical path and, in turn, reducing the inductance.

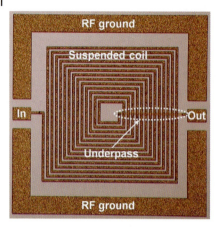

Figure 1.10 Microphotograph of a planar MEMS inductor made of gold in the coplanar waveguide (CPW) configuration. An underpass (buried conductive layer) enables the RF signal to be brought to the output.

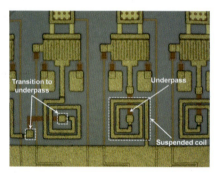

Figure 1.11 Close-up microphotograph taken from an RF-MEMS-based complex network, where spiral suspended inductors are integrated and visible.

1.3.3
Ohmic and Capacitive Switches

The commonest and most well known class of RF-MEMS devices is represented by the microswitches (ohmic and capacitive). The literature reports a large amount of valuable realizations of switches based on MEMS technology for RF applications, as they represent the key components capable, on one hand, of enabling the reconfigurability of the network/platform comprising them, and also presenting, on the other hand, high performance and good characteristics compared with common implementations of relays in standard semiconductor technology. Concerning the mechanical working principle of RF-MEMS microswitches, two main classes can be identified, namely, the clamped–clamped and the cantilever switches. A SEM image of a clamped–clamped switch is reported in Figure 1.12.

1.3 RF-MEMS Lumped Components | 13

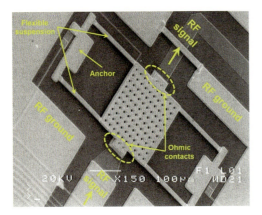

Figure 1.12 SEM microphotograph of a clamped–clamped RF-MEMS series ohmic switch. The central plate is kept suspended with four straight beams, and when the switch pulls in, a low-resistance ohmic contact is established between the input/output RF branches.

In a clamped–clamped (or fixed–fixed) RF-MEMS switch, a metal membrane is suspended and placed transversally over the signal line [48]. The membrane has two anchoring points (i.e., fixed on both ends), and typically is symmetric with respect to the RF line [49], as Figure 1.12 shows, and as was already reported in Figures 1.1 and 1.2. On the other hand, cantilever-type RF-MEMS switches are based on a membrane that is anchored at just one end. The free end realizes the contact with the electrodes underneath (when the RF-MEMS switch is pulled in) and the membrane can be placed transversally [50] on the RF line or can be aligned (i.e., in-line) with it [51]. Figure 1.13 shows a SEM photograph of an RF-MEMS-based network comprising several transversally arranged cantilever-type series ohmic switch-

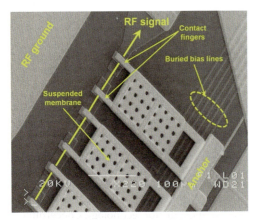

Figure 1.13 SEM microphotograph of cascaded cantilever-type (i.e., single-hinged) RF-MEMS series ohmic switches. The working principle is similar to that of switch reported in Figure 1.12, even though in this case the suspended structure is hinged only at one end, while the metal tips realizing the input/output ohmic short are deployed at the free membrane end.

Figure 1.14 Close-up of a contact finger belonging to one of the cantilever-type switches reported in Figure 1.13. The metal contact pad is visible underneath the suspended finger. When the switch is actuated, the two metal parts touch each other, establishing the ohmic contact.

es, while Figure 1.14 reports a close-up of a contact finger belonging to one of the suspended cantilever-type switches.

Beside the chosen type of geometry that realizes an RF-MEMS switch, microrelays can be divided into two categories depending on the switching function they realize, they being the series or shunt configurations. A series switch passes the RF signal between the input and output, while the shunt switch diverts the RF signal to ground. Both the series and the shunt switches are characterized by two states, namely, on and off. The on state corresponds to an activated switch (i.e., pulled in or actuated), while the off state corresponds to a switch in the rest position (i.e., not biased). The series and shunt switches realize the functions described above when they are in the on state, and they do not realize the functions mentioned when they are in the off state. This means that an RF-MEMS series switch passes the RF signal from the input to the output port when it is in the on state (i.e., actuated), and it isolates the two ports when it is in the off state (i.e., rest position). In other words, looking at the input/output transfer function of an RF-MEMS series ohmic switch, we find it is closed when the microrelay is on, and it is open when the MEMS switch is off. The case of an RF-MEMS shunt switch is the opposite with respect to the case of a series microrelay. A shunt switch diverts the RF signal toward the RF ground when it is in the on state, and it passes the RF signal from the input to the output when it is in the off state. As done previously, looking at the input/output transfer function of an RF-MEMS shunt switch, we find it is closed when the microrelay is off, and it is open when the MEMS switch is on. In the latter case, when the RF signal flows toward the RF ground, it does not reach the device output.

A further distinction that has to be made when referring to RF-MEMS switches (series and shunt) concerns the type of contact they realize with the pads underneath when they are in the on state. When the actuated MEMS metal membrane realizes a metal-to-metal contact with the underlying contact area, it is an ohmic-type microrelay. Differently, if an insulating layer (deposited above the underlying

Table 1.1 Summary of all possible configurations and features a radio frequency (RF) microelectromechanical system (MEMS) switch can exhibit.

Series	Shunt	On	Off	Switch state	Ohmic	Capacitive
X		X		Closed	Low resistance	Low impedance
X			X	Open	High resistance	High impedance
	X	X		Open	High resistance	High impedance
	X		X	Closed	Low resistance	Low impedance

fixed electrode/s) makes a direct metal-to-metal contact impossible, or if the MEMS switch is not meant to touch the underlying electrode also when it is in the on state (i.e., contactless switch), the RF-MEMS microrelay is capacitive. In the first case, referring to a series relay, when the RF-MEMS switch is off, the resistance between the suspended MEMS electrode and the one underneath is very large, virtually infinite, while a low-resistance ohmic contact is established when the switch is on. For RF-MEMS capacitive switches in the off state, referring to a shunt relay, a very small capacitance is realized between the suspended and the underlying fixed parts, that is, a very large impedance. However, such a capacitance becomes significantly larger when the RF-MEMS switch is in the on state, thus establishing a very low impedance path between the movable (actuated) and fixed parts. The possible combinations of switch configurations and features are summarized in Table 1.1. All the characteristics and configurations listed there can be combined in different fashions, resulting in several implementations of RF-MEMS switches. For instance, it is possible to have cantilever-type series ohmic switches [52, 53] as well as RF-MEMS double-hinged capacitive shunt switches [49, 54] and so on [55], each of them with specific advantages and disadvantages concerning performance and characteristics, such as isolation, insertion loss [56], power handling [57], bandwidth, actuation and release voltages, and performance stability versus time [58].

In order to show the typical RF behavior of a few representative typologies of RF-MEMS microswitches, the behavior of measured scattering parameters (S parameters) [59] typical of a few devices is now described. The first example refers to an RF-MEMS series ohmic switch. The S11 parameter (i.e., reflection parameter) for the nonactuated (off) and actuated (on) switch states is reported in Figure 1.15. Concerning the RF-MEMS switch's rest position, S11 is very large as it ranges between 0 and 1.5 dB from 100 MHz to 13.5 GHz. This means that the RF power is almost completely reflected toward its source, and consequently does not reach the output (open switch). The reason why the reflected RF power slightly decreases as the frequency increases is that there is always a small capacitive coupling between the suspended MEMS membrane and the ohmic contact underneath. For higher frequencies, the small series capacitance causes the large impedance to decrease a little between the input and output (due to the two nontouching metal parts), allowing a small fraction of the RF power to flow toward the output port. The value of the S11 parameter for the MEMS switch's actuated position is definitely smaller,

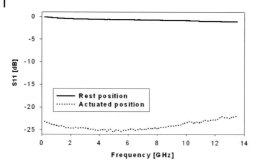

Figure 1.15 Measured reflection parameter (S11) of an RF-MEMS series ohmic switch in the rest and actuated positions. When the switch is not actuated, the RF signal is reflected (open switch), while when it is actuated, the RF signal is not reflected (closed switch).

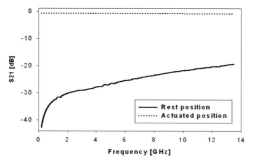

Figure 1.16 Measured transmission parameter (S21) of an RF-MEMS series ohmic switch in the rest and actuated positions. When the switch is not actuated, the RF signal does not reach the output (open switch). In this case, S21 is also referred to as isolation. When the switch is actuated, the RF signal reaches the output (closed switch), and the difference between the amplitude of the RF signal at the RF-MEMS input and output ports is known as loss.

and ranges between 23 and 25 dB in the whole measured frequency range. This is because when the switch is actuated, the RF signal encounters a low-resistance path toward the output, and just a very small fraction is reflected (closed switch). In particular, the match between the characteristic impedance of the RF power source and that of the RF-MEMS switch is better, while the amount of reflected power that bounces back toward the source is lower.

The S21 parameter (i.e., transmission parameter) for a nonactuated and an actuated RF-MEMS series ohmic switch is reported in Figure 1.16.

In the nonactuated configuration (i.e., rest position), S21 is very small and ranges between 42 and 20 dB from 100 MHz to 13.5 GHz. When the switch is open, like in this case, S21 is referred to as isolation, and the fact that isolating the device input from the output gets a little worse when the frequency increases is due to the series capacitive coupling already mentioned when discussing Figure 1.15. On the other hand, when the switch is actuated, the S21 parameter assumes a large value (around 1 dB) over the whole frequency range, proving that, despite some small losses, the RF signal reaches the output port of the microrelay (i.e., closed switch).

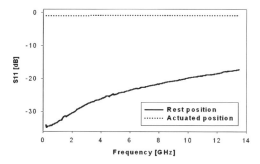

Figure 1.17 Measured reflection parameter (S11) of an RF-MEMS ohmic shunt switch in the rest and actuated positions. When the switch is not actuated, the RF signal is not reflected (closed switch), while when it is actuated, the RF signal is reflected (open switch).

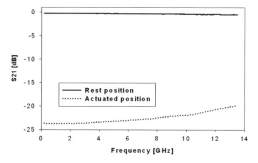

Figure 1.18 Measured transmission parameter (S21) of an RF-MEMS ohmic shunt switch in the rest and actuated positions. When the switch is not actuated, the RF signal reaches the output (closed switch), and the difference between the amplitude of the RF signal at the RF-MEMS input and output ports is known as loss. When the switch is actuated, the RF signal is shorted to ground and does not reach the output (open switch). In this case, S21 is also referred to as isolation.

The next example refers to an RF-MEMS ohmic shunt switch, which is characterized by the behavior of S parameters that is the opposite of that of the series switch just discussed. Figures 1.17 and 1.18 report the S11 and the S21 parameters, respectively, concerning the actuated and nonactuated positions of the microswitch. The reflection (S11 parameter) is very low when the MEMS microrelay is off (and the switch is then closed), while it becomes large when the MEMS microrelay is pulled in (on state), and the RF signal flows to the RF ground (open switch). The characteristic of the shunt switch opposed to that of a series device is also visible in the S21 parameter plot (see Figure 1.18). Indeed, the RF signal travels toward the output when the MEMS microrelay is off, while the input and output ports are isolated when the MEMS microrelay is pulled in (on state).

The next example considered is an RF-MEMS capacitive shunt switch. As mentioned before, since there is no ohmic contact but, instead, capacitive coupling, the RF signal travels toward the output or is shorted to RF ground depending on whether a high-impedance path or a low-impedance path to ground is established,

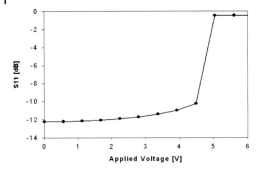

Figure 1.19 Measured reflection parameter (S11) of an RF-MEMS capacitive shunt switch at 10 GHz versus the bias applied to the device. As the switch approaches the pull-in, the capacitance increases and, in turn, so does the fraction of reflected RF signal. When the pull-in occurs, the RF signal in nearly totally reflected, and the output is isolated.

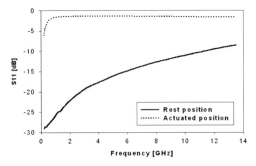

Figure 1.20 Measured reflection parameter (S11) of an RF-MEMS capacitive shunt switch in the rest and actuated positions. When the switch is not actuated, the RF signal is not reflected (closed switch), while when it is actuated, the RF signal is reflected (open switch).

respectively. Figure 1.19 shows the experimental reflection parameter (S11) measured for an RF-MEMS capacitive shunt switch at 10 GHz as a function of the applied controlling bias.

The device exhibits a rather reduced pull-in voltage (lower than 5 V). When no bias is applied, the reflection parameter is around 12 dB, while it tends to increase as the controlling voltage rises. This is because, due to the progressive approach of the suspended membrane toward the underlying fixed electrode, the capacitance increases, and consequently the impedance to ground encountered by the RF signal becomes smaller. The fraction of RF power shorted to ground then increases for larger applied bias levels, until it is almost totally reflected when the RF-MEMS switch pulls in (S11 around 0.5 dB). In this case, the RF-MEMS switch is open.

The S-parameter versus frequency plots for the RF-MEMS capacitive shunt switch are now reported and discussed. In particular, Figure 1.20 shows the S11 parameter measured from 100 MHz to 13.5 GHz, while Figure 1.21 refers to the S21 parameter.

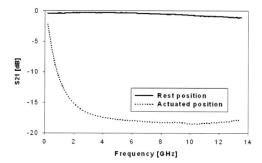

Figure 1.21 Measured transmission parameter (S21) of an RF-MEMS capacitive shunt switch in the rest and actuated position. When the switch is not actuated, the RF signal reaches the output (closed switch), and the difference between the amplitude of the RF signal at the RF-MEMS input and output ports is known as loss. When the switch is actuated, the RF signal is shorted to ground and does not reach the output (open switch). In this case, S21 is also referred to as isolation.

It is useful to compare the behavior of the device in shunt configuration with that of the ohmic shunt previously reported in Figures 1.17 and 1.18, highlighting the difference between an ohmic and a capacitive microrelay. The behavior of the S11 parameter reported in Figure 1.20 for the capacitive shunt switch in the nonactuated position is similar to that of the ohmic shunt switch (in Figure 1.17). However, for the capacitive switch the reflection (i.e., the value of S11) increases more versus frequency than for the ohmic switch. This is because the off-state capacitance of the capacitive switch is larger than the off-state capacitance of the ohmic switch, the overlapped area of the suspended and fixed electrodes being larger for the former device. Actually, in the ohmic shunt switch the capacitive coupling between the suspended membrane and the underlying contacts can be considered as a parasitic capacitance, while in the capacitive shunt switch, it is referred to as off-state capacitance. Consequently, the coupling to ground in a capacitive shunt switch in the off state is larger than that of an ohmic shunt microrelay. Looking now at the S11 parameter in the actuated state of the capacitive shunt switch (Figure 1.20), we find it exhibits lower values in the very low portion of the frequency range, while it reaches a stable, higher value around 2 GHz. This effect is due to the capacitive coupling rather than to the ohmic contact to ground. Despite the capacitance of the switch being much larger in the actuated state than in the rest position, a capacitor tends to behave as an open circuit for very low frequencies, and becomes a short circuit when the frequency increases. Below 2 GHz then, the impedance to RF ground is not small enough to divert the largest part of the RF signal to ground.

The last plot to be discussed refers to the S21 parameter and is reported in Figure 1.21. The transmission parameter for the switch's nonactuated state exhibits very low values, but displays a slight degradation trend for increasing frequency. This effect is due to the increasingly conductive path to RF ground induced by the capacitance in the off state for high frequency. Instead, looking at the S21 parameter for the switch in the on state, we find the isolation exhibits low values below 2 GHz because of the still coarse capacitive coupling to RF ground.

The characteristics presented and discussed in the previous pages refer to the main classes of RF-MEMS switches, depending on their geometry (cantilever-type or clamped–clamped), input/output configuration (series or shunt), and the type of contact (ohmic or capacitive). Given these categories of devices, the scientific literature is rich in several examples of particular realizations of RF-MEMS switches exhibiting specific characteristics. For example, Reines et al. [60] discuss a particular circular geometry of RF-MEMS capacitive switches in order to make them more insensitive to the stress distribution and to the temperature variations. Differently, the focus of [61] is on the design of an RF-MEMS capacitive switch with high power handling. Concerning RF-MEMS ohmic switches, Lee et al. [62] discuss a particular geometry and actuation mechanism in order to improve the isolation of the microrelay in the open state for a very high frequency range (50–110 GHz).

1.4
RF-MEMS Complex Networks

The lumped elements realized in RF-MEMS technology discussed in the previous pages are now to be exploited as building blocks for the realization of more complex functional networks, aiming at the treatment and conditioning of RF signals. There are various types of complex networks whose realization is enabled by RF-MEMS technology, and integration within transceiver and telecommunication platforms can significantly improve their performance and broaden their functionalities. In the following pages, a brief overview of a few of the different classes of complex networks realized in RF-MEMS technology will be given, listing them on the basis of the specific conditioning function they implement.

1.4.1
Reconfigurable Impedance-Matching Networks

RF-MEMS technology is particularly suitable for the realization of impedance-matching networks, able to transform the input impedance into a different one at the output. Such passive networks are necessary to match the characteristic impedance of two functional blocks within a certain RF system or subsystem when it is not the same for the two blocks. A significant example is reported in [63], where a dual-band impedance-matching network (not realized with MEMS technology) is studied in order to match the characteristic impedance of the antenna with the one of the integrated circuit (IC) within an RF identification (RFID) system. In order to make the exploitation of an impedance-matching network as flexible as possible, the network is required to synthesize a large number of different impedance transformations. MEMS technology is capable of enabling a large reconfigurability of an impedance tuner. An example of an impedance-matching network entirely realized with RF-MEMS technology is reported in Figure 1.22.

The network features several metal–insulator–metal capacitors and suspended gold inductors, both visible in Figure 1.22, as well as several RF-MEMS cantilever-

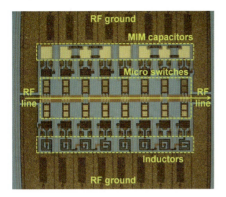

Figure 1.22 Microphotograph of an RF-MEMS-based reconfigurable impedance-matching network. It features several metal–insulator–metal (MIM) capacitors and suspended inductors. Moreover, cantilever-type ohmic microswitches select which reactive elements load the RF line, reconfiguring the impedance transformation realized by the network between the RF input and output.

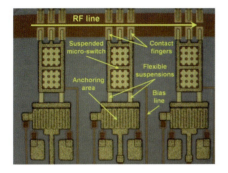

Figure 1.23 Close-up of a few ohmic microswitches in the network shown in Figure 1.22. The RF-MEMS switches, based on a cantilever-type geometry, have suspended contacting fingers at their free ends that realize an ohmic contact with the underlying fixed electrodes when the microrelay is pulled in.

type switches selecting which reactive elements have to load the RF line, changing the impedance transformation realized by the network [64]. Figure 1.23 shows a close-up (microphotograph) of a few cantilever-type ohmic switches in the network and to which the reactive elements are connected. When an RF-MEMS switch is pulled in, the corresponding reactive element loads the RF line. Figure 1.24 shows a 3D image of the whole network acquired with a white light interferometry (WLI) profiling system.

The equivalent lumped element scheme of the reconfigurable impedance-matching network is reported in Figure 1.25. It features four reactive stages, two in series and two in shunt configuration with respect to the RF line. The ohmic switches allow the configuration of the network's scheme (like one or both of the LC-ladder stage/s) as well as the characteristic of each reactive element (capacitive, inductive, or both in parallel). Figure 1.26 shows a Smith chart reporting a few

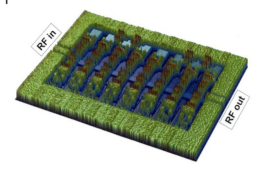

Figure 1.24 Three-dimensional view of the RF-MEMS-based reconfigurable impedance-matching network acquired with a profiling system based on optical (white light) interferometry.

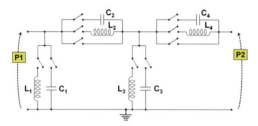

Figure 1.25 Equivalent lumped element network of the RF-MEMS-based reconfigurable impedance-matching network shown in Figure 1.22. RF-MEMS cantilever-type microswitches allow the reactive element/s loading the RF line to be chosen. The whole network realizes two LC-ladder impedance transformation stages.

of the impedance transformations enabled by the matching network. The starting point of each arrow shows the S11 parameter, while the end point indicates the S22 parameter. Consequently, each arrow highlights the impedance transformation between the input and the output of the network, realized at a certain frequency and for each different configuration. The transformations reported in Figure 1.26 refer to a few frequencies up to 25 GHz, and represent just a few of the $2^8 = 256$ different configurations realized by the network.

The scientific literature is populated by several examples of reconfigurable impedance-matching networks based on RF-MEMS technology. A significant proportion of them is based on the concept of a slow-wave distributed MEMS transmission line, like the one realizing $3^8 = 6561$ impedance configurations reported in [65]. Another example of an RF-MEMS-based reconfigurable impedance matching-network, in which the impedance is tuned by changing the length of a few stubs loading the RF line, is reported and discussed in [66]. This solution also enables a good and homogeneous coverage of the impedance Smith chart.

Concerning the integration and exploitation of RF-MEMS-based impedance-matching networks, a relevant example in which it is interfaced to an active power amplifier is reported in [67]. In it, the hybrid CMOS–MEMS functional block (i.e., the power amplifier and the matching network, respectively) is realized in the same technology platform by using a post-CMOS process.

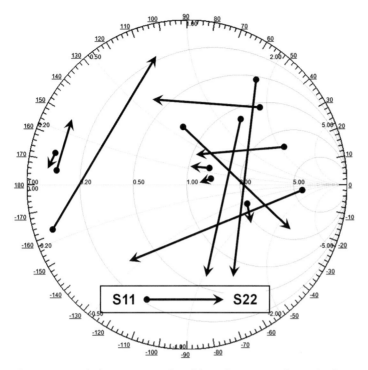

Figure 1.26 Smith chart reporting a few of the impedance transformations (simulated) enabled by the RF-MEMS impedance-matching network shown in Figure 1.22. The transformations refer to a few frequencies up to 25 GHz. The starting and end points of each arrow represent the input (S11) and output (S22) impedance of the network, respectively.

1.4.2 Reconfigurable RF Power Attenuators

Another important class of highly reconfigurable and high-performance complex networks enabled by RF-MEMS technology is represented by RF power attenuators and power dividers. A power divider in microstrip configuration, based on 3 dB couplers that, being reconfigured by several RF-MEMS-based switches, allow several power ratios between the network input and output is discussed in [68, 69]. On the other hand, realizations of RF-MEMS-based reconfigurable power attenuators based on resistive loads are introduced in [70] with a particular focus on radar applications. A general-purpose MEMS-based reconfigurable RF power attenuator is presented in [50]. In this implementation, a few polysilicon resistors load in series the RF line (in a CPW configuration), and can be selectively shorted by pulling in RF-MEMS cantilever-type ohmic switches. If the whole resistance loading the RF line is changed, the attenuation level realized by the network is, in turn, reconfigured. The same working principle is applied to the network reported in [71] and shown in the microphotograph in Figure 1.27.

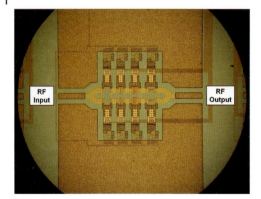

Figure 1.27 Microphotograph of a MEMS-based reconfigurable RF power attenuator. The RF line is loaded by a few resistors, each of which can be shorted by pulling in a cantilever-type ohmic switch, enabling the reconfiguration of the whole resistive load on the RF line and, in turn, the attenuation level realized by the network.

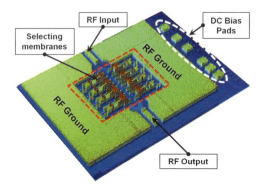

Figure 1.28 Three-dimensional view of the MEMS-based reconfigurable RF power attenuator [71] acquired with a white light interferometry optical profiling system. The RF input/output (in CPW configuration) as well as the pads for biasing the microswitches are labeled. The central part of the device including the RF-MEMS switches is highlighted and magnified in Figure 1.29.

The network features a buried RF line split into two parallel branches, each loaded by the same resistors. In this way, by selecting one or both of the parallel branches, one can double the number of attenuation levels realized by the network. Figure 1.28 shows the 3D topology of the network acquired with WLI, highlighting the pads necessary for the DC biasing of the cantilever-type switches, and the central part of the network where the ohmic switches are.

The latter part of the network is shown in more detail in the close-up (microphotograph) depicted in Figure 1.29. The two parallel RF branches are visible, highlighting the polysilicon resistors and all the switches necessary for shorting them and selecting the branch/es loading the RF line. The 3D close-up of a switch (obtained with WLI) is also depicted in Figure 1.29. Finally, the S21 parameter is measured (up to 30 GHz) for a few attenuation levels realized by the network, and is re-

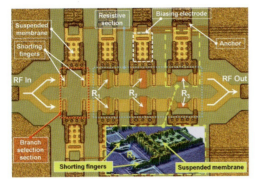

Figure 1.29 Microphotograph (close-up) of the central part of the network reported in Figures 1.27 and 1.28. The RF line shows two parallel branches, each loaded by three resistors with different values (R_1, R_2, R_3). The leftmost switching stage (composed of two microswitches) selects one or both branches, while all the other switches short the resistors, reconfiguring the resistance of the whole network.

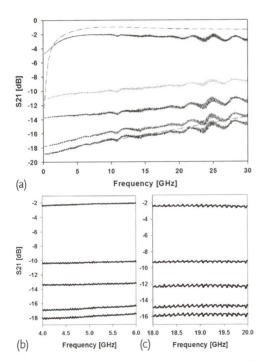

Figure 1.30 (a) Measured transmission parameter (S21 parameter) up to 30 GHz for a few of the attenuation levels realized by the network shown in Figure 1.27. The broken lines refer to the minimum and maximum simulated attenuation levels. (b,c) Close-up of the S21 parameter in the frequency subranges 4–6 GHz (b) and 18–20 GHz (c), emphasizing the flat characteristic of the attenuation levels.

ported in Figure 1.30a. The broken lines refer to the simulated maximum and minimum attenuation levels. The S21 characteristic spans from a few decibels down to 15–19 dB depending on the amount of resistance loading the RF line. Eventually, the S21 characteristic is also rather flat despite the wide frequency range, as highlighted in Figure 1.30b,c.

1.4.3
Reconfigurable Phase Shifters and Delay Lines

A class of complex networks of particular interest concerning possible implementations using RF-MEMS technology is represented by phase shifters. Such devices are needed for phased-array antennas in order to steer the RF beam without the need for rotating physical parts. The advantages are significant, as the dimensions and the complexity (i.e., hardware redundancy) of the systems can be drastically reduced, resulting in the possibility to integrate the whole radar system in a very small area, as well as to have better control of the steering of the beam. A significant example of an RF-MEMS-based reconfigurable phase shifter is that proposed in [72]. It features three bits, that is, three stages of two-state shunt capacitors realized with RF-MEMS technology, and shows a rather flat phase up to 30 GHz that is independent of the frequency. Another relevant implementation of a reconfigurable phase shifter based on RF-MEMS technology for the Ka-band is reported in [73]. It is a four-bit implementation that uses capacitive switches to reconfigure the phase shift, and exhibits a very reduced insertion loss for all the configurations. A more recent realization of a multistate phase shifter based on RF-MEMS technology, intended for K-band and W-band (24 and 77 GHz, respectively), is presented in [74]. A comparison of different phase shifter designs is given in [75], having as an objective the robustness of the network itself in terms of power handling (i.e., capability of managing significant power levels).

Another type of network that is rather similar to a phase shifter is a delay line. Of particular interest for wireless applications are true-time delay (TTD) networks, which are basically able to introduce a certain delay on the RF signal that is independent of the frequency or, in other words, a constant time delay over a wide frequency range. An example of a TTD network implemented with RF-MEMS technology and realizing several delay configurations is discussed in [76]. It features RF-MEMS ohmic switches for reconfiguring the network and has been tested up to 40 GHz. A similar implementation of a TTD line based on RF-MEMS technology with a very reduced phase error is reported in [77].

1.4.4
Reconfigurable Switching Matrices

Another field of application in which RF-MEMS technology exhibits great potential is the switching and rerouting of RF signals in telecommunication platforms (both terrestrial and satellite). Given the very reduced dimensions of RF-MEMS switches compared with their traditional (electromechanical) counterparts, it is possible in

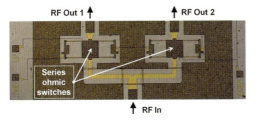

Figure 1.31 Microphotograph of an RF-MEMS based single pole, double throw switch realized in a surface micromachining process.

principle to increase the complexity of the switching unit, that is, increase the number of channels that can be managed by a single switching unit, and at the same time drastically reduce the area occupied. The basic component for the realization of more complex switching and commutation schemes can be considered to be the single pole, double throw (SPDT) switch, which is a T-like path, in which the input RF signal can be diverted to one of the two output branches, to both of them, or to neither of them. The microphotograph in Figure 1.31 shows an RF-MEMS-based SPDT switch realized with surface micromachining technology. The single RF input and the double RF output (selected by means of two RF-MEMS ohmic switches) are visible. An additional example of an SPDT switch realized with RF-MEMS technology, also explaining the working principle of such a switching device, is reported in [78]. Another realization of an SPDT switch based on RF-MEMS technology exhibiting low insertion loss and good isolation in the Ku-band is described in [79].

The switching capability of SPDT switches can be raised by increasing the number of output branches on which the input RF signal can be commuted. In a single pole, four throw (SP4T) switch, for instance, the signal is switched on four output channels [80]. More generally, the number of commutation channels can be increased rather freely depending on the requirements of the specific application, leading to the so-called single pole, multiple throw (SPMT) switch [81]. The configuration of a switching unit can be flexibly modified, for instance, by increasing also the number of input channels. An example in this direction is represented by double pole, double throw (DPDT) commutation devices [82]. The composition of the basic switching elements, mentioned briefly above, enables the realization of actual switching matrices, entirely realized with RF-MEMS technology, and presenting various orders of complexity. An example of a 2 × 2 RF-MEMS-based switching matrix is reported and discussed in [83]. Differently, higher-order switching matrices based on RF-MEMS technology are reported in [84]. In conclusion, new concepts of switching units to build up higher-order commutation matrices also represent a focus for the research community, as reported in [85].

1.5
Modeling and Simulation of RF-MEMS Devices

This section discusses the state of the art in the modeling and simulation of RF-MEMS devices and networks, reporting the most diffused approaches pursued by the scientific community.

1.5.1
The Finite Element Method Approach

The correct prediction of the behavior of complex systems via the use of appropriate simulation software tools is not a simple task. This becomes more critical when dealing with MEMS devices, as they require the coupling of different physical domains (e.g., electrical and mechanical). A valuable approach is represented by the use of finite element method (FEM) based software. Briefly, such a method divides the 2D/3D geometry of the problem into subelements (e.g., in the shape of tetrahedrons) connected to each other by means of nodes. Within each of these small elements, the problem equations are approximated and solved locally, like Maxwell equations for the electromagnetic problem or elasticity theory relations for the mechanical behavior [86]. Finally, the relations among the set of elements are defined by imposing suitable boundary conditions. Commercial FEM tools able to simulate and couple the device characteristics belonging to multiple physical domains are available on the market. In particular, concerning the simulation of MEMS and RF-MEMS, a few of the most diffused FEM tools are CoventorWare® [87], ANSYS Multiphysics™ [88], and COMSOL Multiphysics® [89].

Despite FEM tools allowing very accurate predictions of the RF-MEMS electromagnetic and mechanical behavior, sometimes their use is not a reasonable choice. Indeed, they require a considerable amount of memory as well as simulation time. For this reason, when a structure with a complex geometry (e.g., an RF-MEMS network featuring several switches) has to be optimized to meet certain requirements, a reasonably accurate and fast tool rather than a very accurate but slow tool would be preferable. This is the main reason why efforts are being expended by the scientific community in the development of fast simulation methods to be used beside FEM tools, widening the choice the designer has depending on his/her needs and requirements.

1.5.2
Compact Modeling of RF-MEMS

The approach in pursuing a simplified description of certain structures (RF-MEMS) to be analyzed is well known in the literature as *compact modeling*. Basically, compact models can be obtained and derived according to different methods and approaches. One of the most diffused is analytical modeling, which can be further split into two categories depending on the approaches through which compact models are obtained. These are the so called *bottom-up* and *top-down* approaches [90, 91]. In the first one, the starting point is the accurate analysis of entire RF-MEMS devices, including all the mechanical and electrical effects, performed, for example, by means of FEM-based tools. Hence, through suitable methods, a simplified description of the system is derived and its geometry is divided into elementary models with limited loss of accuracy within certain analysis ranges. This method is called model order reduction and will be described briefly below. The other method in the compact modeling of RF-MEMS, the opposite with respect to the one just mentioned, is the top-down approach. In this case, the analysis commences at the system level. The system, for instance, an RF transceiver (i.e., transmitter and receiver) subblock, is described by means of lumped elements based on analytical models (lumped analytical modeling). When dealing with a single physical domain, that is, electrical, a network of lumped elements, such as capacitors and inductors, is able to describe the behavior of the entire system with reasonable accuracy. Nevertheless, when multiple physical domains are involved, as happens with MEMS technology, different strategies must be employed to adapt the lumped elements approach to such a generic problem. One valuable solution is well known in the literature and is based on equivalent electrical circuits [92]. The approach that will be discussed in detail throughout this book, however, is the development of a mixed domain simulation environment based on the concept of hierarchical structural analysis.

Model Order Reduction Method
One of the most effective methods to simplify the initial MEMS/RF-MEMS structure employing a bottom-up approach is the model order reduction method [93]. In the FEM simulation of an entire structure (e.g., a MEMS switch), the number of internal nodes is generally very high (e.g., 50 000). Nonetheless, within certain ranges, many of them are not critical to determine the structure's behavior. Consequently, very limited sets of nodes, for example, 10 or 20, are sufficient to model each part of the initial topology with small losses in accuracy but, on the other hand, with very low computational complexity and reduced simulation time. The MOR method allows one to identify which nodes are more critical than others in achieving the correct prediction of the structure's behavior. Suitable software has been developed for this purpose [94]. Such tools interact with the matrices generated as output by the FEM simulator and, by following certain criteria, reduce their order down to a few tens of nodes. After the screening of matrices is complete, the more significant nodes (or degrees of freedom) in the determination of the structure's behavior, within defined ranges, are figured out. On the other hand, all the other degrees of freedom are simply disregarded [95].

Equivalent Electrical Circuits
The top-down approach based on equivalent electrical circuits was proposed and explained very well in detail by Senturia [92]. The elementary models, or lumped elements, use the analogy existing between the mechanical and the electrical physical domains. It is possible to determine an equivalent electric component for each

mechanical effect of interest. For instance, a spring defines a certain relationship between its deformation and the consequent restoring force (function of the spring constant) in the mechanical domain. This is analogous to the relationship imposed by a capacitor between the applied voltage and the current in the electrical domain. Following the same approach, other mechanical effects, such as inertia and viscous damping, are *translated* into their electric equivalents. Eventually, a complete RF-MEMS device is represented by its equivalent electrical lumped element network and can be easily simulated in a commercial circuit simulator working in the electrical domain [96].

1.5.3
Mixed-Domain Electromechanical Simulation Environment

The top-down approach for compact modeling of RF-MEMS, on which this book is focused, is different from the approaches described thus far. It is based on a multidomain simulation environment, which means that the electrical and mechanical physical domains are kept separated. In other words, mechanical magnitudes are not transformed into their electrical equivalents.

The implemented simulation environment allows *hierarchical structural analysis*. The entire system is divided into elementary parts that, according to the concept of hierarchy, can be connected together in several fashions in order to assemble structures with more complex topologies [97, 98]. For instance, an RF-MEMS switch based on a suspended electrostatic transducer can be decomposed into two parts. One includes the mechanical spring effects and the other includes the actual electromechanical transduction. When elementary models implementing the first and the second effects are available, such as deformable beams and suspended rigid plate electrostatic transducers, respectively, their interconnection allows one to reconstruct the behavior of the entire RF-MEMS switch.

Elementary components are based on mathematical models able to characterize their behavior in certain physical domains. For instance, the behavior of flexible structures comes from the theory of elasticity and structural mechanics. Differently, the electromechanical transduction model can be easily derived from an electrostatic problem. Such models are valid in rather large ranges, ensuring a certain flexibility in using basic components. This results in a broad variety of complete configurations of RF-MEMS devices that can be successfully simulated and characterized following such an approach.

Each component has a limited number of nodes through which it is connected to the other elements forming a complex structure. Nodes also represent the input/output points by which the mathematical model imposes the relations among the physical magnitudes involved. For example, the deformation applied in input to one node of a flexible structure causes a reaction force, defined by the internal model, which is the output of the same node [99]. The coexistence of parameters associated with different physical domains (i.e., electrical and mechanical) is managed by the simulator. Indeed, tools whose simulation machines are based on the resolution of the Kirchhoff flow law and the Kirchhoff voltage law are suitable for multidomain analysis. In other words, two parameters are associated with each node: a *through* and an *across* value, allowing a generalization of the physical domain they belong to in the simulator [100]. In the electrical domain, the through value is the current flowing, while the across value is the applied voltage. Similarly, in the mechanical domain, the through value is the force applied to a certain node, while the across value is the displacement of that very node. This is the approach chosen for the development of the mixed-domain simulation environment presented in this work. The simulator used is Spectre®, which is available in the Cadence IC development framework, and the language chosen for the implementation of the compact models is Verilog-A [101, 102].

Beside the above-mentioned implementation in the Cadence development framework, on which there will be more focus throughout this book, the mixed-domain compact modeling approach is also used in a few other commercial tools. The two most significant examples are ARCHITECT™ from Coventor [103] and SYNPLE™ from IntelliSense [104].

1.6
Packaging of RF-MEMS

In developing a solution for the fabrication of a protective cap to be applied to certain microdevices, one must take several considerations into account. First of all, the main task of the package is to offer adequate protection to the devices from harmful factors, such as mechanical shocks as well as moisture, dust particles, and various contaminants. Of course, depending on the type of devices and on the environment in which they are meant to operate, the capping part should be particularly robust with respect to the more harmful external factors [105]. The type of application for which the particular devices are conceived also plays an important role in defining the packaging solution [106].

Concerning issues related to the fabrication of the capping part, different materials are employed for this purpose, such as silicon, glass, and ceramic compounds [107]. Depending on the particular material chosen for the realization of the capping, one can apply different techniques to set up the electrical signal interconnection scheme as well as the necessary space to accommodate the devices to be packaged.

A packaging solution for RF-MEMS based on a glass substrate was demonstrated in [108]. Through-vertical vias are opened in the cap by means of sandblasting. These are then filled with gold to redistribute the electrical signals from the device wafer to the outside world. Another viable solution in order to package RF-MEMS switches was demonstrated in [109]. A benzocyclobutene ring is patterned around each device in order to provide the necessary vertical space to house the MEMS itself, as well as to isolate it from external dangerous factors. The electrical signals are brought out of the package by means of CPWs placed on the device wafer and passing underneath the benzocyclobutene protective ring. Differently, a packaging solution for RF-MEMS switches in which vertical vias are etched in a silicon substrate and filled with conductive material prior to the fabrication of the MEMS devices themselves is presented and discussed in [110]. After a silicon capping wafer has been bonded onto the switches, the electrical signals are accessible on the bottom face of the packaged device wafer.

The final utilization of RF-MEMS devices and networks implies their integration and interfacing with other functional blocks in a certain system. The packaging represents a delicate step also in the sense of making easier the final mounting of a chip including MEMS devices onto a board. Within the scientific community, the definitions of *zero-level* and *first-level* packaging have been introduced [111]. Zero-level packaging refers to a packaging solution performed at the device level after the manufacture of MEMS/RF-MEMS structures (i.e., encapsulation of naked devices). The electrical interconnect scheme is made available for the final mounting of the MEMS chip. This is based, for instance, on a ball grid array, a dual in-line package, a pin grid array, or a leadless chip carrier [112]. On the other hand, in first-level packaging, a chip within which a packaged MEMS/RF-MEMS has already been included is packaged/integrated in a more complex system, and so on, obtaining increasing architectural and functional complexity of the subsystem [113].

Among the various zero-level packaging solutions for MEMS and RF-MEMS devices, a widely known one is referred to as wafer-level packaging. This means that an entire device wafer has to be bonded to the package. The wafer-to-wafer bonding can be performed, for instance, by means of solder reflow, anodic bonding [114], or *via* the use of adhesive materials such as SU-8 [115, 116]. After the bonding step, the capped wafer must be singulated, which means sawing it into dies of smaller size (e.g., $1 \times 1\,\text{cm}^2$). Finally, the single die must be made ready for final on-board mounting, for example, by wire-bonding it to a carrier chip provided with leads for a dual in-line package or solder balls for flip-chip mounting [117]. In other words, a zero-level solution is an intermediate packaging stage which requires additional steps to get a chip ready for standard surface-mount technologies, such as the ones mentioned above. This is necessary when dealing with MEMS devices as they need to be protected immediately after fabrication (e.g., during wafer handling and singulation). Another valuable solution to fabricate a package at zero level is represented by so-called thin film capping. In this case, the package is not based on a second substrate that has to be processed first and then wafer-bonded to the device wafer, but is processed directly onto the MEMS/RF-MEMS device to be protected, by means of additional fabrication steps performed after the MEMS device processing (postprocessing) [118, 119].

Beside this, still referring to the integration of RF-MEMS devices, it is worth investigating additional functionalities that the package might provide. One of the most interesting ideas is surely the possibility to interface CMOS circuitry with a passive RF-MEMS part directly on-chip by exploiting the cap to provide a proper housing for the CMOS chip, along with an interconnect scheme for interfacing the signals. This allows one to get a complete functional block (e.g., an oscillator based on a MEMS resonator and a CMOS sustaining circuitry) packaged and ready for surface-mount technologies. Because the RF-MEMS and CMOS parts are obtained via different incompatible technologies, this solution is called hybrid packaging [120, 121].

Another important characteristic can be enabled by the package. This is the hermeticity and the vacuum sealing which are possible depending on the particular solution chosen for the bonding of the capping part to the MEMS device. It has been demonstrated that a MEMS lateral comb-drive resonator exhibits a Q factor of 27 when it is operated in air [122]. The Q factor for the same device rises to about 50 000 when it is operated in a vacuum. Typically, it is very difficult to maintain the vacuum condition in the packaged cavity over a long period of time, as leakages through the sealing may occur as outgassing of the materials within the chamber (e.g., the sealing material or the MEMS device itself) may take place. This leads to a drift (i.e., degradation) over time of the performance of the capped MEMS devices. An effective solution leading to a drastic reduction of this problem is offered by the use of so-called *getters* [123, 124]. Such materials exhibit very selective absorption properties with respect to certain species which might jeopardize the vacuum condition in the package. Getter materials are usually arranged in thin films, so they can be easily accommodated in the cavity in which the vacuum condition must be enhanced and maintained. They are obtained by sintering different materials to achieve the desired sensitivity to the gaseous species that must be trapped [125]. The effectiveness of getters has been demonstrated, proving that the Q factor of a MEMS resonator operated in vacuum conditions is stabler over time when a getter layer is added within the sealed cavity hosting the device [126, 127].

1.7
Brief Overview of Exploitation of RF-MEMS in RF Systems

The purpose of this section is to provide a brief insight into the exploitation of the simple and complex RF-MEMS devices discussed in the previous pages in modern RF components and systems. As mentioned before, RF-MEMS technology enables the realization of a wide range of passive components with different levels of complexity (i.e., from switches and varactors to complex networks), all being characterized by large reconfigurability and very good performance in terms of high Q factor, low losses, high isolation, and good linearity. Table 1.2 lists several devices and applications, characterized by an increasing level of complexity, that benefit from the availability of RF-MEMS components. For each item it is reported if it is a passive or an active device/application, and if its complexity is low, high, or at the system level. Furthermore, the rightmost columns indicate which RF passive component(s) or network(s) are included in each device/application mentioned on the left.

Among the applications listed in Table 1.2, the cell phone (i.e., a system-level application) is now taken as an example to show in more detail in which of its components and subsystems RF-MEMS technology can be profitably exploited. The high-level block diagram of a GSM (Global System for Mobile Communications) cell phone architecture, as reported in [128], is depicted in Figure 1.32.

Table 1.2 The most relevant RF and microwave applications of the devices and networks discussed thus far. The level of complexity of each application is divided into three classes, namely, low (L), high (H), and system-level (S) complexity when referring to a complete RF/microwave system. The letter X indicates which RF passive devices and networks are used in each application mentioned in the leftmost column.

Device/applications	Passive component	Active component	Complexity	Variable capacitors (varactors)	Variable inductors	Switches and relays	Impedance tuners	Tunable filters	Resonators	Phase shifters and delay lines	Power splitters and attenuators	Switching units and matrices
Resonator	X		L			X			X			
LC tank	X		L	X	X	X						
Impedance tuner	X		H	X	X	X						
Tunable filter	X		H	X	X	X						
Phase shifter and delay line	X		H	X	X	X	X					
Power splitter and attenuator	X		H	X			X	X				
Switching unit	X		H			X						
Switching matrix	X		H			X						X
Coupler	X		H	X	X	X						
Circulator	X		H	X	X	X						
Mixer	X		H	X	X	X						X
Cell phone		X	S	X	X	X	X	X	X			
Smartphone		X	S	X	X	X	X	X	X			
Radio receiver		X	S	X	X	X	X	X	X	X		
Radio transmitter		X	S	X	X	X	X	X	X	X		
Radar system		X	S	X	X	X	X	X	X	X	X	
Satellite for telecommunications		X	S	X	X	X	X	X	X	X	X	X
Wireless sensor network		X	S	X	X	X	X	X	X			
RF identification systems		X	S	X	X	X	X	X	X			

Referring to such an architecture, we can briefly describe the function of each block reported in Figure 1.32 as follows:

CPU The CPU controls the functionality of all the system, it being interfaced with the memory unit, the peripheral drivers, and the analog to digital converter/digital to analog converter (ADC/DAC).

Memory The memory contains both the routines and programs necessary for the cell phone operations, as well as the user-defined data (such as the list of contacts, text messages, multimedia content, etc.).

1.7 Brief Overview of Exploitation of RF-MEMS in RF Systems | **35**

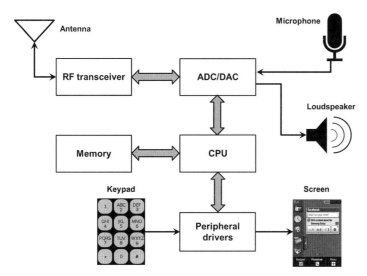

Figure 1.32 High-level block diagram of a typical cell phone's architecture. The block controlling all the others is the central processing unit (CPU), it being interfaced with the memory, peripheral drivers, analog to digital converter/digital to analog converter (ADC/DAC), and RF transceiver (through the ADC/DAC) functional blocks.

- **Peripheral drivers** The peripheral drivers constitute the low-level interface necessary to control and exchange information with the input and output devices used in the cell phone, namely, the keypad (input) and the screen (output).
- **ADC/DAC** The ADC/DAC block performs the transformation of the input analog signal, that is, the human voice collected by the microphone, into a digital signal, as well as the opposite operation, that is, the transformation of the received digital signal (detected by the cell phone) into the analog signal feeding the loudspeaker.
- **RF transceiver** The RF transceiver (i.e., transmitter/receiver) transforms the digital codification of the speaker's voice collected by the microphone from a baseband signal into an RF signal (i.e., a radio signal) that is then broadcasted through the surrounding space by means of the antenna. The RF transceiver also performs the opposite function, transforming the RF signal collected by the antenna from a radio signal into a baseband signal, which is subsequently converted into an analog signal (delivered by the loudspeaker to the listener).

The RF transceiver in Figure 1.32 is the functional block of a cell phone where RF-MEMS components find room to be integrated and exploited. The typical architecture of an RF transceiver front end is based on the superheterodyne scheme, discussed in [129, 130], and described in the block diagram in Figure 1.33.

The RF transceiver employs two branches, one for receiving the signal (the upper one in Figure 1.33, from the antenna to the ADC/DAC), and one for the transmission (the bottom one in Figure 1.33, from the ADC/DAC to the antenna). In superheterodyne transceivers, the received RF signal is down-converted to a frequency

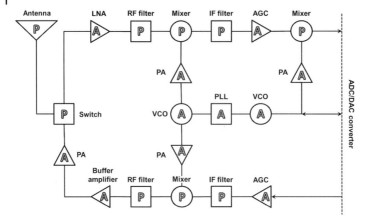

Figure 1.33 Detailed block diagram of the RF transceiver used in the cell phone architecture shown in Figure 1.32. It is composed of both active and passive components, labeled A and P, respectively, and is based on the super-heterodyne transmitter/receiver scheme. The labels AGC, IF, LNA, PA, PLL, and VCO represent automatic gain control, intermediate frequency, low-noise amplifier, power amplifier, phase-locked loop, and voltage-controlled oscillator, respectively.

band which is intermediate between that of the radio signal and that of the baseband signal (referred to as intermediate frequency, IF), as it eases the treatment of the signal (i.e., filtering, gain control, etc.). The same signal treatment is performed on the baseband signal to be transmitted, as it is up-converted to the IF band, before being transformed into a radio signal to be broadcasted by the antenna. An RF transceiver is composed of both passive and active elements. Such elements are listed as follows:

Active elements

Low-noise amplifier The low-noise amplifier realizes a low-noise amplification of the RF signal received from the antenna, as it is rather attenuated and needs to be regenerated without amplifying also the noise and the interference (i.e., spurious signals) detected by the antenna.

Automatic gain control The automatic gain control is an amplification stage internal to the RF transceiver that adjusts the signal level at the interface of the transceiver with the other blocks of the cell phone architecture.

Power amplifier The power amplifier realizes an amplification of the signal, as required before being broadcasted, as well as before being provided as input to the mixer stages.

Buffer amplifier The buffer amplifier realizes an amplification of the signal up to a standard level, it being an intermediate stage before the subsequent amplification realized by the power amplifier.

Voltage-controlled oscillator The voltage-controlled oscillator is a circuit generating a frequency carrier, which is necessary in the transceiver to perform the frequency up-conversion and down-conversion between the radio signal, the IF signal, and the baseband signal.

Phase-locked loop The phase-locked loop is an active circuit capable of generating a frequency signal with a fixed correlation of the phase with that of a reference signal. In this case the phase-locked loop is used to control the VCO and the frequency of the carrier it generates.

Passive elements

Switch The role of the switch is to select whether the RF transceiver transmitting branch or the RF transceiver receiving branch is loading the antenna, the latter being exploited for both broadcasting and detecting RF signals.

RF filter The RF filter filters the signal (i.e., it passes a certain frequency range and blocks other frequency ranges) in the RF range.

IF filter The IF filter filters the signal (i.e., it passes a certain frequency range and blocks other frequency ranges) in the IF range.

Mixer The mixer performs the frequency up-conversion or down-conversion of the input signal by mixing it with the reference signal generated by the voltage-controlled oscillator (i.e., carrier frequency).

Having in mind all the passive elements reported in the block diagram of an RF transceiver (Figure 1.33), we can basically replace all of them by implementations of basic and complex devices based on RF-MEMS technology.

Eventually, it is worth discussing a few final considerations in order to understand the potentialities of RF-MEMS technology. As stressed up to now, the implementation of passive RF components based on MEMS technology enables the targeting of very high performance. Consequently, the replacement of switches, varactors, and inductors, realized in standard semiconductor technology, with their implementation in RF-MEMS technology in an RF transceiver or subsystem can lead to better performance of the whole wireless apparatus. Nonetheless, a further step enabled by RF-MEMS technology should be carefully kept in mind, it being the large complexity achievable with such solution. As mentioned in the previous pages, the availability of very high performance lumped passive components allows the realization of highly reconfigurable complex networks, such as tunable filters, reconfigurable impedance tuners and phase shifters, and high-order switching matrices. Such functional subsystems based on RF-MEMS technology are not only components with better performance to replace standard ones in an RF circuit, but also pave the way for rethinking the architecture of an RF front end, it being possible to integrate more and more functions within the same block, while reducing hardware redundancy of the wireless system itself. A straightforward vision of this scenario was provided by Nguyen [131] in 2001 by discussing the impact of RF-MEMS technology on the architecture of a superheterodyne RF receiver. In particular, the scheme reported in Figure 1.34 shows a typical transceiver architecture, and the exploitation of RF-MEMS technology for the replacement of standard passive elements was already discussed in Figure 1.33. On the other hand, relying on

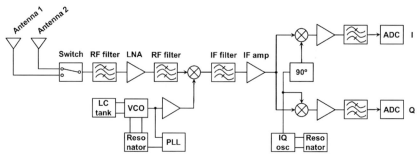

Figure 1.34 Block diagram of the standard architecture of a superheterodyne RF receiver.

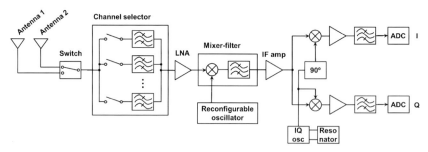

Figure 1.35 Block diagram of a superheterodyne RF receiver whose architecture is partially redesigned by exploiting complex RF-MEMS functional subblocks.

the availability of very low loss passive elements, like in the case of RF-MEMS technology, one can integrate more functionalities within the same block, leading to the modified architecture reported in Figure 1.35. As seen, the number of channels that can be selected is increased by grouping multiple switchable filters in one unique block, while the mixing and filtering functions, performing the down-conversion to the IF, are integrated within the same functional component. The architecture in Figure 1.35 is less complex in terms of the number of blocks with respect to that in Figure 1.34, and exhibits at the same time improved characteristics in terms of reconfigurability and hardware redundancy reduction. For instance, among the various benefits arising from the availability of RF-MEMS technology, the need for a very limited number of amplification stages to regenerate the signal is one of the more relevant, as it leads to a significant reduction of hardware redundancy and power consumption.

1.8
Conclusions

The chapter provided a comprehensive overview of the state of the art in RF-MEMS technology. After the discussion of some aspects concerning the history of RF-MEMS, mainly related to the development of the technology, the most relevant classes of RF passive components whose manufacturing is enabled by RF-MEMS technology were presented. The devices, together with their performance and characteristics, were reported according to an increasing complexity trend, starting from lumped elements (e.g., variable capacitors and microrelays) and then moving to networks characterized by a significantly increased complexity, such as reconfigurable phase shifters and impedance tuners. The chapter also treated two important aspects related to the field of RF-MEMS technology, they being the modeling and the packaging (i.e., the encapsulation of naked devices). The overview of RF-MEMS technology and applications provided in this chapter is necessary in order to correctly understand the content of the following chapters. Other important fields related to RF-MEMS technology, such as reliability and experimental testing, were not reported in this first, introductory chapter, as they will not be treated in the following chapters.

2
The Book in Brief

Abstract: This chapter contains information that is the basis for the following chapters, and it is divided into two parts. First of all, fundamental details on the technology platform exploited for the fabrication of the radio frequency (RF) microelectromechanical system (MEMS) devices discussed throughout the book are provided. The specific process is a surface micromachining technology based on gold metallizations. Subsequently, all the RF-MEMS device examples that will be studied in detail in the next chapters are introduced. Such a set of devices comprises lumped RF-MEMS elements, such as switches and variable capacitors, as well as complex networks, like, for instance, a reconfigurable RF power attenuator and an impedance-matching network. Three-dimensional schematics for all the RF-MEMS case studies presented are reported in this chapter, also highlighting the configuration of the buried layers when necessary to help understand their working principles.

2.1
Introduction

In this chapter, a few radio frequency (RF) microelectromechanical system (MEMS) devices and networks will be introduced, providing some details concerning their design, working principles, and specifications. These devices will be discussed throughout the book, focusing on different aspects related, for example, to the design, modeling, and simulation, depending on the aim of each chapter. All the devices that will to be described are fabricated with the RF-MEMS surface micromachining technology available at Fondazione Bruno Kessler (FBK; Italy). A suitable label is assigned to each device (e.g., the first RF-MEMS device that is presented in the following pages is named Dev A) as it will help in referencing each component in the next chapters. However, before the devices are described, a brief introduction to the technology with which they are realized is presented in order to provide the reader with the necessary information to understand the topology of such MEMS components.

2.2
A Brief Introduction to the FBK RF-MEMS Technology

The technology exploited for the realization of the RF-MEMS devices and networks discussed in this work is available at FBK in Italy [132]. It is based on a surface micromachining process on silicon or quartz 4-inch wafers. Two buried layers are available, namely, a high-resistivity/low-resistivity polysilicon layer (depending on the implanted boron dose amount), suitable for DC biasing lines and electrodes, and a multimetal layer based on aluminum, suitable for RF signals [133]. Two electroplated gold layers realize the actual MEMS devices, and a sacrificial layer (i.e., photoresist) is used to define the suspended movable parts [50]. The standard eight-mask FBK RF-MEMS process is briefly described in Figures 2.1–2.11.

Figure 2.1 A 4-inch silicon substrate is employed for the fabrication.

Figure 2.2 A silicon oxide layer is deposited onto the wafer.

Figure 2.3 High-resistivity or low-resistivity polysilicon is implanted and patterned on the wafer (first mask).

Figure 2.4 A silicon oxide layer covers the whole wafer, and holes are opened where electrical contact with the underlying polysilicon is required (second mask).

Figure 2.5 An aluminum-based multimetal layer is sputtered onto the wafer and subsequently patterned (third mask).

Figure 2.6 A silicon oxide layer covers the whole wafer, and holes are opened where electrical contact with the underlying multimetal layer is required (fourth mask).

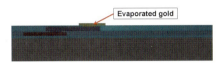

Figure 2.7 A thin gold layer is evaporated onto the contact openings above the multimetal layer (fifth mask).

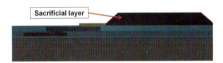

Figure 2.8 A sacrificial layer (photoresist) is patterned where microstructures are meant to be suspended (sixth mask).

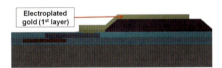

Figure 2.9 A first gold layer is electroplated, defining lines, pads, anchors, and the suspended deformable structures (seventh mask).

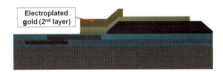

Figure 2.10 A second gold layer (thicker than the first one) is electroplated, and defines the parts of the microstructures that are meant to be more rigid (eighth mask).

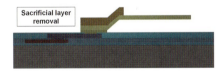

Figure 2.11 The sacrificial layer is removed, releasing the suspended microelectromechanical system (MEMS) structures.

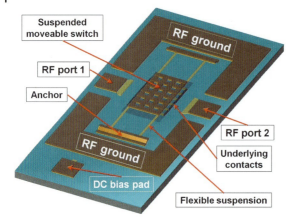

Figure 2.12 Three-dimensional schematic of the radio frequency (RF) MEMS series ohmic switch generated with the MEMS Pro software tool. The device consists of a suspended gold membrane that can close or open the electrical path between RF ports 1 and 2 depending on its position, that is, the rest or pulled-in position.

2.3
An RF-MEMS Series Ohmic Switch (Dev A)

The first device presented is an RF-MEMS series ohmic switch. The 3D schematic obtained with the MEMS Pro® software tool from SoftMEMS [134, 135] is reported in Figure 2.12.

The actual switch is surrounded by a coplanar waveguide (CPW) frame, allowing both ground–signal–ground probing, and integration of the RF-MEMS switch with waveguide-based blocks and systems. The pads of RF ports 1 and port 2, highlighted in Figure 2.12, are connected to a buried conductive layer that is not continuous underneath the central suspended MEMS membrane [136]. When no DC bias is applied between the suspended plate and the buried controlling electrode (not visible in Figure 2.12), the MEMS membrane is in the rest position, and the switch is off (i.e., open). Differently, when a DC bias is applied between the fixed and the movable plate, the electrostatic attraction force causes the suspended MEMS part to move downward, toward the fixed electrode, as the four straight beam suspensions are flexible. When the DC bias reaches a critical level called the pull-in voltage [23], the plate collapses onto the contacts underneath, realizing a short circuit between RF ports 1 and 2 (switch on or closed).

Since the RF-MEMS series ohmic switch is the first example of a device presented in this book, the most significant fabrication steps will be described briefly in the next few pages by means of 3D schematics.

2.3 An RF-MEMS Series Ohmic Switch (Dev A)

Step 1 The first step consists of growing a silicon oxide layer on top of a silicon wafer as shown in Figure 2.13.

Figure 2.13 Step 1: Silicon oxide layer on top of a silicon wafer.

Step 2 A high-resistivity polysilicon layer is deposited onto the oxide, as reported in Figure 2.14. The polysilicon defines the fixed electrode controlling the vertical position of the MEMS movable membrane, depending on the applied DC bias. Such an electrode is interdigitated, as reported in Figure 2.14. A bias line connects

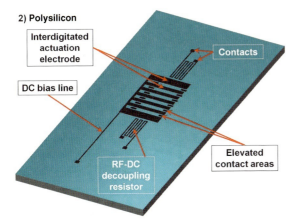

Figure 2.14 Step 2: The high-resistivity polysilicon layer realizes the fixed electrode, the RF–DC ground decoupling resistors, and elevated areas to enhance the ohmic contact of the two RF lines with the pulled-in switch.

all the fingers of the electrode. The polysilicon also defines two resistors (serpentine shaped) that are needed to decouple the RF signal from the DC ground when the MEMS switch is actuated. The DC ground applied to the suspended gold membrane is necessary to define the voltage drop when the fixed electrode is biased.

Nonetheless, without these resistors, the RF signal would be shorted to ground when the MEMS switch is pulled in. Finally, the polysilicon is also used to define small elevated areas upon which the two RF line contacts will be deployed. This solution makes possible the realization of better ohmic contacts between the suspended plate and the underlying areas.

Step 3 Another oxide layer is grown onto the wafer, and openings are etched that correspond with the RF line contacts, decoupling resistors, and bias electrode contacts (see Figure 2.15).

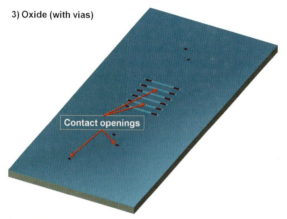

Figure 2.15 Step 3: Additional oxide layer with openings for the RF and DC contacts.

Step 4 A conductive multimetal layer is patterned on top of the oxide. Such a layer defines the two RF branch contacts underneath the suspended gold membrane, and redistributes the ohmic contact from the buried polysilicon to the top surface of the oxide for the resistors and the fixed electrode (see Figure 2.16).

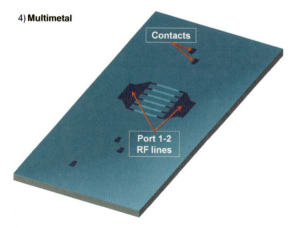

Figure 2.16 Step 4: The multimetal layer defines the two RF branch contacts and redistributes the DC contacts.

Step 5 Another oxide layer covers the multimetal thin film, and openings are etched for the redistribution of the multimetal thin film contacts on the top of the insulating layer (see Figure 2.17).

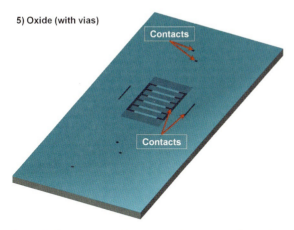

Figure 2.17 Step 5: An additional oxide layer covers the multimetal thin film and etched openings enable the redistribution of the electrical signal on the top of the insulator.

Step 6 An evaporated gold layer establishes the ohmic contact with the buried multimetal thin film through the openings etched in the oxide layer (see Figure 2.18).

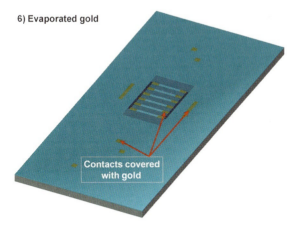

Figure 2.18 Step 6: A gold thin film is evaporated onto the openings in the oxide layer to establish an ohmic contact with the underlying multimetal thin film.

Step 7 A sacrificial layer (photoresist) is patterned where the gold membrane (still to be deposited) is meant to be suspended, as depicted in Figure 2.19.

7) Sacrificial layer

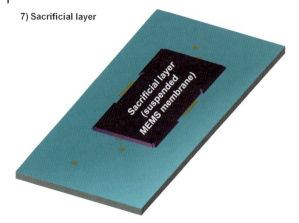

Figure 2.19 Step 7: Patterning of the sacrificial layer defining the suspended MEMS switch structure.

Step 8 Electrodeposition of the first gold layer (see Figure 2.20). The gold layer defines the suspended part (i.e., MEMS switch flexible suspensions), the anchors of the suspended membrane, the surrounding CPW frame and the pad for imposing the DC controlling voltage on the fixed electrode.

8) First gold layer

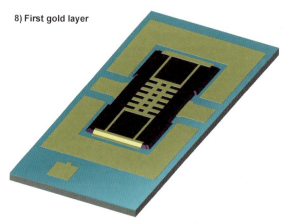

Figure 2.20 Step 8: Deposition of the first electroplated gold layer.

Step 9 A second gold layer, thicker than the first one, is selectively electroplated onto it. For this particular device, it is deposited everywhere but on the flexible suspensions, as seen in Figure 2.21. This is because a stiffer gold membrane is needed on the surrounding pads and especially in the anchoring parts of the suspended MEMS movable switch, but not for the straight flexible suspensions, in order not to increase their elastic constant and, in turn, the pull-in voltage of the device.

9) Second gold layer

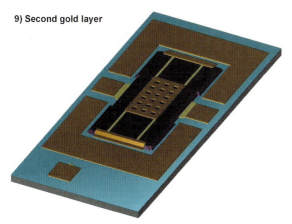

Figure 2.21 Step 9: Deposition of the second electroplated gold layer.

2.4 RF-MEMS Capacitive Switches/Varactors

In this section, two different designs for a capacitive switch realized with RF-MEMS technology are presented. The large capacitance is reached when the suspended MEMS membrane collapses onto the substrate, realizing a low impedance path to ground for the RF signal (i.e., open switch) [49]. However, such devices can also be used as varactors (variable capacitors), as the capacitance can be tuned continuously in the DC bias range from no voltage applied up to the pull-in level [137].

In the description of the following devices, fewer details on the fabrication process will be reported with respect to the RF-MEMS ohmic switch discussed in the previous section, as the technology platform employed is the same.

2.4.1
Design 1 (Dev B1)

The first varactor design proposed is shown in Figure 2.22. The movable electrode is a suspended gold membrane (in thick gold), and the four flexible suspensions (thin gold) are shaped like a meander in order to lower the elastic constant and, in turn, the activation voltage of the device [138]. In this case, there is no dedicated polysilicon electrode to apply the DC controlling bias, but the multimetal fixed capacitor electrode is also used to control the movable electrode (see Figure 2.23). This means that the DC bias and ground have to be superimposed on the RF signal and RF ground, respectively.

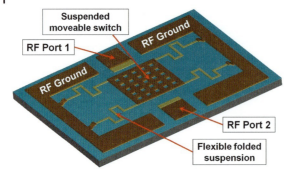

Figure 2.22 RF-MEMS capacitive switch/varactor with the movable gold electrode suspended by means of four folded beams.

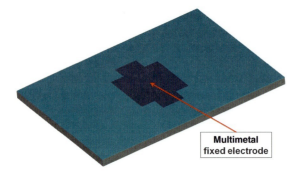

Figure 2.23 The multimetal fixed capacitor electrode. The controlling DC bias is superimposed on the RF signal.

2.4.2
Design 2 (Dev B2)

The second proposed design, schematically shown in Figure 2.24, is still a capacitive switch (varactor) with a central suspended gold electrode.

However, in this case there are two rather than four suspensions, they are serpentine shaped, and are realized in thick gold (two gold layers) [139]. The fixed electrode is realized in multimetal and, as for the previous design (Dev B1), the DC bias has to be superimposed on the RF signal.

2.4.3
RF-MEMS Ohmic Switch with Microheaters (Dev C)

The device introduced in this section is an RF-MEMS ohmic switch with an active self-release mechanism meant to improve reliability. RF-MEMS devices can suffer from poor reliability, stiction being a common source of malfunctioning [140]. Stiction is, by definition, the missed release of an actuated (i.e., pulled-in) RF-MEMS switch when the DC controlling voltage is removed. Two of the most frequent caus-

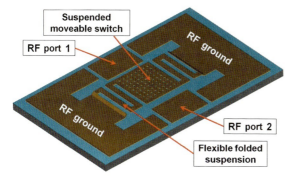

Figure 2.24 Three-dimensional complete schematic of the capacitive switch/varactor with two serpentine-shaped flexible suspensions.

es of stiction are charge entrapment within the insulating layer underneath the MEMS suspended membrane [141], and the occurrence of microwelding [142]. In the first case, the charge within the insulator generates a spurious bias that, if it is larger than the release voltage (i.e., pull-out), keeps the plate in the down position. On the other hand, when large RF signals are flowing through the MEMS device, the elevated current density can induce the formation of welded points on the ohmic contacts between the MEMS membrane and the in/out pads. Such joints are capable of keeping the plate in the down position even if the DC bias is removed [143]. In both just discussed cases, the RF-MEMS switch is not operable anymore. The solution to overcome such an issue, proposed by the design discussed here, is based on polysilicon microheaters embedded underneath the actual MEMS device [144]. Figure 2.25 shows a 3D schematic of the structure. The MEMS switch is realized by a typical gold membrane connected to four straight flexible suspensions within a CPW frame. The pads for activating the buried heaters are also visible and are highlighted in Figure 2.25. Figure 2.26 shows the polysilicon layer, realizing both the MEMS controlling electrode and the resistors for heating. The same polysilicon layer is characterized by two different resistivities, as, after the first boron low-dose implantation step, it is selectively exposed to a second high-

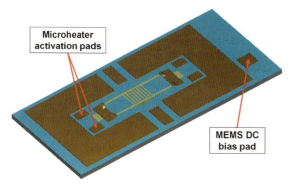

Figure 2.25 Complete 3D schematic of the RF-MEMS ohmic switch with microheaters.

52 | *2 The Book in Brief*

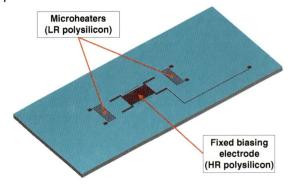

Figure 2.26 The polysilicon layer realizing the DC electrode and bias lines (high-resistivity, HR, polysilicon), and the microheaters (low-resistivity, LR, polysilicon).

dose implantation. As a result, the polysilicon of the biasing electrode and biasing lines has a sheet resistance of about 1.5 kΩ/□ (high-resistivity polysilicon), while the serpentine-shaped microheaters are implemented in low-resistivity polysilicon, exhibiting a sheet resistance in the range of 150–300 Ω/□ [50]. When the heater is activated by means of an electric current, the heat generated causes thermal expansion of the gold membrane, inducing shear and restoring forces on the welded contacts [145]. Moreover, the temperature increase speeds up the dispersion of the charge entrapped within the oxide, helping the recovery of the stuck RF-MEMS switch [146].

2.4.4
MEMS-Based Reconfigurable RF Power Attenuator (Dev D)

The device discussed in this section is the first example of a complex network based entirely on RF-MEMS technology. It is a reconfigurable RF power attenuator that uses different resistive loads, realized in low-resistivity polysilicon, to lower the amplitude of the RF signal passing through the device [71]. A complete schematic

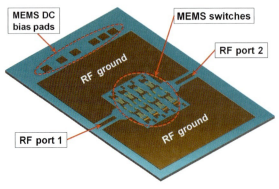

Figure 2.27 Complete 3D schematic of the RF-MEMS-based reconfigurable power attenuator.

of the network is depicted in Figure 2.27. The resistive loads can be inserted on the RF line or can be shorted, depending on the state (actuated or not actuated) of several series ohmic switches, as reported in the close-up in Figure 2.28. Each MEMS switch (hinged only at one end) has two gold fingers that, when the switch is actuated, short the buried polysilicon resistor by touching the two contact areas placed underneath them [50, 71]. Figure 2.29 shows the whole scheme of the polysilicon, while the close-up in Figure 2.30 focuses on more details, including both the biasing electrodes and lines (high-resistivity polysilicon) and the resistive loads (low-resistivity polysilicon). The high-resistivity polysilicon also realizes resistors to decouple RF from DC signals, while the low-resistivity polysilicon is also used to implement microheaters placed underneath the gold areas anchoring the suspended switches. These heaters are meant for reliability purposes, and are activated in the case of stiction of the corresponding MEMS switch, according to the principle presented and discussed in the previous section. Finally, Figure 2.31 shows the multimetal layer, and Figure 2.32 reports a close-up where the RF lines, connecting the polysilicon resistors to the device ports, as well as the microheater contacts are visible.

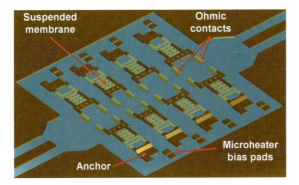

Figure 2.28 Close-up of the MEMS switches (single hinged) used to select or short the resistive loads on the RF line.

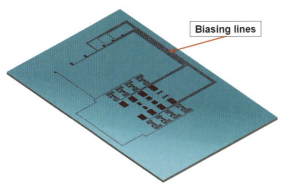

Figure 2.29 Complete view of the polysilicon layer scheme.

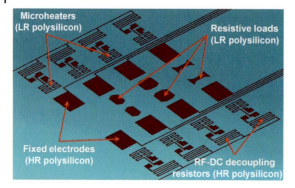

Figure 2.30 Close-up of the polysilicon layer in relation to the MEMS switches. The biasing electrodes and RF–DC decoupling resistors are realized in HR polysilicon. The resistive loads and the microheaters are realized in LR polysilicon.

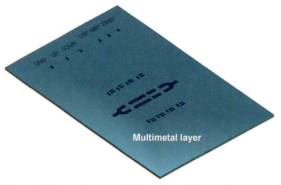

Figure 2.31 Complete view of the multimetal layer scheme.

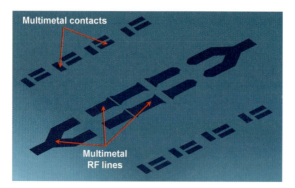

Figure 2.32 Close-up of the multimetal layer in relation to the MEMS switches. The RF lines connecting the polysilicon resistive loads to the two ports of the network are visible, as are the multimetal contacts of the microheaters.

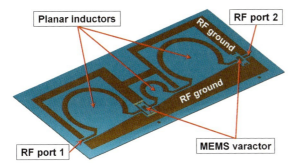

Figure 2.33 Complete 3D schematic of the GSM impedance matching network discussed in this section.

2.4.5
MEMS-Based Reconfigurable Impedance-Matching Network (Dev E)

The second RF-MEMS-based complex network discussed in this section is a reconfigurable impedance-matching network for GSM applications. The network is designed to match the output $12\,\Omega$ impedance of a CMOS power amplifier to the $50\,\Omega$ impedance of the antenna [147, 148]. The network uses LC ladders to match the impedance, and is composed of three planar inductors and a couple of two-state capacitors, reconfigured by means of two MEMS switches [149]. A schematic of the network is reported in Figure 2.33, and the two-state capacitors are meant to perform the impedance conversion at both GSM frequencies, namely, 900 MHz and 1.8 GHz. The two capacitors are realized in multimetal, similarly to the varactors previously discussed in this chapter.

2.5
Conclusions

In this chapter, relevant information to understand the content of the following chapters was provided. First of all, the specific technology platform exploited for the fabrication of the RF-MEMS devices discussed and analyzed throughout the book was briefly detailed. Subsequently, all the devices employed as case studies throughout the whole book were introduced. This set of devices comprises both basic passive components, such as switches and variable capacitors, and complex networks based on RF-MEMS technology, such as a reconfigurable power attenuator and an impedance-matching tuner. All the devices were described by showing 3D schematics that clarify their configuration and working principles.

3
Design

Abstract: This chapter focuses on the design of radio frequency microelectromechanical system (RF-MEMS) devices, and this topic is treated according to a rather practical approach. The concept of a set of design rules is introduced, stressing their close link to the specific technology employed for the development of an RF-MEMS device. Furthermore, the concept of a set of design rules is particularized for the specific technology platform chosen for the discussion of all the RF-MEMS examples treated in this book. For this purpose, the design of an RF-MEMS series ohmic microswitch from scratch is presented in detail and step-by-step. A specific software tool for the realization of such a design is chosen, but the operations performed are presented stressing the concept lying behind them, rather than focusing on how such operations are performed in a specific software environment. In this way, the reader who wants to design an RF-MEMS device using a different tool will find this information on how to perform such a task in an efficient way useful. The design developed will stress also the way design rules are complied with and the important concept of hierarchy, as well as other useful tricks that will be introduced to reduce the effort expended by the designer of RF-MEMS devices.

3.1
Introduction

In this chapter, the important aspects of the design will be addressed. The design phase, despite it perhaps seeming like a mere task the designer has to accomplish, is, in fact, much more than that. In fact, it represents the link between theory and practice, or in other words, between the results of the optimization of the characteristics of device and the subsequent manufacturing step. Given a certain set of specifications the new microelectromechanical system (MEMS)/radio frequency (RF) MEMS device or network has to comply with, a significant number of constraints are defined concerning the geometry and the topology of such a component. For example, if it is required to lower the pull-in voltage (i.e., activation) of an RF-MEMS switch, the elastic constant of the flexible suspensions should be lowered as well. This means designing suspending beams with a small section and as long as pos-

Practical Guide to RF-MEMS, First Edition. Jacopo Iannacci.
© 2013 WILEY-VCH Verlag GmbH & Co. KGaA. Published 2013 by WILEY-VCH Verlag GmbH & Co. KGaA.

sible. However, the set of specifications and trends imposed on the geometry by the specifications have to match the constraints imposed by the specific technology platform that has been chosen for the realization of the RF-MEMS devices. The set of technology-imposed constraints is known as design rules, and the phase through which the specification-oriented geometry meets the design rules is the design. It is clear that the designer needs to have rather good knowledge of the set of design rules for a certain technology. However, good knowledge of the technology represents, for the designer, significant added value, as it enables him/her to find the optimum solution when some geometry requirements are not compatible with the possibilities offered by the technology. This is true especially if, as happens in most cases, the designer is not the same person who performed the preliminary study on the device characteristics and came up with the geometry constraints to satisfy the specifications. Moreover, the person who translates the specifications into trends and constraints on the geometry/topology of the device does not necessarily know the technology platform with which the RF-MEMS devices will be manufactured. Consequently, very likely there will be topological requirements that cannot be realized in practice, and the person in charge of seeking the best compromises between the individual who conceived the design and the clean-room technicians is once again the designer.

3.2
Design Rules of the Fondazione Bruno Kessler RF-MEMS Technology

The design rules are a set of constraints imposed by the specific technology employed that have to be respected by the designer in designing the layout of new devices and structures, in order to ensure a satisfactory yield concerning the fabricated samples. The design rules are always concerned with minimum or maximum geometrical features and distances between layers, and are determined both by the minimum resolution of the lithography employed in the process and by the specific fabrication steps. For instance, if the minimum size that can be managed by lithography is 2 µm, assuming that a folded line is designed within a certain layer with a width of 1.5 µm and a lateral spacing between two adjacent branches of the serpentine of 1 µm, it most probably will not be fabricated as designed. Because of this, for that specific layer realized with the available lithography, two design rules must be introduced; that for the minimum width of the layer that can be reasonably fixed to 3 µm, and that for the minimum spacing between two adjacent edges of the layer of at least 3 µm. Complying with these two rules in the design of new devices will ensure their proper physical manufacturing. Apart from the design rules related to the minimum/maximum geometrical feature sizes of a certain layer, other design rules are dedicated to the minimum overlap of two or more layers to allow the realization, for example, of an ohmic contact between two stacked layers even when the maximum misalignment due to lithography occurs. Finally, further design rules can be necessary to prevent the problems that might occur while performing actual process steps. For example, certain metals might

3.2 Design Rules of the Fondazione Bruno Kessler RF-MEMS Technology | 59

Figure 3.1 The Fondazione Bruno Kessler (FBK) radio frequency (RF) microelectromechanical system (MEMS) technology masks. The number of each mask corresponds to the lithography sequence during the process. The name of each mask is reported as it is employed in the layout editor, as is the GDS number and the color per layer.

need to be always covered by another layer in order not to be removed or damaged during an etching step.

Figure 3.1 shows the sequence of masks of the whole Fondazione Bruno Kessler (FBK) surface micromachining process. Increasing mask number corresponds to the actual sequence of lithography steps performed during the process. For each layer, the name is reported (as it is used in the layout editor) as well as the GDS number (whose meaning will be discussed later). Mask 1a is optional, as the basic process relies on eight masks and gives the polysilicon layer a unique resistivity value. However, as will be discussed further in this book, it is possible to add an additional mask after the POLY definition in order to obtain, for the same layer, two values of resistivity (i.e., a double-dose implantation step). Table 3.1 shows the sequence of all the layers involved in the process with their physical properties, that is, electrical and mechanical, as well as their thickness. The levels that correspond to actual process masks (i.e., patterning or removal of layers) are also reported. On the other hand, Table 3.2 summarizes all the design rules of the FBK RF-MEMS surface micromachining technology platform. All the letters reported in the table refer to the design rules shown in detail in Figures 3.2–3.7.

Table 3.1 Physical properties of the layers in the Fondazione Bruno Kessler (FBK) radio frequency microelectromechanical system technology process. List of the superscripts included in the table.

Physical layer Name	Mask name and GDS number	Thickness (nm)	Sheet resistance ($\Omega/\square$)	Dielectric constant (F/m)	Stress (MPa)	Stress gradient (MPa/μm)	Young's modulus (GPa)
Substrate		525 ± 15 [a)h)]	> 5000 [d)]	11.9			
Field oxide		1000 ± 30		3.94	−274		
Polysilicon	POLY, 1	630 ± 15	1584 ± 52		−260		
	POLY HR, 1				36		
TEOS	CONHO, 2	300 ± 10		3.94			
TiN		80 ± 5 [b)]					
Aluminum	TIN, 3	440 ± 25	0.0654 ± 0.0011 [e)]				
TiN		110 ± 15 [b)]					
LTO	VIA, 4	100 ± 5 [i)]		3.94	150		
Flomet	FLOMET, 8	150 ± 5 [f)]	0.126 ± 0.010				
Spacer	SPACER, 5	3000 ± 200					
Bridge	BRIDGE, 7	1800 ± 200 [c)]	0.022 ± 0.004		120	6.95 ± 0.4 [e)]	98.5 ± 6
CPW	CPW, 7	1800 ± 400 [c)]	0.0055 ± 0.0008		120		98.5 ± 6

a) Micrometer.
b) Titanium (30 nm) as an adhesion layer.
c) Chrome (2.5 nm) as an adhesion layer.
d) Centimeter.
e) Applies to the metal multilayer.
f) Chrome (5 nm) as an adhesion layer.
g) A beam bends down (toward the wafer).
h) Also possible is 200 ± 15 μm.
i) This is an effective value; for the stress values a positive value indicates tensile stress and a negative value indicates compressive stress.

3.3
Design of an RF-MEMS Series Ohmic Switch (Dev A)

In this section the design of the RF-MEMS series ohmic switch introduced in Chapter 2 (referred to as Dev A) will be presented. The design is performed with the L-Edit software tool, which is dedicated to layout creation and editing, and is distributed by Tanner Tools [150]. However, the philosophy followed and described in the example that will be discussed is applicable regardless of the specific software tool the designer uses. The steps and operations that will be discussed in the following pages are available and implemented in most commercial layout editors presently available, as well as in several open source tools, even though in the

3.3 Design of an RF-MEMS Series Ohmic Switch (Dev A)

Table 3.2 All the design rules to be complied while designing with the FBK technology. Each design rule is identified by a letter that is associated with a geometrical feature reported in Figures 3.2–3.7.

Mask level (GDS Number)	Layer name	Minimum features (μm)	Minimum space (μm)	Overlap (μm)	Comments
1	POLY	3 (a)	4 (b)	–	Usually first mask
9	POLY (HR)	–	5	5	Overlap on the POLY
2	CONHO	4 (c)	4 (d)	3 (e)	Polysilicon overlap around contact
3	TIN	5 (f)	5 (g)	3 (h)	Metal overlap around contact
4	VIA	4 (i)	4 (j)	4 (k)	TIN overlap around via
4	VIA	4 (i)	4 (j)	6 (l)	Distance to polysilicon
8	FLOMET	5 (m)	5 (n)	3 (o)	FLOMET overlap around via
8	FLOMET	5 (m)	5 (n)	10 (p)	FLOMET to SPACER distance
5	SPACER	8	8	6 (q)	Distance to VIA
5	SPACER	8	8	10 (r)	Overlap around VIA
5	SPACER	8	8	10 (s)	Overlap around FLOMET
6, 7	BRIDGE & CPW	10 (t)	10 (u)	–	(1) Minimum air hole size 10 μm (2) Maximum spacing between adjacent air holes 10 μm (3) Maximum distance between air holes and borders 15 μm
6, 7	BRIDGE & CPW	10 (t)	10 (u)	4 (w)	Metal overlap around VIA
6, 7	BRIDGE & CPW	10 (t)	10 (u)	4 (x)	Distance to CONHO
6, 7	BRIDGE & CPW	10 (t)	10 (u)	10 (y)	BRIDGE/SPACER overlap

latter case some limitations are possible. Another important aspect that has to be stressed concerns the particular sequence of operations that are described in the following pages, that is, what has just been referred to as the design philosophy. Such a procedure is not the only one that exists, and cannot even be considered the best one. This is because the effectiveness of the procedure chosen is always related to what are the targets one, in this case the designer, wants to achieve. To give an idea of possible different directions that can be followed, one specification could be to complete the design of a certain device or network in the shortest possible time. In this case it might be sensible to perform the design all at once. On the other hand, if the device has to be designed in multiple versions with slight modifications, or if several devices and/or networks have to be designed, effort should

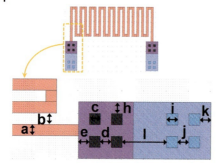

Figure 3.2 Design rules concerning the minimum features and distances of POLY, CONHO, and VIA, and overlap with POLY and TIN.

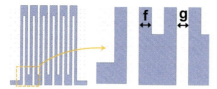

Figure 3.3 Design rules concerning the minimum features and distances of the TIN.

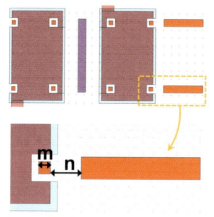

Figure 3.4 Design rules concerning the minimum features and distances of the FLOMET.

be expended in order to identify parts and subsections of the design that can be reused (as they are) within the topology of other devices. In this case, the generation of the so-called cells and their instantiation during the layout editing becomes crucial to avoid duplication of the design work. The RF-MEMS series ohmic switch that will be presented as an example is arranged according to the latter case and, for this reason, the definition of cells might appear redundant. However, in the upcoming example, cells are used not only to avoid redesign of certain blocks, but also to make easier their placement in the layout itself. All the above considerations will appear clearer to the reader after the presentation of the steps in the next

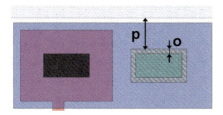

Figure 3.5 Design rules concerning the FLOMET overlap around VIA and the FLOMET to SPACER minimum distance.

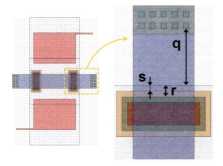

Figure 3.6 Design rules concerning the minimum distance between SPACER and VIA, the minimum overlap of the SPACER above the FLOMET, and the minimum SPACER overlap around VIA.

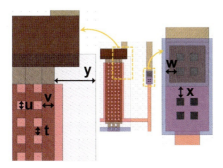

Figure 3.7 Design rules concerning the minimum distance between BRIDGE/CPW and SPACER and the minimum size of the openings and spacing in the suspended gold membranes. Design rules concerning the BRIDGE/CPW overlap around VIA and distance to CONHO.

pages. The only concept the designer should keep in mind is that a certain procedure he/she might adopt as his/her standard might not be the most efficient in all the situations he/she will have to face during daily work.

First of all, all the layers of the RF-MEMS switch are incrementally visualized, according to the fabrication sequence of the FBK technology, in Figures 3.8 and 3.9 as follows:

64 | *3 Design*

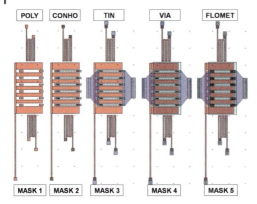

Figure 3.8 Progressive superimposition of the first five FBK RF-MEMS technology masks for the series ohmic switch discussed in this section (Dev A).

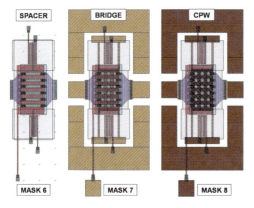

Figure 3.9 Progressive superimposition of the last three FBK RF-MEMS technology masks for the series ohmic switch discussed in this section (Dev A).

Mask 1 The high-resistivity polysilicon (POLY) defines the fixed buried biasing electrode (interdigitated) connected to the biasing line, two RF–DC decoupling resistors, as well as the elevated areas for the ohmic contact pads between RF ports 1 and 2.

Mask 2 The contact openings through the oxide (CONHO) define contacts between the polysilicon and the conductive multimetal layer above.

Mask 3 The multimetal layer (TIN) defines the buried RF lines, the contact pads underneath the movable MEMS membrane, and the contacts with the buried polysilicon.

Mask 4 The contact openings through the oxide (VIA) ensure electric contact to the buried multimetal layer. The VIA is also defined on the transverse interdigitated polysilicon fingers of the biasing electrode. This is meant to remove the oxide also on top of the biasing electrode and, in turn, mitigate charge accumulation within the oxide layer during the biasing of the device.

Mask 5 The floating metal (FLOMET), that is, a thin gold evaporated layer, is placed on top of all the ohmic contacts.

Mask 6 The photoresist sacrificial layer (SPACER) is designed where the MEMS structures (still to be designed) are meant to be suspended.

Mask 7 The first gold electroplated layer (BRIDGE) defines the RF-MEMS suspended ohmic switch, the surrounding CPW frame, as well as the biasing pad.

Mask 8 The second gold electroplated layer (CPW) is superimposed on the first layer (BRIDGE). The two gold layers are present everywhere, with the exception of flexible suspending beams connected to the RF-MEMS switch. This is because a double gold suspension would have a larger elastic constant (i.e., would be stiffer) and, in turn, the RF-MEMS switch actuation voltage would significantly increase.

Figure 3.10 shows a screenshot of the main window of L-Edit. The subwindow reporting all the technology layers is highlighted on the left of the screen, and is also placed in foreground in the center of Figure 3.10 to make it more readable. The technology file, where proper names, GDS numbers, and colors are assigned to the layers depending on the specific fabrication process chosen, has to be defined by the designer prior to the actual design phase. On the left of the subwindow, the layer names (in this case those of the FBK technology) are listed in the desired sequence, while on the right the corresponding GDS numbers are visible. The latter identify each layer when the layout is exported in the GDS format, which is a universal file format for the layout that enables consistent data import and export between different layout editors and computer platforms [151]. In the GDS numbers list in Figure 3.10, note that the FLOMET does not follow the process sequence (the GDS number is 8 while the FLOMET is the fifth mask). This is because the FLOMET layer was not present in the initial FBK RF-MEMS technology, which was originally a seven-mask process. In order to maintain consistency with the GDS numbers of

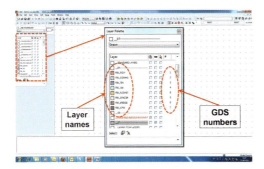

Figure 3.10 Main window of the L-Edit layout editor. The subwindow reporting the layer names and the corresponding GDS numbers is visible in the center of the image.

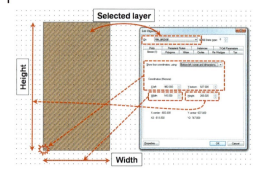

Figure 3.11 Definition of the footprint of the suspended MEMS switch (BRIDGE layer). By means of the properties window, we can modify the dimensions of the block, as well as other characteristics, after it has been generated in the editor.

all the other masks, the FLOMET has been assigned to the first available number, following the seven-mask process, that is, 8. Maintaining consistency in the GDS numbers assigned to the layers is fundamental in order not to generate misunderstandings and mistakes when exchanging layouts with other designers or with mask shops.

Figure 3.11 shows the design of the first block of the RF-MEMS switch discussed here. The design does not necessarily have to start from the first technology layer (in this case the POLY), and also the sequence of levels in the design does not have to be followed as it is in the manufacturing process. Let us say that once the designer has in mind how the device should look like, he/she can choose the best strategy to make easy its design.

This means that, for instance, a layer could be more suitable than others to begin the building of the structure. Also, the use of temporary fictitious layer/s could help in this process. Here, we decided to define the footprint of the suspended RF-MEMS switch first. For this reason, the first step, as depicted in Figure 3.11, is the definition of a BRIDGE rectangle with dimensions of 140 µm × 260 µm, as required by the previously defined specifications. The BRIDGE rectangle can be generated with proper dimensions at once, that is, by moving the mouse pointer on the screen while keeping the left button pressed, and maintaining reference with the grid visible in the editor window. However, when the block to be designed has considerable dimensions, like in this case, the method described is rather difficult to perform. The alternative solution we propose, reported in Figure 3.11, is to design a small block without minding the dimensions. Subsequently, we open the properties windows for that block (in L-Edit, once a block is selected by clicking on it, its properties are recalled with the command Ctrl+E). In the properties windows we can perform several actions on the block, as will be shown in the next few pages. In this case, we are interested only in assigning proper dimensions to the BRIDGE block. For this reason, we type 140 in the Width field, and 260 in the Height field. The selected coordinates mode is "Bottom left corner and dimensions" (see Figure 3.11). This means that, since we did not modify the fields identifying the origin

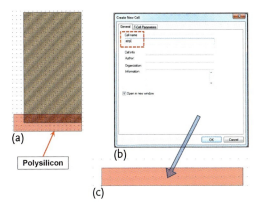

Figure 3.12 (a) Generation of the first polysilicon actuation electrode finger. (b) Window for the definition of a new cell within the current hierarchy. (c) Design of the cell just defined in a new empty layout editor window.

of the block, the bottom left corner of the block will not move, but only the length of it edges will change on the basis of the number we specified.

Figure 3.12a shows the generation of the first polysilicon finger of the biasing electrode (width and height of 165 and 40 μm, respectively), defined according to the same procedure previously described for the MEMS switch footprint. Subsequently, it is necessary to generate five fingers of dimensions of 165 μm × 20 μm to be placed underneath the suspended switch, and separated by 20 μm from the first finger (just designed) and with a spacing of 20 μm between them. This is a good opportunity to show how the definition of cells can be used to make easier the design. Instead of designing five fingers, we create a new cell. Figure 3.12b shows the pop-up window after we choose the option to generate a new cell, and the field where we specify the new cell's name is highlighted. After we click OK, an empty new layout editor window opens, and we design in it a polysilicon block with the proper dimensions (i.e., 165 μm × 20 μm), as Figure 3.12c depicts.

After saving this new cell (named strip), we close the window, and return to the main layout (Figure 3.13a). Now, we instance the cell we just generated in the main layout window. In L-Edit, the instance window can be opened by looking for it in the palette (menu Cell→ Instance), or just by pressing the I key. Afterward, the window reported in Figure 3.13b pops up, and in it we can see all the hierarchy of cells contained inside the top cell, the latter being the one associated with the main layout window. In this case, since we just generated one cell in the hierarchy, we can only see the strip cell below the top cell, named ohmic_sw_ser_CELL. We select the strip cell and we press OK. The polysilicon will appear in the main layout window, and we move it reaching the proper alignment and spacing (20 μm) with the already existing polysilicon finger, like Figure 3.13c shows.

Subsequently, we open the properties window for the just instanced object (Ctrl+E), and we select the Instances palette (see Figure 3.14). Here, we can visualize the reference system. In the lower part of the window there are some fields to

68 | 3 Design

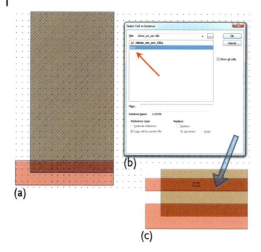

Figure 3.13 (a) Main layout window after saving and closing the new cell generated in the previous step. (b) Instance window to select the proper name linked to the cell to be instanced in the main layout. (c) Placement and alignment of the cell instanced in the main layout.

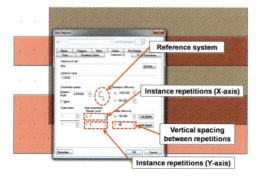

Figure 3.14 Once the cell has been instanced, the properties window is opened in order to specify the direction, number, and spacing of repetitions to form at once the interdigitated polysilicon actuation electrode (see Figure 3.15).

define the repetitions of the instance along the X and Y axes. We want to duplicate the finger five times along the Y direction; therefore, we type 5 in the Y field, while we leave 1 in the X field, as we do not want any repetition along the horizontal axis. We also want a spacing of 20 µm between one finger and the next one. Because the finger has a height of 20 µm, we specify $20 + 20 = 40$ µm in the Delta field for the Y axis.

After pressing OK, we obtain the result reported in Figure 3.15a. We then complete the polysilicon fingers by adding another 165 µm × 40 µm block above the just instanced five thinner fingers, as Figure 3.15b reports.

Now, we face the definition of the contacts between the two RF lines underneath the RF-MEMS switch. Since the contacts have to be replicated within the gaps of

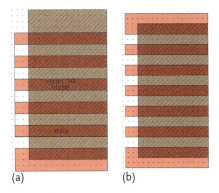

Figure 3.15 (a) The repetition specification discussed for Figure 3.14 enables the generation, at once, of the five biasing fingers. (b) The biasing electrode is completed by adding the last wider finger on top of the ones just generated.

the biasing fingers (see the previous steps), we will follow once again the approach based on the definition of a new cell.

First of all, in the empty editor window of a new cell we create a block of a contact hole within the oxide layer above the polysilicon (i.e., CONHO mask) of dimensions 11 µm by 165 µm (see Figure 3.16a). This is because we also want to remove the oxide layer above the polysilicon (the one above will be removed by the VIA, defined everywhere on the electrode, as previously shown in Figure 3.8). However, the oxide layer above the polysilicon will remain on the biasing finger (i.e., no

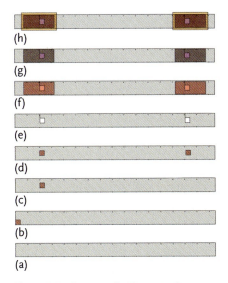

Figure 3.16 Sequence for the generation of the RF in/out fixed contact. (a) The starting point is the definition of a CONHO strip. (b–e) Two dimples are subtracted from it in order to generate small elevated areas made of oxide. The contact is completed by adding (f) POLY elevated pads, (g) TIN contacts, and (h) FLOMET overlapped areas.

CONHO will be defined there), otherwise the polysilicon of the biasing electrode will be totally exposed to the suspended MEMS membrane. In the case of contact between the two electrodes, when the pull-in is reached, touching of gold of the suspended membrane with the polysilicon would generate an undesired short circuit that would zero the DC bias, and most probably also weld the two electrodes together, making the switch not operable anymore.

The next step consists in the removal of small portions of the CONHO in order to create dimples that correspond with the ohmic contacts for the RF lines. Where the CONHO is not defined, the oxide is not removed, and all the other layers deployed above that area will feel its more elevated vertical position. We use the POLY as a temporary layer for the definition of such holes within the CONHO strip. A 4 μm × 4 μm POLY square is defined in the bottom-left corner of the CONHO strip, as depicted in Figure 3.16b.

The next action is to position this object in the proper place, that is, at the center of the ohmic RF contacts. To do so, we use the move function, which, once a block has been selected, enables the definition of an offset along the X and Y axes for its displacement. The displacement we want to impose is 20 μm along X and 3.5 μm along Y, and, after selecting the block to be moved, we specify these numbers in the Move window, as indicated in Figure 3.17a. The result of this operation is reported in Figure 3.16c.

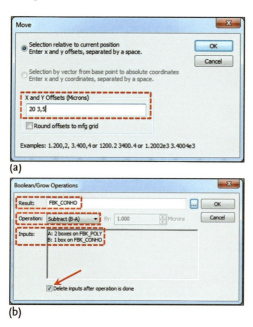

Figure 3.17 (a) Move window, where it is possible to specify the offset along the X and Y axes for displacing the selected object/s. (b) Boolean operation window. In this case the selected operation is the subtraction of two POLY blocks from a CONHO strip. The layer on which the result will be put is the CONHO, and the original blocks involved in the operation are deleted.

3.3 Design of an RF-MEMS Series Ohmic Switch (Dev A)

Following the same procedure, we define another POLY block placed symmetrically close to the right edge of the CONHO strip (see Figure 3.16d).

The subsequent step is to subtract the two POLY dimples from the CONHO strip, and to do so we have to perform a Boolean operation. After selecting the CONHO strip and the two POLY blocks, we open the Boolean/Grow Operations window reported in Figure 3.17b. In the Inputs field we can see the current selection of objects, and, in more detail, the two POLY blocks are identified by the letter A, while the CONHO strip is identified by the letter B. What we have to perform, then, is the B−A operation, as we chose in the Operation field. We also specify the layer on which we want to generate the result of the subtraction, which in our case is the CONHO, as specified in the Result field. Finally, we check the option Delete inputs after operation is done, as we are not interested in keeping the three original blocks, but are only interested in the result of the operation, which is displayed in Figure 3.16e.

Now we define the polysilicon elevated pads, where the RF contacts will be deployed. For this purpose, two POLY blocks with dimensions of 25 µm by 10 µm are defined and aligned with the CONHO strip by means of the operations previously described, as reported in Figure 3.16f.

The two POLY blocks now have to be duplicated in the multimetal layer (TIN). For this purpose, they are copied and pasted first. Afterward, while it is still selected, the properties window (Ctrl+E) is opened and the layer is modified from POLY to TIN, as Figure 3.18 shows. The result is reported in Figure 3.16g.

The final step for the completion of the contacts is the definition of FLOMET pads on top of the POLY elevated area and of the TIN RF lines. This operation is

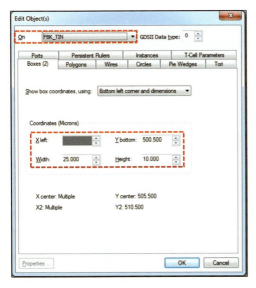

Figure 3.18 The properties window is used to transform two blocks from the POLY layer to the TIN layer.

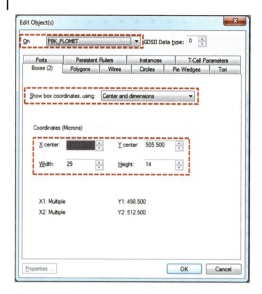

Figure 3.19 The properties window is used to transform two blocks from the POLY or TIN layer to the FLOMET, and also to increase their dimensions, keeping their center fixed.

similar to the previous one, as we can select the two POLY or TIN pads, duplicate and align them, and then modify the layer to FLOMET in the properties window. However, since the FLOMET, according to the design rules, has to have a certain overlap with respect to the underlying TIN, we increase the dimension of the starting blocks by 4 µm each side, that is, 29 µm × 14 µm. Moreover, if we select `Center and dimensions` in the `Show box coordinates, using:` field (see Figure 3.19), the center of the two FLOMET blocks will not move, while their edges will expand by 2 µm in all four directions.

The complete contact cell is reported in Figure 3.16h, while the placement and repetition of the contact cell in the main layout editor window is reported in Figure 3.20. The initial BRIDGE footprint is intentionally not displayed.

The next step in the design of the RF-MEMS switch is the definition of the multimetal to gold transitions, for RF ports 1 and 2, which have to be connected to the RF contact pads defined in the previous steps, which are the areas where the low-impedance ohmic contact is established when the MEMS switch pulls in. The first step consists in the definition of a 10 µm × 210 µm TIN rectangle, reported in Figure 3.21a. Then, a rectangle with perpendicular edges of 40 µm (along X) and 45 µm (along Y) is aligned with the TIN rectangle (Figure 3.21a), and is duplicated and mirrored on its lower edge (Figure 3.21b). Another TIN rectangle is generated (65 µm × 120 µm) and is placed between the two triangles (Figure 3.21c). Finally, all the TIN blocks are selected and merged together in order to create a unique object from them (Figure 3.21d). Subsequently, the transition to the gold level above is designed by adding a VIA onto the TIN (Figure 3.21e), and this is then covered and overlapped with a FLOMET rectangle (Figure 3.21f). The first

3.3 Design of an RF-MEMS Series Ohmic Switch (Dev A) | **73**

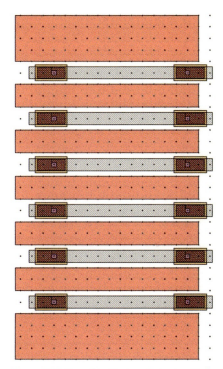

Figure 3.20 Complete biasing electrode with the RF contacts instantiated.

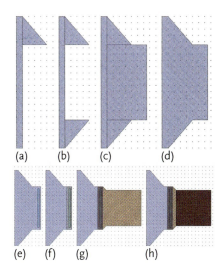

Figure 3.21 (a–d) Definition of the TIN geometry for the in/out RF buried lines. (e–h) Definition of the TIN to gold transition and deployment of the RF pad for probing and testing.

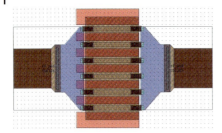

Figure 3.22 Instantiation and alignment of the RF multimetal to gold transition beside the central actuation electrodes and in/out contacts.

gold layer (BRIDGE) is then defined (Figure 3.21g) following steps similar to the ones described for the TIN layer, and the RF pad (for probing) is reinforced with the second gold layer (CPW) (Figure 3.21h). The transitions just described, designed as a cell in the layout hierarchy, are then instanced in the main layout window, and are aligned with the elevated contacts placed between the biasing fingers, as reported in Figure 3.22.

The next cell we define includes flexible suspensions to be connected to the RF-MEMS switch and the anchoring area where they are fixed. The beams are 10 µm × 165 µm (in BRIDGE), and are linked to a gold rectangular area (BRIDGE + CPW) of dimensions 220 µm × 40 µm. The sacrificial layer (SPACER) is placed underneath the two flexible beams, and is overlapped by the anchoring area for 10 µm. The block just described is depicted in Figure 3.23a, and its instantiation in the main layout window is reported in Figure 3.23b. In the latter case, a SPACER block has been added also in the area where the actual RF-MEMS movable switch has to be defined.

It is now time to define the geometry of the actual suspended RF-MEMS gold switch. Starting from the switch footprint chosen at the beginning of the design (i.e., 140 µm × 260 µm BRIDGE rectangle), we now have to define openings on its surface. Such openings are required by the technology process in order to enable effective removal of the sacrificial layer underneath the switch. The openings have to be defined by means of a Boolean operation, that is, subtraction, as was already described in the previous pages.

Figure 3.24a shows the BRIDGE block on which we defined two rectangles of dimensions 40 µm × 20 µm, with a square hole (side of 20 µm) spaced 20 µm horizontally and from the bottom edge. The temporary layer exploited is the CPW, and the strip just described is repeated along the Y direction with a spacing of 20 µm between each occurrence (see Figure 3.24b). Once the subtraction of the BRIDGE box and the CPW objects has been performed, we have the BRIDGE layer of the RF-MEMS switch with the desired openings, as depicted in Figure 3.24c. To obtain the CPW layer for the RF-MEMS switch, we copy the BRIDGE layer that has just been generated and we add two rectangles of dimensions 20 µm × 260 µm at the lateral edges in order to get a matrix of 20 µm × 20 µm openings (see Figure 3.24d). We then merge such objects and we modify the layer from BRIDGE to CPW. Finally,

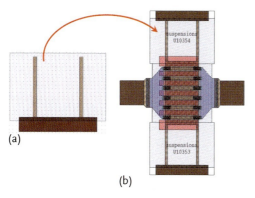

Figure 3.23 (a) Cell comprising the two flexible suspensions, the anchoring area, and the sacrificial layer defining the suspended structures. (b) Instantiation and alignment of the cell with suspensions in the main layout window.

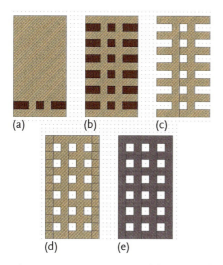

Figure 3.24 (a–c) Generation of the BRIDGE layer for the suspended RF-MEMS switch. (d,e) Generation of the CPW layer. Once the two layers are aligned, the suspended part is complete.

we align the BRIDGE and CPW layers, getting the final geometry of the suspended RF-MEMS switch, as reported in Figure 3.24e.

The following step consists in the definition of the RF ground frame that has to surround the actual RF-MEMS switch in order to build a CPW-like structure. Such a frame is realized in BRIDGE + CPW gold, and has suitable dimensions to surround the central device and realize the desired gap between the RF lines and the ground (i.e., 50 μm). The frame is reported in Figure 3.25a, and its instantiation in the main layout is shown in Figure 3.25b.

The very last step consists in the inclusion of RF–DC decoupling resistors to be applied between the suspended MEMS switch and the RF ground lines, and of a DC biasing line connecting the polysilicon interdigitated electrode underneath the

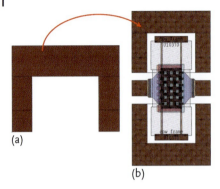

Figure 3.25 (a) Definition of the RF ground frame that wraps the actual MEMS device in double gold layer. (b) Instantiation of the RF ground frame around the switch in the main layout window.

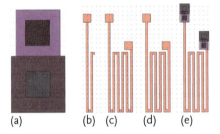

Figure 3.26 (a) Definition of the polysilicon to gold transition cell. (b–e) Steps for the definition of the DC–RF decoupling POLY resistor. It is based on a folded serpentine and exploits the POLY to FLOMET transitions.

MEMS suspended membrane. First we define a cell for the transition between the polysilicon and the gold in such a way that it can be used for the two reasons just described.

The complete cell of the transition is shown in Figure 3.26a. The top part consists of a POLY square of side 18 µm, with a square of CONHO of side 10 µm centered within it. The POLY square is then duplicated in TIN and superimposed on it. The lower part consists of a TIN square of side 22 µm. A FLOMET square is superimposed and aligned with it, while a square VIA of side 10 µm is centered within it. Figure 3.26b–e shows the definition of the POLY decoupling resistor based on a serpentine (width of 5 µm). The number of squares of the entire serpentine is roughly 100, and since each POLY square has a sheet resistance of about 1.5 kΩ/□, the resistance of the folded serpentine is around 150 kΩ.

The RF-MEMS series ohmic switch design is completed by placing two instances of the polysilicon serpentine just described and by designing a POLY biasing line connecting the fingers underneath the MEMS switch to a gold biasing pad for probing, as reported in the screenshot of the complete layout shown on the right in Figure 3.9. The 3D measured profile of a physical RF-MEMS ohmic switch is reported in Figure 3.27.

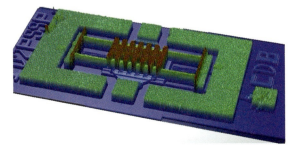

Figure 3.27 Three-dimensional measured profile of the RF-MEMS series ohmic switch whose design was discussed in the previous pages.

3.4
Generation of 3D Models Starting from the 2D Layout

In this section a brief description of the 3D functionality incorporated in the MEMS Pro tool [135] (also featuring the L-Edit layout editor) from Tanner Tools is provided. This extension enables the generation of 3D models corresponding to the 2D layout of RF-MEMS devices and networks, after the proper characteristics and features of the technology employed have been defined. Such additional functionality is important because it allows one to export 3D structures in file formats compatible with 3D finite element method based simulation tools, such as the Standard ACIS Text (SAT) format [152]. As will be discussed in the chapter focusing on modeling using the finite element method (Chapter 4), the possibility to import the 3D geometry of an RF-MEMS structure to be simulated significantly makes easier and speeds up the design phase.

In the following program code, a list of all the layers necessary for the generation of 3D models in the FBK RF-MEMS technology in MEMS Pro is given. In this list, to the standard eight-mask process an additional mask is added, concerning the second level of doping that is possible for the polysilicon layer, as reported in Figure 3.1. All nine active layers, that is, the actual layers the designer uses in designing RF-MEMS structures, are reported at the top of the list in the following program code (from `FBK_POLY_LOW_RES` to `FBK_CPW`). Subsequently, a second block defines additional layers, optionally used in the layout, to define the dimensions of the substrate area to be included in the complete 3D model (from `FBK_SILICON_SUB` to `FBK_AIR`). Finally, service layers are also defined in order to support the translation of active layers into specific fabrication steps they are linked to in the technology process, for instance, etching or deposition steps. Such layers are those from `FBK_LOW_POLY_MASK` to `FBK_CPW_MASK`.

In the following program code, the lines beginning with a double underscore are comment lines. After the list of layers, each of them is discussed in the following pages, specifying if it is an active or a service layer, and also its geometry characteristics and the options selected in MEMS Pro in order to associate each of them with the proper technology step are discussed. The order in which the layers are described in the following pages is the same as the order in the physical devices, starting from the silicon substrate and finishing with the second gold metallization. The tool described in this section enabled the generation of all the images reported in Chapter 2, starting from 2D layouts of RF-MEMS and test structures.

```
__FBK TECHNOLOGY
__&&&&&&&&&&&&&&&&&&&&&&&&&&&&&&&&&&
__--------------------
__STANDARD LAYERS
__--------------------
FBK_POLY_LOW_RES
FBK_POLY_HIGH_RES
FBK_CONHO
FBK_TIN
FBK_VIA
FBK_FLOMET
FBK_SPACER
FBK_BRIDGE
FBK_CPW
__--------------------
__&&&&&&&&&&&&&&&&&&&&&&&&&&&&&&&&&&
__--------------------
__LAYERS FOR THE
__DEFINITION OF THE
__WORKING PLANE
__--------------------
FBK_SILICON_SUB
FBK_FIELD_OXIDE
FBK_TEOS
FBK_LTO
__--------------------
__&&&&&&&&&&&&&&&&&&&&&&&&&&&&&&&&&&
__--------------------
__SERVICE LAYER
__NOT TO BE USED
__--------------------
FBK_LOW_POLY_MASK
FBK_HIGH_POLY_MASK
FBK_TIN_MASK
FBK_FLOMET_MASK
FBK_SPACER_MASK
FBK_BRIDGE_MASK
FBK_CPW_MASK
__--------------------
__&&&&&&&&&&&&&&&&&&&&&&&&&&&&&&&&&&
__END
```

Layer 1: FBK_SILICON_SUB This is a layer defining the silicon substrate. It does not necessarily have to be included in the layout. When designed in the layout, it defines the in-plane dimensions of the silicon block; otherwise it is automatically defined by L-Edit.

Command type: Wafer.
Thickness: 500 µm.

Layer 2: FBK_FIELD_OXIDE This is a layer defining the field oxide layer on top of the silicon substrate. It does not necessarily have to be included in the layout. When designed in the layout, it defines the in-plane dimensions of the field oxide; otherwise it is automatically defined by L-Edit. It has to be used in combination with FBK_SILICON_SUB.
Command type: Deposit.
Deposit type: Conformal.
Face: Top.
Thickness: 1 µm.

Layer 3: FBK_LOW_POLY_MASK This is a service layer defining the low-resistivity polysilicon over the whole design plane. It must not be used in the layout.
Command type: Deposit.
Deposit type: Fill.
Face: Top.
Thickness: 630 nm.

Layer 4: FBK_POLY_LOW_RES This is an active layer used in the actual layout to define the low-resistivity polysilicon.
GDS number: 1.
Command type: Etch.
Etch type: Dry.
Face: Top.
Depth: 3 µm.
Angle: 90°.
Undercut: 0.5.
Mask type: Outside.
Etch removes: FBK_LOW_POLY_MASK.

Layer 5: FBK_HIGH_POLY_MASK This is a service layer defining the high-resistivity polysilicon over the whole design plane. It must not be used in the layout.
Command type: Deposit.
Deposit type: Fill.
Face: Top.
Thickness: 630 nm.

Layer 6: FBK_POLY_HIGH_RES This is an active layer used in the actual layout to define the high-resistivity polysilicon.
GDS number: 9.
Command type: Etch.
Etch type: Dry.
Face: Top.
Depth: 3 µm.
Angle: 90°.
Undercut: 0.5.

Mask type: Outside.
Etch removes: FBK_HIGH_POLY_MASK.

Layer 7: FBK_TEOS This is a service layer defining the silicon oxide (tetraethylorthosilicate) over the whole design plane. It must not be used in the layout.
Command type: Deposit.
Deposit type: Fill.
Face: Top.
Thickness: 300 nm.

Layer 8: FBK_CONHO This is an active layer used in the actual layout to define the opening of the vias through the TEOS. It is necessary to ensure electrical interconnection to the underlying polysilicon.
GDS number: 2.
Command type: Etch.
Etch type: Dry.
Face: Top.
Depth: 3 µm.
Angle: 90°.
Undercut: 0.5.
Mask type: Inside.
Etch removes: FBK_TEOS.

Layer 9: FBK_TIN_MASK This is a service layer defining the multimetal over the whole design plane. It must not be used in the layout.
Command type: Deposit.
Deposit type: Fill.
Face: Top.
Thickness: 630 nm.

Layer 10: FBK_TIN This is an active layer used in the actual layout to define the multimetal conductive layer.
GDS number: 3.
Command type: Etch.
Etch type: Dry.
Face: Top.
Depth: 3 µm.
Angle: 90°.
Undercut: 0.5.
Mask type: Outside.
Etch removes: FBK_TIN.

Layer 11: FBK_LTO This is a service layer defining the silicon oxide (low-temperature oxide) over the whole design plane. It must not be used in the layout.
Command type: Deposit.
Deposit type: Fill.
Face: Top.
Thickness: 100 nm.

Layer 12: FBK_VIA This is an active layer used in the actual layout to define the opening of the vias through the low-temperature oxide to ensure the electrical interconnection to the underlying multimetal layer.
GDS number: 4.
Command type: Etch.
Etch type: Dry.
Face: Top.
Depth: 3 µm.
Angle: 90°.
Undercut: 0.5.
Mask type: Inside.
Etch removes: FBK_LTO.

Layer 13: FBK_FLOMET_MASK This is a service layer defining the floating metal (thin evaporated gold layer) over the whole design plane. It must not be used in the layout.
Command type: Deposit.
Deposit type: Fill.
Face: Top.
Thickness: 150 nm.

Layer 14: FBK_FLOMET This is an active layer used in the actual layout to define the floating metal layer. It is necessary to improve the ohmic contact with the underlying multimetal layer, in relation to the vias opened through the LTO.
GDS number: 8.
Command type: Etch.
Etch type: Dry.
Face: Top.
Depth: 3 µm.
Angle: 90°.
Undercut: 0.5.
Mask type: Outside.
Etch removes: FBK_FLOMET.

Layer 15: FBK_SPACER_MASK This is a service layer defining the sacrificial layer (photoresist defining the suspended RF-MEMS parts) over the whole design plane. It must not be used in the layout.
Command type: Deposit.
Deposit type: Fill.
Face: Top.
Thickness: 3 µm.

Layer 16: FBK_SPACER This is an active layer used in the actual layout to define the sacrificial layer. It is necessary to release the suspended movable parts of the RF-MEMS device.
GDS number: 5.
Command type: Etch.
Etch type: Dry.

Face: Top.
Depth: 6 µm.
Angle: 40°.
Undercut: 0.5.
Mask type: Outside.
Etch removes: FBK_SPACER_MASK.

Layer 17: FBK_BRIDGE_MASK This is a service layer defining the first electroplated gold layer (BRIDGE) over the whole design plane. It must not be used in the layout.
Command type: Deposit.
Deposit type: Snowfall.
Face: Top.
Thickness: 1.8 µm.

Layer 18: FBK_BRIDGE This is an active layer used in the actual layout to define the first electroplated gold thin film (BRIDGE), which is a structural layer defining the actual RF-MEMS movable parts, the anchors, the nonsuspended parts and the signal pads.
GDS number: 6.
Command type: Etch.
Etch type: Dry.
Face: Top.
Depth: 1.8 µm.
Angle: 90°.
Undercut: 0.5.
Mask type: Outside.
Etch removes: FBK_BRIDGE_MASK.

Layer 19: FBK_CPW_MASK This is a service layer defining the second electroplated gold layer (CPW) over the whole design plane. It must not be used in the layout.
Command type: Deposit.
Deposit type: Snowfall.
Face: Top.
Thickness: 3 µm.

Layer 20: FBK_CPW This is an active layer used in the actual layout to define the second electroplated gold layer (CPW), which is a structural thin film defining actual (parts of the) RF-MEMS movable membranes, the anchors, the nonsuspended parts and the signal pads.
GDS number: 7.
Command type: Etch.
Etch type: Dry.
Face: Top.
Depth: 3 µm.
Angle: 90°.
Undercut: 0.5.

Mask type: Outside.
Etch removes: FBK_CPW_MASK.

3.5 Conclusions

This chapter introduced the important aspects in the design of RF-MEMS devices and components. The topic was treated according to a rather practical approach. First, the concept of design rules and their link to the technology were introduced. Subsequently, the complete design phases of an RF-MEMS device employed as a case study, that is, an RF-MEMS series ohmic microrelay, were presented and discussed in detail. In the development of such a design, compliance with the set of design rules was stressed and useful tricks to make easier the work of the designer of RF-MEMS devices were also provided.

4
Simulation Techniques (Commercial Tools)

Abstract: This chapter focuses on the use of commercial simulation tools for the prediction of the behavior of radio frequency microelectromechanical system (RF-MEMS) devices and components. As stated in Chapter 1, RF-MEMS devices are characterized by a multiphysical behavior that involves the electrical, mechanical, and electromagnetic domains. For this reason, the use of different simulation tools is necessary for the prediction and observation of the whole behavior of RF-MEMS devices, and this is the approach followed in this chapter. Different behavioral aspects of a few devices belonging to the set of examples previously introduced in the book are analyzed by means of commercial simulation tools. Like in Chapter 3, the approach followed will be methodological, that is, not linked to the use of a specific software tool.

4.1
Introduction

This chapter reports on the coupled multiphysical simulations of radio frequency (RF) microelectromechanical system (MEMS) devices based on the use of commercial finite element method (FEM) software tools. The FEM approach to solve such problems is very effective, as it enables both the possibility of coupling different physical domains (such as the electrical and mechanical ones) and also the gathering of accurate results and predictions of the behavior of MEMS/RF-MEMS. On the other hand, FEM analysis might require significant computational resources for it to be performed properly, and this aspect has an impact on the simulation time. Moreover, because of this consideration, FEM tools enable the coupling of different physical domains, but do not allow the easy simulation, at once, of entire systems, for example, an entire RF transceiver, comprising the RF-MEMS passive part and the CMOS active circuitry. For these reasons, the FEM approach, as well as any other means or resource, has to be used within the context where it is possible to maximize the benefits it introduces, and should not be referred to as the best way to solve a certain problem in all possible cases. Throughout this book, the philosophy that is adopted is that of developing and acquiring a strategy to solve a certain problem related to RF-MEMS technology. Concerning the specific topic treated in

Practical Guide to RF-MEMS, First Edition. Jacopo Iannacci.
© 2013 WILEY-VCH Verlag GmbH & Co. KGaA. Published 2013 by WILEY-VCH Verlag GmbH & Co. KGaA.

this chapter, that is, the simulation, a good strategy is not based on excluding all possible approaches and favoring the one considered the best one, but rather on combining the benefits of different approaches. More specifically, in this book the two simulation approaches considered are the FEM one (the focus of this chapter) and compact modeling based on on-purpose developed tools (Chapter 5). The latter is used for the fast and fairly accurate simulation of RF-MEMS devices and networks when the influence of numerous degrees of freedom on the performance has to be assessed, aiming to identify trade-offs. The compact modeling approach is also suitable for development of a mixed-simulation domain, enabling the fast simulation of entire hybrid RF-MEMS/CMOS systems, such as a transceiver for telecommunications. On the other hand, the FEM-based approach is very important when a high level of accuracy is needed, for example, when the layout of a new RF-MEMS device has to be optimized for the fine-tuning of its characteristics and performance. Moreover, FEM tools also allow the inclusion of second-order physical effects which can affect the properties of a certain structure, and that are rather complicated to implement in compact models relying on specific analytical or semiempirical models.

4.2
Static Coupled Electromechanical Simulation of the RF-MEMS Ohmic Switch (Dev A) in ANSYS Multiphysics™

In this section, the use of an FEM-based tool is introduced by simulating the static coupled electromechanical behavior of the RF-MEMS series ohmic switch discussed in Chapter 2, and referred to as Dev A. The commercial software tool employed is ANSYS Multiphysics [153]. It is a powerful and flexible simulation framework that can be used according to two different approaches, namely, through the graphical interface, or through the ANSYS Parametric Design Language (APDL). The latter is a programming language that enables one to perform all the operations that can be conducted through the graphical user interface (GUI) of ANSYS Multiphysics. The APDL is the most convenient approach, as it easily enables the parameterization of the simulations, as well as their splitting into modules, which can be combined together and extended. Figure 4.1 shows the GUI of ANSYS Multiphysics. The central window is the design space, where the structure to be simulated is visualized. The subwindow on the left shows all the groups of commands necessary to set up the simulation (e.g., design phase, boundary conditions assignment, material properties).

In this book, we chose to use the APDL because of the flexibility it introduces, as already mentioned. However, regardless of the specific FEM tool adopted and regardless of the analysis to be performed (e.g., electromechanical, thermomechanical, electromagnetic), the operations to be performed to set up the simulation can always be grouped in the same way. First of all, the structure to be simulated has to be defined (i.e., designed) in the FEM tool, and can be 1D, 2D, or 3D depending on its geometry and also on the analysis to be performed. In this case, the structure

4.2 Static Coupled Electromechanical Simulation of the RF-MEMS Ohmic Switch ...

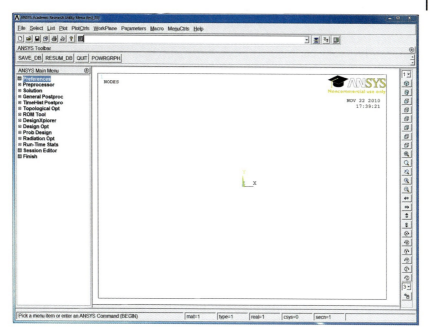

Figure 4.1 Graphical user interface of the ANSYS Multiphysics finite element method based simulation tool. The central window (in this case blank) displays the structure to be simulated and the results of the simulation, while the sub-window on the left contains all the operations to be used to define the proper simulation.

to be analyzed is the RF-MEMS ohmic switch (Dev A), which is a 3D structure. Moreover, the properties of the material/s from which the structure is made have to be defined depending on the type of analysis to be performed. For example, if the analysis is structural, the mechanical properties of the materials, such as Young's modulus and Poisson's ratio, have to be defined. If the analysis also includes thermoelectric effects, the electrical and thermal properties of the materials also have to be specified, such as the resistivity, the thermal expansion coefficient, and the heat capacity. Subsequently, the structure has to be properly meshed, that is, divided into subelements, in order to perform the simulation. The generation of the mesh is a delicate step, as it has a significant impact on the complexity of the simulation in terms of the required computational effort and simulation time, as well as on the accuracy of the results produced. For these reasons, the refinement of the mesh must be carefully considered. Moreover, the geometrical features of the structure to be simulated also influence the type of mesh chosen. Typical mesh elements are bricks (i.e., parallelepiped elements) and tetrahedrons. In the latter case, for example, the mesh density is determined by the minimum feature of a certain 3D layer to be meshed. This is because tetrahedrons grow with the same density along the X, Y, and Z axes. Consequently, for layers with one dimension much smaller than the other two (e.g., a very thin metal or oxide layer), the mesh complexity can be reduced by choosing brick rather than tetrahedron-based elements. After the mesh has been defined, all the proper boundary conditions have to be applied to the structure. Typical boundary conditions in the mechanical domain are constraints (e.g., the anchoring points of a suspended MEMS membrane) and loads (e.g., applied pressure and force). Finally, the type of analysis has to be defined, for example, mechanical, electromechanical, thermal, or electromagnetic, with all its features, such as the number of steps in which a load has to be applied. The final step consists in performing the simulation and visualizing the results once the computation phase has been successfully completed. In summary, the steps to be addressed in setting up a simulation in an FEM-based tool are as follows:

Step 1 Design of the geometry to be simulated.
Step 2 Definition of the properties of the materials.
Step 3 Meshing of the structure.
Step 4 Definition of the boundary conditions and loads.
Step 5 Definition of the simulation properties and characteristics.
Step 6 Simulation execution.
Step 7 Visualization of the simulation results.

In the example discussed in this section, the analysis consists in the observation of the static pull-in/pull-out characteristic of the RF-MEMS ohmic switch (Dev A). Consequently, the type of simulation is a coupled electromechanical one in which the applied voltage is swept from zero up to a certain level in order to detect the pull-in voltage (switch activation), and is then swept back to zero to detect the pull-out voltage (switch deactivation). In the following pages, the whole command files for the static pull-in/pull-out simulation of the RF-MEMS switch based on the APDL will be listed and divided into a few parts in accordance with the simulation steps previously reported. Each block of commands will be introduced and discussed in detail before being reported. Moreover, the code itself contains brief comments that always start with an exclamation mark (in APDL the execution of what follows such a symbol is ignored till the next carriage return).

4.2.1
Block 1: Definition of the Geometry and Properties of the Material

The list of commands (based on the APDL) for the definition of the geometry and properties of the material is reported below. Each APDL command has a certain number of fields and options that are to be specified. Such fields are always separated by commas following the command itself. The user guide helps one understand what the fields are for each specific command, and what their order is [153]. After a few commands devoted to the initialization of the simulator, to the definition of the Cartesian system in the window, and so on, the system of units is defined to be the uMKS system. It is a system of units available in ANSYS Multiphysics that is specific for microsystems, being the standard MKS system scaled down by six orders of magnitude (10^{-6}). This means that in specifying, for instance, the length of a structure as 100 µm, we do not have to type "100×10^{-6}"; instead we just have to type "100." Care must be taken in the proper conversion of all the magnitudes from the uMKS system to the MKS system, and checking the conversion tables, available in the ANSYS Multiphysics guidelines, helps avoid errors [153]. The element type is then specified in the code by issuing the ET (element type) command, it being SOLID186. ANSYS Multiphysics allows one to choose between a wide variety of 1D, 2D, and 3D elements, which have to be associated with the geometry. In other words, when the structure to be simulated is meshed, it will be *made of* the elements we chose. Consequently, the dimensions of the elements (1D, 2D, or 3D) depend on the size of the structure designed. Moreover, each element has certain properties. For example, there are elements describing just the mechanical behavior of the structure they are associated with. Other elements manage the thermomechanical coupling, while others deal with the magnetic behavior, and so on. SOLID186 is a 3D element that is chosen in this example to describe the mechanical properties (stiffness and deformation) of the gold membrane defining the RF-MEMS movable part of Dev A [153]. In our case, we do not need to define further elements to account for additional physical effects other than the mechanical properties. SOLID186 is then identified with the number 1. Subsequently, the properties of the material are defined by means of the MP command. The MP command has different fields that have to be specified. One of them is the element type the property refers to (in this case it is always element number 1). The code below reports the MP command several times. Every time it is issued, a different property of the material is defined. In our case, the material of interest is gold, so Young's modulus (option EX), Poisson's ratio (option PRXY), and the density (option DENS) are set. Other properties (electric and thermal) are also specified for demonstration purposes, despite their not being necessary for this specific simulation. The following block of commands defines a few parameters, such as air gaps, substrate thickness, and tolerances, that will be used and explained later. The definition of the parameters enables the parameterization of the simulations, as it will be necessary to change the value of one of them in order to modify the value in all the occurrences throughout the whole APDL command file. The final part of the block of commands under discussion concerns the generation of the 3D model. ANSYS Multiphysics features several commands for the definition of blocks, surfaces, and arched shapes, plus several Boolean operations to manipulate them (e.g., subtraction and merging). However, the philosophy pursued in this book is to integrate as much as possible the software tools adopted. For this reason, we generated a complete 3D model of the RF-MEMS ohmic switch (Dev A) in SAT format by using the 3D generation tool available in MEMS Pro, as described in Section 3.4 (Chapter 3). The SAT file of Dev A is then imported into ANSYS Multiphysics by issuing the SATIN command, which concludes the first block of commands. A 3D view of the switch imported into ANSYS Multiphysics is shown in Figure 4.2.

```
!!----------------------------------------!!
FINISH ! Exits normally from a processor
/CLEAR ! Clears the database
/PREP7 ! Enters the model creation preprocessor
/VIEW,1,1,1,1 ! Defines the viewing direction for the display
/VUP,1,Z ! Specifies the global Cartesian coordinate
  ! system reference orientation
/TITLE, Pull-In/Pull-Out ! Defines a main title
/COM, uMKS SYSTEM ! Places a comment in the output
/ESHAPE,1 ! Displays elements with shapes determined from
  ! the real constants or section definition
/PNUM,TYPE,1 ! Controls entity numbering/coloring on plots
/NUM,1 ! Specifies whether numbers, colors,
  ! or both are used for displays
/ICSCALE,1,250 ! Scales the icon size for elements
  ! supported in the circuit builder
/RGB,INDEX,100,100,100, 0 ! Specifies the RGB color values for
  ! indexes and contours
/RGB,INDEX, 80, 80, 80, 13
/RGB,INDEX, 60, 60, 60, 14
/RGB,INDEX, 0, 0, 0, 15
/PREP7
/UNITS,uMKS ! Annotates the database with
  ! the system of units used
ET,1,SOLID186 ! Element type selection

!**** GOLD ****! ! Material properties of Gold
MP,DENS,1,19300e-18 ! Density of Gold
MP,KXX, 1,315e+06 ! Electric conductivity
MP,RSVX,1,2.44e-14 ! Electric resistivity
MP,EX, 1,76e+3 ! Young's modulus
MP,PRXY,1,0.42 ! Poisson's ratio
MP,ALPX,1,14.2 ! Thermal expansion coefficient
MP,C,1,130.5e+12 ! Specific heat

!**** PARAMETERS ****! ! Definition of parameters
D1 = 2.0e-02 ! X coordinate of the substrate
D2 = 2.4e-02 ! Y coordinate of the substrate
TOL = 1e-4 ! Tolerance
GAPMIN = 1e-4 ! Minimum gap
TBLK = 300 ! Substrate thickness
VLT = 25 ! Max voltage
GAP1 = 3.53 ! Air gap 1
GAP2 = 3 ! Air gap 2

!**** MODEL ****!
! Transfers a .SAT file into the ANSYS program
~SATIN,'OHMIC_SW_SER_4_ANSYS_CELL','SAT',,SOLIDS,0,1
!!----------------------------------------!!
```

4.2 Static Coupled Electromechanical Simulation of the RF-MEMS Ohmic Switch ...

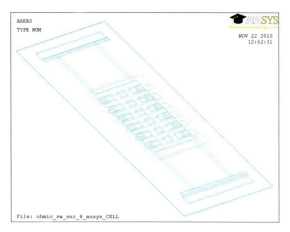

Figure 4.2 ANSYS Multiphysics 3D view of the switch (Dev A) as imported in the SAT format after being generated in MEMS Pro (see Section 3.4) starting from the 2D layout.

Since the only part of the RF-MEMS ohmic switch necessary for the static pull-in/pull-out simulation is the movable suspended membrane, all the other volumes are removed from the model in Figure 4.2. After the unnecessary volumes have been identified (operation performed via the GUI, and not shown here), their labels are selected with the VSEL command, and they are deleted with the VDELE command. The remaining volumes are then glued (command VGLUE) to avoid inconsistencies between them (e.g., their physical separation during the simulation). The updated structure after these additional operations is reported in Figure 4.3.

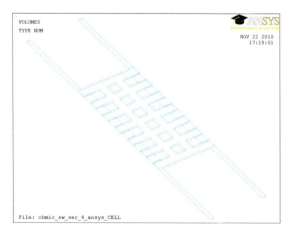

Figure 4.3 Three-dimensional schematic of a radio frequency (RF) microelectromechanical system (MEMS) switch (reported in Figure 4.2) after all the volumes not necessary for the simulation discussed in this section have been removed.

```
!!----------------------------------------!!
VSEL,U,,,30,35,5 ! Selects a subset of volumes
VDELE,ALL,,,1 ! Deletes unmeshed volumes
ALLSEL ! Selects all entities with a single command
VGLUE,ALL ! Generates new volumes by "gluing" volumes
!!----------------------------------------!!
```

4.2.2
Block 2: Meshing of the Structure

The subsequent block of APDL commands concerns the generation of the mesh. The structure of interest for the simulation (reported in Figure 4.3) is composed of two volumes, namely, volume 1, corresponding to the BRIDGE of the Fondazione Bruno Kessler (FBK) technology, and volume 2, corresponding to the CPW of the FBK technology, both being made of the same gold electroplated in two different lithography steps. The mesh is generated separately for the two volumes, as reported in the command lines below. The command MSHAPE specifies that the meshed elements have to be of 3D type and with tetrahedral shape. The SMRTSIZE command enables automatic meshing of the selected volume with a refinement defined by the number 10. The value of this parameter can range from 1 (fine mesh) to 10 (coarse mesh). Finally, the command VMESH executes the actual generation of the mesh. Figure 4.4 shows the entire meshed 3D structure (i.e., two volumes), while in Figure 4.5 only volume 1 (i.e., BRIDGE) is displayed. Despite the mesh refinement level is the same for both layers, the BRIDGE shows a finer mesh density compared to the CPW. This difference is due to the minimum geometrical feature of the two layers, already introduced in the previous pages. The thickness of the BRIDGE is 1.8 µm, and therefore its mesh based on tetrahedral elements is denser compared with that of the CPW, since the latter is characterized by a thickness of 3 µm.

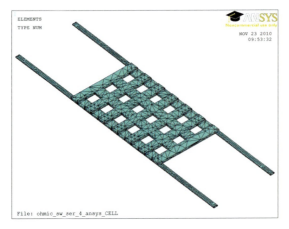

Figure 4.4 ANSYS Multiphysics 3D schematic of the meshed RF-MEMS ohmic switch to be simulated. The mesh density of the top layer (BRIDGE) is coarser than that of the bottom layer (CPW).

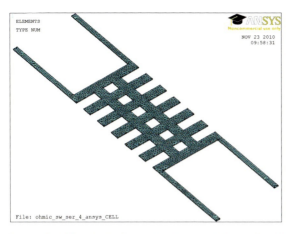

Figure 4.5 ANSYS Multiphysics 3D schematic of the meshed RF-MEMS ohmic switch where only the BRIDGE volume is displayed, highlighting the higher density of its mesh.

```
!!----------------------------------------!!
TYPE,1 ! Sets the element type attribute pointer
MAT,1 ! Sets the element material attribute pointer
ALLSEL

!**** MESH OF VOLUME 1 ****!
VSEL,S,,,1
MSHKEY,0 ! Specifies whether free meshing or mapped
 ! meshing should be used to mesh a model
MSHAPE,1,3D ! For elements that support multiple shapes,
 ! specifies the element shape
 ! to be used for meshing
SMRTSIZE,10 ! Specifies meshing parameters for automatic
 ! (smart) element sizing
VMESH,all ! Generates nodes and volume
 ! elements within volumes
ALLSEL

!**** MESH OF VOLUME 2 ****!
VSEL,S,,,2
MSHKEY,0
MSHAPE,1,3D
SMRTSIZE,10
VMESH,ALL
ALLSEL
!!----------------------------------------!!
```

4.2.3
Block 3: Generation of the Elements for the Electromechanical Coupling

The coupling between the mechanical and electrical domains is performed by means of the TRANS126 element, that is, a lumped parallel plate capacitor asso-

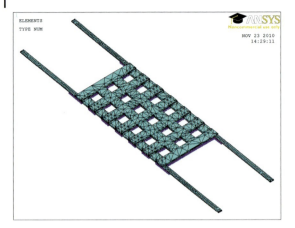

Figure 4.6 ANSYS Multiphysics schematic of the RF-MEMS ohmic switch under test, after the TRANS126 elements have been generated.

ciated with each node of the mesh structure, on the area facing the underlying controlling electrode. Each TRANS126 element applies a force on the mesh node with which is associated when a voltage drop is imposed between the two nodes of the small capacitor. First of all, the nodes of the mesh above the polysilicon biasing electrode are selected by ranges of coordinates through the command NSEL (node selection). Then, the selection is grouped into a component, identified with a label, and this operation is issued by the command CM. Finally, the generation of the TRANS126 elements, each of which is attached to a node of the selection, is performed by the command EMTGEN. The parameters required by the EMTGEN command include the name of the selection, the direction along which the TRANS126 elements have to be generated, and the gap between the two plates of the capacitors, corresponding to the air gap of the RF-MEMS switch. The TRANS126 elements also detect the contact between the suspended plate and the substrate when pull-in occurs, specified by the air gap. Figure 4.6 shows the structure to be simulated in ANSYS Multiphysics after the TRANS126 elements have been generated. They are visible underneath the meshed gold structure. In order to show more clearly the deployment of the TRANS126 elements, a different perspective of the structure showing its bottom surface is reported in Figure 4.7. The distribution of the small capacitors corresponds to the disposition of the polysilicon fingers underneath the suspended RF-MEMS switch, as reported in Figure 3.15b. Details of how the TRANS126 elements are displayed in ANSYS Multiphysics are reported in the close-up shown in Figure 4.8.

```
!!----------------------------------------!!
! Selects a subset of nodes
NSEL,S,LOC,Y,367.28-TOL,406.28+TOL
NSEL,A,LOC,Y,427.28-TOL,446.28+TOL
NSEL,A,LOC,Y,467.28-TOL,486.28+TOL
```

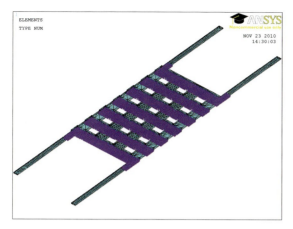

Figure 4.7 View of the bottom surface of the RF-MEMS switch in order to highlight the portions where the TRANS126 elements are deployed. The distribution corresponds to the polysilicon finger scheme, as reported in Figure 3.15b.

```
NSEL,A,LOC,Y,507.28-TOL,526.28+TOL
NSEL,A,LOC,Y,547.28-TOL,566.28+TOL
NSEL,A,LOC,Y,587.28-TOL,606.28+TOL
NSEL,A,LOC,Y,627.28-TOL,666.28+TOL
NSEL,R,LOC,Z,  6.56-TOL,  6.56+TOL

! Groups geometry items into a component
CM,ACTIVE_PLATE_AREA,NODE

! Generates a set of TRANS126 elements
EMTGEN,'ACTIVE_PLATE_AREA','ACTIVE_PLATE_AREA_ELEM','GND_ELECTRODE','UZ',-GAP1,0.53,1
ALLSEL
!!---------------------------------------!!
```

4.2.4
Block 4: Definition of the Mechanical Boundary Conditions

The next operation consists in defining the mechanical constraints of the suspended structure. The whole RF-MEMS switch geometry was simplified, as reported in Figures 4.2 and 4.3. Consequently, the mechanical constraints related to the anchoring parts of the suspended structure are to be applied to the four free ends of the beams. At first, all the mesh nodes belonging to the ends of the beams are selected. Subsequently, the mechanical constraints (i.e., zero displacement) along the X, Y, and Z axes are applied by issuing the D command. The symbols indicating the applied mechanical constraints are then visible at the ends of the beams in the schematic reported in Figure 4.9.

4 Simulation Techniques (Commercial Tools)

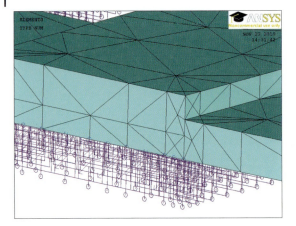

Figure 4.8 Close-up of the ANSYS Multiphysics schematic showing how the TRANS126 elements are displayed.

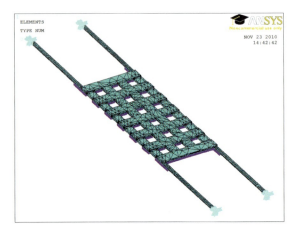

Figure 4.9 ANSYS Multiphysics schematic of the RF-MEMS ohmic switch after the mechanical constraints have been applied to the four free ends of the suspending beams.

```
!!----------------------------------------!!
NSEL,S,LOC,Y,222.38-TOL,222.38+TOL
NSEL,A,LOC,Y,811.18-TOL,811.18+TOL

! Defines degree-of-freedom constraints at nodes
D,ALL,UX,0,,,,UY,UZ
ALLSEL
!!----------------------------------------!!
```

4.2.5
Block 5: Definition of the Simulation

The next block concerns the definition of the simulation type and of all its features and characteristics. The following code is split into three subsections. In the first one, the general simulation settings are defined. The simulation type is chosen to be transient (ANTYPE command), despite the fact that the voltage will be swept slowly enough in order not to observe any dynamic (inertial) effect in the pull-in/pull-out characteristic. The simulation setting part is followed by a section in which the mechanical and electrical constraints for the element grouping the TRANS126 transducers (i.e., the actuation electrode) are defined. The initial applied voltage is zero.

```
!!----------------------------------------!!
/SOLU ! Enters the solution processor
ANTYPE,TRANS ! Specifies the analysis type and restart status
NLGEOM,ON ! Includes large-deflection effects in a static
 ! or full transient analysis
OUTRES,ALL,ALL ! Controls the solution data
 ! written to the database
RESCONTROL,,NONE ! Controls file writing for multi-frame restarts
TRNOPT,FULL ! Specifies transient analysis options
TIMINT,OFF ! Turns on transient effects
SSTIF,ON ! Activates stress stiffness effects
 ! in a nonlinear analysis
TIME,.0001 ! Sets the time for a load step
NSUBST,1 ! Specifies the number of sub-steps
 ! to be taken for this load step
KBC,1 ! Specifies stepped or ramped
 ! loading within a load step
ALLSEL
CNVTOL,U,1,0.005,L2 ! Sets convergence values for nonlinear analyses

!**** ELECTRIC BCs FOR THE ACTIVE_PLATE_AREA COMPONENT ****!
CMSEL,S,ACTIVE_PLATE_AREA,NODE
D,ALL,VOLT,0
ALLSEL

!**** ELECTRIC AND MECHANICAL BCs FOR THE GND_ELECTRODE COMPONENT ****!
CMSEL,S,GND_ELECTRODE,NODE
D,ALL,VOLT,0
D,ALL,UX,0,,,,UY,UZ
ALLSEL

LSWRITE ! Writes load and load step option data to a file
!!----------------------------------------!!
```

The following subblock of commands defines the voltage up-ramp of the sweep, the maximum bias being defined by the VLT parameter (see the first section of the code on previous pages), whose value is 25 V. This step of the simulation will enable the detection of the pull-in voltage.

```
!!----------------------------------------!!
TIMINT,ON
TIME,1
NEQUIT,200 ! Specifies the maximum number of equilibrium
 ! iterations for nonlinear analyses
NSUBST,60,120,60
AUTOTS,ON ! Specifies whether to use automatic
 ! time stepping or load stepping
KBC,0
NLGEOM,ON
OUTRES,ALL,ALL
CNVTOL,U,1,0.005,L2

!**** ELECTRIC BCs FOR THE ACTIVE_PLATE_AREA COMPONENT ****!
CMSEL,S,ACTIVE_PLATE_AREA,NODE
D,ALL,VOLT,VLT
ALLSEL

!**** ELECTRIC AND MECHANICAL BCs FOR THE GND_ELECTRODE COMPONENT ****!
CMSEL,S,GND_ELECTRODE,NODE
D,ALL,VOLT,0
D,ALL,UX,0,,,,UY,UZ
ALLSEL

LSWRITE
!!----------------------------------------!!
```

The last of the three subblocks concerns the down-ramp of the sweep. In this case, the electrical boundary condition applied to the TRANS126 transducers is 0 V. This means that, starting from the final condition of the previous simulation step, which was 25 V, the bias is decreased to 0 V in order to detect the pull-out voltage.

```
!!----------------------------------------!!
! Text
TIMINT,ON
TIME,2
NEQUIT,200
NSUBST,60,120,60
AUTOTS,ON
KBC,0
NLGEOM,ON
OUTRES,ALL,ALL
CNVTOL,U,1,0.005,L2

!**** ELECTRIC BCs FOR THE ACTIVE_PLATE_AREA COMPONENT ****!
CMSEL,S,ACTIVE_PLATE_AREA,NODE
D,ALL,VOLT,0
```

```
ALLSEL

!**** ELECTRIC AND MECHANICAL BCs FOR THE GND_ELECTRODE COMPONENT ****!
CMSEL,S,GND_ELECTRODE,NODE
D,ALL,VOLT,0
D,ALL,UX,0,,,,UY,UZ
ALLSEL

LSWRITE
SAVE ! Saves all current database information
!!---------------------------------------!!
```

4.2.6
Block 6: Simulation Execution

The three simulation steps just defined (i.e., initialization, bias up-ramp, and bias down-ramp) are now executed with the LSSOLVE command.

```
!!---------------------------------------!!
LSSOLVE,1,3,1 ! Reads and solves multiple load steps
FINISH
!!---------------------------------------!!
```

4.2.7
Block 7: Postprocessing and Visualization of Results

The very last group of commands are devoted to the visualization of the data produced by the completed simulation. In this case, a plot is visualized where the X axis reports the applied voltage (during the up-ramp and down-ramp) and the Y axis shows the vertical height of the suspended bridge, indeed identifying the static pull-in/pull-out characteristic reported in Figure 4.10. The pull-in occurs at 18 V, while the pull-out occurs at about 8.5 V. In conclusion, Figure 4.11 shows the deformed shape of the RF-MEMS switch when the pull-in occurs. The color scale refers to the extent of the vertical displacement.

```
!!---------------------------------------!!
/POST26 ! Enters the time-history results post-processor
KSEL,S,,,444 ! Selects a subset of keypoints or hard points
NSLK,S ! Selects those nodes associated
 ! with the selected keypoints
N_DETECT=NDNEXT(0) !
NSOL,2,N_DETECT,VOLT,,VOLTAGE ! Specifies nodal data to be stored
 ! from the results file
NSOL,3,N_DETECT,U,Z
ALLSEL
XVAR,2 ! Specifies the X variable to be displayed
/AXLAB,X,VOLTAGE ! Labels the X and Y axes on graph displays
/AXLAB,Y,DISPLACEMENT (UM)
```

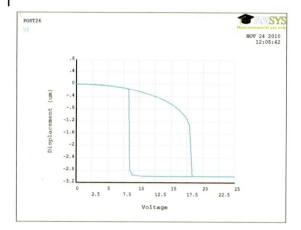

Figure 4.10 Pull-in/pull-out characteristic resulting from the ANSYS Multiphysics simulation of the RF-MEMS ohmic switch (Dev A) discussed here.

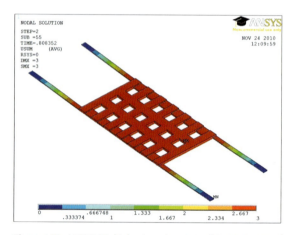

Figure 4.11 ANSYS Multiphysics schematic of the RF-MEMS ohmic switch deformed after the pull-in occurs. The color scale refers to the magnitude of the vertical deformation distributed on the movable membrane surface.

```
PRVAR,3 ! Lists variables vs. time (or frequency)
PLVAR,3 ! Displays up to ten variables
 ! in the form of a graph
FINISH
!!----------------------------------------!!
```

4.3
Modal Analysis of the RF-MEMS Capacitive Switch (Dev B2) in ANSYS Multiphysics

In this section, the modal analysis of RF-MEMS suspended structures is covered. The device chosen is the capacitive shunt switch referred to as Dev B2 in Chapter 2. The modal analysis allows one to determine the dynamic properties of a mechanical structure under vibrational excitation [154]. The modal analysis does not require any stimulus to be applied to the device analyzed, as the characteristic resonant frequencies (or modes) of a structure depend on the structure itself, and not on the particular excitation applied. Moreover, an excitation applied to a certain structure might, depending on its characteristics, excite a few modes of the structure rather than all of them. Differently, the modal analysis that will be performed highlights all the eigenmodes of the structure [155]. The only features to be defined, apart from the geometry and the material characteristics, are the mechanical constraints applied to the structure to be analyzed.

The list of commands (based on the APDL) for setting and performing the modal analysis of device Dev B2 in ANSYS Multiphysics is reported below. Most parts of the commands and operations are similar to those in the previous example (i.e., pull-in/pull-out characteristic of an RF-MEMS ohmic switch) and for this reason, they will not be commented on and described again in the following command lines.

```
!!----------------------------------------!!
FINISH
/CLEAR
/PREP7
/VIEW,1,1,1,1
/VUP,1,Z
/TITLE, Modal analysis of CVAR

/ESHAPE,1
/PNU,TYPE,1
/NUM,1

/ICS,1,250

/RGB,INDEX,100,100,100, 0
/RGB,INDEX, 80, 80, 80,13
/RGB,INDEX, 60, 60, 60,14
/RGB,INDEX,  0,  0,  0,15
/PREP7

ET, 1,MESH200, 7
ET, 2,SOLID226,111

!**** GOLD ****!
MP,DENS,1,19300e-18
MP,KXX, 1,315e+06
MP,RSVX,1,2.44e-14
```

```
MP,EX, 1,76e+3
MP,PRXY,1,0.42

!**** PARAMETERS ****!
D1 = 2
D2 = 3
ZERO_LEVEL = 30
TOL = 1e-3
GAP = 3

!**** MODEL ****!
~SATIN,'CVAR_3_MOD_SIMPLE_CELL','SAT',,SOLIDS,0,1

VDELE,1,2

ASEL,U,LOC,Z,ZERO_LEVEL-TOL,ZERO_LEVEL+TOL
ADELE,ALL,,,1
ALLSEL
NUMCMP,ALL

!**** MESH ****!
LESIZE,ALL,15
AMESH,ALL

ALLSEL
!!----------------------------------------!!
```

The result of the commands just listed is reported in Figure 4.12, where the meshed schematic of the RF-MEMS capacitive shunt switch is shown.

The following list of commands completes the modal analysis of the device studied in this section.

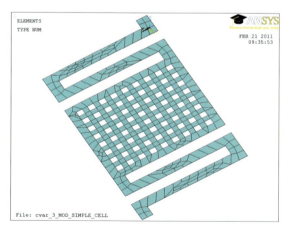

Figure 4.12 ANSYS Multiphysics schematic of the RF-MEMS capacitive shunt switch after the definition of the geometry and the mesh phase, and before the modal analysis is performed.

4.3 Modal Analysis of the RF-MEMS Capacitive Switch (Dev B2) in ANSYS Multiphysics

```
!!----------------------------------------!!
TYPE,2
ESIZE,,1
VEXT,1,,,,D1 ! Generates additional volumes by extruding areas
VEXT,2,,,,D2

!**** BOUNDARY CONDITIONS ****!
NSEL,S,LOC,Y, 0.8-TOL, 0.8+TOL
NSEL,A,loc,Y,379.8-TOL,379.8+TOL
D,ALL,UX,,,,UY,UZ

ALLSEL

!**** SIMULATION PHASE ****!
/SOLU
ANTYPE,MODAL ! Specifies the analysis type and restart status
 ! Performs a modal analysis.
 ! Valid for structural and fluid degrees of freedom.

MODOPT,UNSYMM,25 ! Specifies modal analysis options
 ! Unsymmetrical matrix. This option cannot be
 ! followed by a subsequent spectrum analysis.
 ! The number of modes to extract is specified.

MXPAND,,,,YES ! Specifies the number of modes to expand and
 ! write for a modal or buckling analysis

SOLVE

*GET,F1,MODE,1,FREQ ! Retrieves a value and stores it as a scalar
 ! parameter or part of an array parameter
*GET,F5,MODE,5,FREQ
*GET,F10,MODE,10,FREQ
*GET,F15,MODE,15,FREQ
*GET,F20,MODE,20,FREQ
*GET,F25,MODE,25,FREQ

!**** OUTPUT SECTION ****!
! This section defines which are the simulation results, in terms of modal
! frequencies, to be displayed in the Ansys schematic window
/ANN,DELE
/TLA,-0.35, 0.90,NATURAL FREQUENCIES OF CVAR
/TLA,-0.10, 0.85, 1ST MODE:
/TLA,-0.10, 0.80, 5TH MODE: %f5% Hz
/TLA,-0.105,0.75,10TH MODE:
/TLA,-0.105,0.70,15TH MODE: %f15% Hz
/TLA,-0.105,0.65,20TH MODE:
/TLA,-0.105,0.60,25TH MODE: %f25% Hz

FINISH

/POST1
SET,LIST ! Defines the data set to be
```

```
    ! read from the results file
    ! Scans the results file and lists a
    ! summary of each load step
FINISH
!!----------------------------------------!!
```

The frequencies for all 25 resonant modes predicted by the ANSYS Multiphysics modal simulation are listed in Table 4.1. Moreover, the ANSYS Multiphysics 3D schematics for the deformed structure corresponding to the first, second, and tenth resonant modes are reported in Figures 4.13–4.15, respectively.

4.4
Coupled Thermoelectromechanical Simulation of the RF-MEMS Ohmic Switch with Microheaters (Dev C) in ANSYS Multiphysics

The problem discussed in this section is more complex from the physical point of view with respect to the previously reported ones. This is because in the current problem three different physical domains are involved and coupled, namely, the

Table 4.1 Frequencies for all 25 resonant modes predicted by the ANSYS Multiphysics modal simulation described in the previous pages.

Mode 1	Mode 2	Mode 3	Mode 4	Mode 5
8.8 kHz	12.4 kHz	23.6 kHz	32.7 kHz	41.2 kHz

Mode 6	Mode 7	Mode 8	Mode 9	Mode 10
41.7 kHz	77.3 kHz	89.5 kHz	106.4 kHz	106.5 kHz

Mode 11	Mode 12	Mode 13	Mode 14	Mode 15
118.5 kHz	118.8 kHz	185.8 kHz	186.1 kHz	207 kHz

Mode 16	Mode 17	Mode 18	Mode 19	Mode 20
212.3 kHz	230.4 kHz	300.3 kHz	306.7 kHz	381 kHz

Mode 21	Mode 22	Mode 23	Mode 24	Mode 25
405.7 kHz	406.5 kHz	428 kHz	467.2 kHz	533.7 kHz

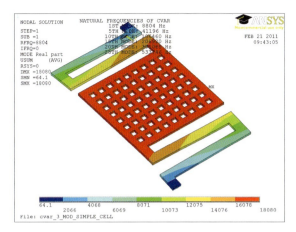

Figure 4.13 ANSYS Multiphysics 3D schematic of the deformed structure corresponding to the first resonant mode (8.8 kHz as reported in Table 4.1).

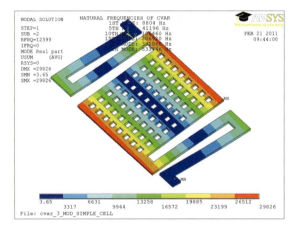

Figure 4.14 ANSYS Multiphysics 3D schematic of the deformed structure corresponding to the second resonant mode (12.4 kHz as reported in Table 4.1).

electrical, thermal, and mechanical ones. The device under test is the RF-MEMS switch previously mentioned in Chapter 2 and referred to as Dev C. It is a series ohmic switch based on a clamped–clamped suspended movable gold membrane. The peculiarity of this device is represented by the two polysilicon serpentines deployed underneath the anchoring areas. The structures just mentioned behave as microheaters to be activated in case the RF-MEMS switch remains stuck in the on position because of stiction [146]. Indeed stiction, that is, the missed release of the switch to the rest position when the DC bias is removed, can occur in RF-MEMS ohmic microrelays mainly for two reasons, namely, charge entrapment within the insulating layer and the formation of microwelding joints in relation to the input/output RF ohmic pads when a large signal flows through the device [156]. The

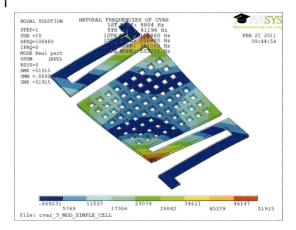

Figure 4.15 ANSYS Multiphysics 3D schematic of the deformed structure corresponding to the tenth resonant mode (106.5 kHz as reported in Table 4.1).

activation of the polysilicon microheaters with an electric current is expected to increase the temperature of the whole device. The consequent thermal expansion of gold will induce shear and restoring forces on the microwelding points, indeed easing the recovery phase of the malfunctioning microswitch [146]. The aim of the FEM analysis that is reported in this section is the observation of the sample temperature increase and the study of the thermally induced mechanical deformation when one of the two polysilicon serpentines is activated with a DC current.

The RF-MEMS ohmic switch with embedded microheaters discussed in this section was also characterized experimentally, and the tested sample is included in a piece of the silicon wafer significantly larger than the switch itself (silicon wafer piece with a size of a few centimeters). In order to have simulated results consistent with the measurements, the same shape and dimensions of the silicon piece are reproduced in the ANSYS Multiphysics model that is going to be described. Indeed, the increase in temperature induced by the activation of the heater strongly depends on how the heat diffuses into the substrate (i.e., heat conduction), and it is critical to model properly such a characteristic. As in the previous examples, the ANSYS Multiphysics command lines are split into a few blocks, corresponding to the completion of the main simulation phases, such as the model definition, mesh construction, applications of boundary conditions, and the final simulation execution and visualization of the results.

The definition of the 3D geometry and of the material properties is reported in the next set of commands, while the resulting ANSYS Multiphysics schematics after their execution are shown in Figures 4.16 and 4.17.

```
!!----------------------------------------!!
fini
/clear
/PREP7
/VIEW,1,1,1,1
```

4.4 Coupled Thermoelectromechanical Simulation of the RF-MEMS Ohmic Switch... | 107

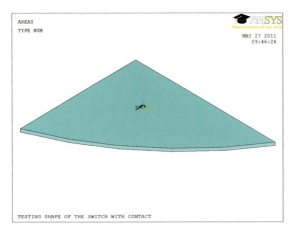

Figure 4.16 Schematic of the 3D model in ANSYS Multiphysics, comprising the silicon wafer piece onto which the RF-MEMS ohmic switch with microheaters is placed.

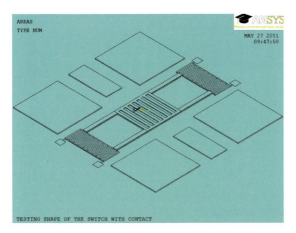

Figure 4.17 Close-up of the central part of the ANSYS Multiphysics 3D schematic in Figure 4.16, highlighting the structure of the RF-MEMS ohmic switch.

```
/VUP,1,Z
/TITLE, TESTING SHAPE OF THE SWITCH WITH CONTACT
/COM the thickness of the central plate is 2.2 micron

/ESHAPE,1
/pnu,type,1
/num,1

/ics,1,250

/RGB,INDEX,100,100,100, 0
/RGB,INDEX, 80, 80, 80,13
/RGB,INDEX, 60, 60, 60,14
/RGB,INDEX,  0,  0,  0,15
```

```
/PREP7

ET, 1,MESH200, 7
ET, 2,SOLID226,111

!* POLY
MP,DENS,1,2300
MP,KXX, 1,150 ! CONDUCTIVITY
MP,RSVX,1,113e-6 ! Electrical resistivity, Ohm-m
MP,EX, 1,175e+9 ! Young's modulus, Pa
MP,PRXY,1,0.36 ! Poisson's ratio
MP,ALPX,1,2.9e-6

!* SILI
MP,DENS,2,2330
MP,KXX, 2,153 ! CONDUCTIVITY
MP,RSVX,2,2.3e+3 ! Electrical resistivity, Ohm-m
MP,EX, 2,169e+9 ! Young's modulus, Pa
MP,PRXY,2,0.064 ! Poisson's ratio
MP, C,2,700
MP,ALPX,2,2.6e-6

!* SiO2
MP,DENS,3,2650
MP,KXX, 3,1.4 ! CONDUCTIVITY
MP,RSVX,3,10e+10 ! Electrical resistivity, Ohm-m
MP,EX, 3,66e+9 ! Young's modulus, Pa
MP,PRXY,3,0.17 ! Poisson's ratio
MP, C,3,705.5
MP,ALPX,3,0.55e-6

!* GOLD
MP,DENS,4,19300
MP,KXX, 4,315 ! CONDUCTIVITY
MP,RSVX,4,2.44e-8 ! Electrical resistivity, Ohm-m
MP,EX, 4,76e+9 ! Young's modulus, Pa
MP,PRXY,4,0.42 ! Poisson's ratio
MP,ALPX,4,14.2e-6
MP, C,4,130.5

!* PARAMETERS
d1 = 2.0e-02 ! X coordinate of substrate
d2 = 2.4e-02 ! Y coordinate of substrate
tol = 1e-8
gapmin = 1e-7
Tblk = 300
Vlt = 80

!* MODEL
!*serpentine
rectng, 180e-06, 205e-06, 437.5e-06, 462.5e-06
rectng, 180e-06, 205e-06, 607.5e-06, 632.5e-06
```

```
rectng, 205e-06, 285e-06, 447.5e-06, 452.5e-06
rectng, 205e-06, 285e-06, 617.5e-06, 622.5e-06

rectng, 215e-06, 285e-06, 457.5e-06, 462.5e-06
agen,16,5,,,,10e-06

rectng, 280e-06, 285e-06, 452.5e-06, 457.5e-06
agen,9,21,,,,20e-06

rectng, 215e-06, 220e-06, 462.5e-06, 467.5e-06
agen,8,30,,,,20e-06

aglue,all
numcmp,all

arsym,x,all,,,,0,0
asel,u,loc,x,180e-06-tol,285e-06+tol
agen,2,all,,,1000e-06,,,,1
allsel

numcmp,all

!* Contact area
rectng, 425e-06, 575e-06, 460e-06, 610e-06
rectng, 290e-06, 425e-06, 460e-06, 475e-06
rectng, 290e-06, 425e-06, 595e-06, 610e-06
rectng, 575e-06, 710e-06, 460e-06, 475e-06
rectng, 575e-06, 710e-06, 595e-06, 610e-06

!* The device boundary (sector)
k,1145,-2*d1/7,-2*d2/7
k,1146,-2*d1/7, 5*d2/7
k,1147, 5*d1/7,-2*d2/7
k,1148, , ,

larc,1146,1147,1148,50e-03
l,1145,1146
l,1145,1147

al,321,322,323

boptn,keep,no
aptn,all
aglue,all

numcmp,all

!* switch
rectng,215e-06, 285e-06,447.5e-06, 622.5e-06
rectng,715e-06, 785e-06,447.5e-06, 622.5e-06

!* squares
rectng,215e-06, 415e-06,215e-06, 415e-06
```

```
rectng,215e-06, 415e-06,655e-06, 855e-06
rectng,585e-06, 785e-06,215e-06, 415e-06
rectng,585e-06, 785e-06,655e-06, 855e-06

!* rf-lines
rectng,465e-06, 535e-06,215e-06, 385e-06
rectng,465e-06, 535e-06,685e-06, 855e-06

aptn,all
numcmp,all

!* serpentines element size
lsel,s,loc,y,437.5e-06-tol,632.5e-06+tol
lsel,u,loc,x,-tol,tol
lsel,u,loc,x,1000e-06-tol,1000e-06+tol
lsel,u,loc,x, 290e-06-tol, 710e-06+tol
lesize,all,5e-06
allsel

!* contact area element size
lsel,s,loc,x, 290e-06-tol, 710e-06+tol
lsel,r,loc,y, 460e-06-tol, 610e-06+tol
lesize,all,15e-06
allsel

! element size of the wafer piece edges
lsel,s,loc,x,-2*d1/7-tol,-2*d1/7+tol
lsel,a,loc,y,-2*d2/7-tol,-2*d2/7+tol
lesize,all,2e-03
allsel

lsel,s,,,303
lesize,all,2e-03

!* Gold squares element size
lsel,s,loc,y,215e-06-tol,215e-06+tol
lsel,a,loc,y,415e-06-tol,415e-06+tol
lsel,a,loc,y,655e-06-tol,655e-06+tol
lsel,a,loc,y,855e-06-tol,855e-06+tol

!*rf-lines element size
lsel,a,loc,y,385e-06-tol,385e-06+tol
lsel,a,loc,y,685e-06-tol,685e-06+tol

!* Gold squares element size
lsel,a,loc,x,215e-06-tol,215e-06+tol
lsel,a,loc,x,415e-06-tol,415e-06+tol
lsel,a,loc,x,585e-06-tol,585e-06+tol
lsel,a,loc,x,785e-06-tol,785e-06+tol

!*rf-lines element size
lsel,a,loc,x,465e-06-tol,465e-06+tol
lsel,a,loc,x,535e-06-tol,535e-06+tol
```

```
lesize,all,25e-06
allsel

type,1
mat,2
amesh,all

type,2
mat,2
esize,,3
vext,all,,,,,500e-06

asel,s,loc,z,500e-06-tol,500e-06+tol
esize,,1
mat,3
vext,all,,,,1e-06

asel,s,loc,z,501e-06-tol,501e-06+tol
vext,all,,,,0.63e-06

asel,s,loc,z,501.63e-06-tol,501.63e-06+tol
vext,all,,,,0.3e-06

asel,s,loc,z,501.93e-06-tol,501.93e-06+tol
asel,r,loc,y, 215e-06-tol, 855e-06+tol
asel,u,loc,x, -tol, 215e-06-tol
asel,u,loc,x, 785e-06+tol, 1000e-06-tol
asel,u,,,2054
asel,u,,,2034,2044,10
asel,u,,,2039,2049,10
mat,4
vext,all,,,,3e-06

vsel,s,,,249,318,1
vsel,a,,,324,325,1
vsel,a,,,360,361,1
vsel,a,,,368,371,1
eslv,s
mat,1
emodif,all
allsel

nummrg,node,tol/2

block, 425e-06, 575e-06, 460e-06, 610e-06, 504.93e-06, 507.13e-06
*do,i,0,6,1
block, 435e-06+i*20e-06,445e-06+i*20e-06,470e-06,600e-06, 504.93e-06, 507.13e-06
*enddo

vsbv, 607, 608,,dele,dele
vsbv, 615, 609,,dele,dele
vsbv, 607, 610,,dele,dele
```

```
vsbv, 608, 611,,dele,dele
vsbv, 607, 612,,dele,dele
vsbv, 608, 613,,dele,dele
vsbv, 607, 614,,dele,dele

numcmp,volu

k,2000,285e-06, 460e-06, 504.93e-06
k,2001,285e-06, 460e-06, 507.13e-06
k,2002,285e-06, 475e-06, 504.93e-06
k,2003,285e-06, 475e-06, 507.13e-06
k,2004,425e-06, 460e-06, 504.93e-06
k,2005,425e-06, 460e-06, 507.13e-06
k,2006,425e-06, 475e-06, 504.93e-06
k,2007,425e-06, 475e-06, 507.13e-06

v,2000,2001,2003,2002,2004,2005,2007,2006

k,2008,715e-06, 460e-06, 504.93e-06
k,2009,715e-06, 460e-06, 507.13e-06
k,2010,715e-06, 475e-06, 504.93e-06
k,2011,715e-06, 475e-06, 507.13e-06
k,2012,575e-06, 460e-06, 504.93e-06
k,2013,575e-06, 460e-06, 507.13e-06
k,2014,575e-06, 475e-06, 504.93e-06
k,2015,575e-06, 475e-06, 507.13e-06

v,2008,2009,2011,2010,2012,2013,2015,2014

k,2016,715e-06, 595e-06, 504.93e-06
k,2017,715e-06, 595e-06, 507.13e-06
k,2018,715e-06, 610e-06, 504.93e-06
k,2019,715e-06, 610e-06, 507.13e-06
k,2020,575e-06, 595e-06, 504.93e-06
k,2021,575e-06, 595e-06, 507.13e-06
k,2022,575e-06, 610e-06, 504.93e-06
k,2023,575e-06, 610e-06, 507.13e-06

v,2016,2017,2019,2018,2020,2021,2023,2022

k,2024,285e-06, 595e-06, 504.93e-06
k,2025,285e-06, 595e-06, 507.13e-06
k,2026,285e-06, 610e-06, 504.93e-06
k,2027,285e-06, 610e-06, 507.13e-06
k,2028,425e-06, 595e-06, 504.93e-06
k,2029,425e-06, 595e-06, 507.13e-06
k,2030,425e-06, 610e-06, 504.93e-06
k,2031,425e-06, 610e-06, 507.13e-06

v,2024,2025,2027,2026,2028,2029,2031,2030

vsel,s,,,607,611,1
vglue,all
```

4.4 Coupled Thermoelectromechanical Simulation of the RF-MEMS Ohmic Switch...

```
allsel

numcmp,all

lsel,s,,,3759,3761,2
lsel,a,,,3780,3781,1
lsel,a,,,3766,3768,2
lsel,a,,,3784,3785,1
lsel,a,,,3773,3775,2
lsel,a,,,3788,3789,1
lsel,a,,,3752,3754,2
lsel,a,,,3776,3777,1
lesize,all,15e-06
allsel
!!----------------------------------------!!
```

The subsequent batch of commands for meshing the 3D structure just defined in ANSYS Multiphysics is reported below. The results of the meshing phase are shown in Figures 4.18 and 4.19. In particular, Figure 4.18 shows the whole 3D structure (i.e., silicon wafer piece and RF-MEMS ohmic switch), highlighting the division of the 3D volume into elements. On the other hand, Figure 4.19 shows a close-up of the elements into which the RF-MEMS ohmic microrelay is divided. The mesh density is clearly finer (i.e., denser) in relation to the minimum feature sizes of the structure, such as the RF-MEMS switch, when compared with the whole 3D wafer piece and the anchoring area with the underlying polysilicon microheaters (not visible here).

```
!!----------------------------------------!!
type,1
mat,4
amesh,2648
amesh,2662
amesh,2678
amesh,2672

type,2
vsweep,607,2648,2679
vsweep,608,2662,2683
vsweep,609,2672,2687
vsweep,610,2678,2691

vsel,s,,,611
aslv,s
lsla,s
lesize,all,10e-06
type,1
amesh,2697
mat,4
vsweep,611,2697,2698

allsel
!!----------------------------------------!!
```

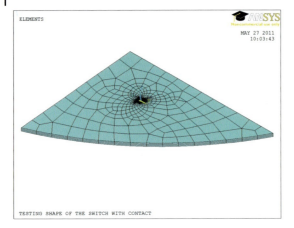

Figure 4.18 Schematic of the meshed 3D structure comprising the silicon wafer piece and the RF-MEMS ohmic switch with microheaters.

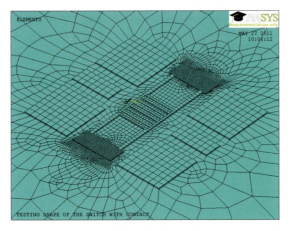

Figure 4.19 Close-up of the meshed 3D schematic corresponding to the central RF-MEMS ohmic switch with microheaters.

In the following batch of commands all the proper boundary conditions are applied to the 3D structure. The boundary conditions refer to the mechanical, thermal, and electrical physical domains involved. The mechanical boundary conditions refer to physical constraints applied to the bottom of the silicon wafer piece, in order not to allow any movement of the structure. On the other hand, the thermal constraints refer to the initial temperature condition (room temperature, 300 K) and to the thermal conduction and convection through the silicon. Finally, the electrical boundary conditions concern the imposition of a voltage drop (80 V) on the polysilicon microheater terminals, in order to drive a heating current through it. Figure 4.20 shows the thermal boundary conditions, while Figure 4.21 shows the mechanical constraints.

4.4 Coupled Thermoelectromechanical Simulation of the RF-MEMS Ohmic Switch ... | 115

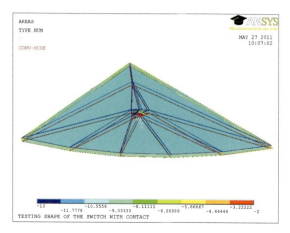

Figure 4.20 ANSYS Multiphysics 3D schematic of the structure highlighting the thermal boundary conditions applied.

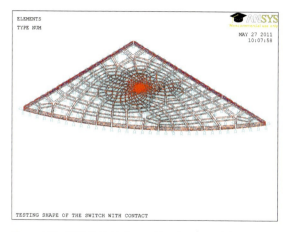

Figure 4.21 ANSYS Multiphysics 3D schematic of the structure highlighting the mechanical boundary conditions (i.e., constraints) applied.

```
!!----------------------------------------!!
vsel,s,,,607,611,1
aslv,u
asel,r,loc,z,504.93e-06-tol,504.93e-06+tol
mat,4
vext,all,,,,,2.2e-06
allsel

aclear,2648
aclear,2662
aclear,2678
aclear,2672

nummrg,node,9*tol
```

```
nummrg,kp,9*tol

allsel

!* LOADING
asel,s,loc,z,0-tol,0+tol
aclear,all
da,all,UX,0
da,all,UY,0
da,all,UZ,0
allsel

! Thermal Boundary Conditions
MPTEMP,1,300,500,700,900,1100,1300
MPTEMP,7,1500

allsel

!* Face up
asel,s,,,3101,3116,5
sfa,all,,RDSF,0.018,1

! AMBIENT TEMPERATURE
spctemp,1,Tblk

! STEFAN-BOLTZMANN RADIATION CONSTANT, J/(K)4(M)2(S)
stef,5.6704E-8
sfa,all,,conv,-4,Tblk
mpdata,hf,4,1,11.87,40.38,44.14,46.07,47.27,48.20
mpdata,hf,4,7,48.96

asel,s,loc,z,507.13e-06-tol,507.13e-06+tol
asel,r,loc,y,447.5e-06-tol,622.5e-06+tol
asel,a,,,3121,3126,5
asel,u,,,2681,2693,4
asel,u,,,2698
sfa,all,,RDSF,0.018,1
sfa,all,,conv,-6,Tblk
mpdata,hf,6,1,15.44,52.49,57.39,59.90,61.45,62.66
mpdata,hf,6,7,63.65

asel,s,,,2681,2693,4
sfa,all,,RDSF,0.018,1
sfa,all,,conv,-9,Tblk
mpdata,hf,9,1,22.69,77.15,84.36,88.04,90.32,92.10
mpdata,hf,9,7,93.55

asel,s,,,2698
sfa,all,,RDSF,0.018,1
sfa,all,,conv,-12,Tblk
mpdata,hf,12,1,25.11,85.38,93.35,97.43,99.96,101.9
mpdata,hf,12,7,103.5
```

```
asel,s,,,2189
sfa,all,,RDSF,0.7,1
sfa,all,,conv,-13,Tblk
mpdata,hf,13,1,4.465,15.18,16.60,17.33,17.77,18.13,
mpdata,hf,13,7,18.41

!* Face down
asel,s,loc,z,0+tol,0-tol
sfa,all,,RDSF,0.7,1
sfa,all,,conv,-2,Tblk
mpdata,hf,2,1,4.876,14.60,16.12,17.04,17.70,18.25
mpdata,hf,2,7,18.73

asel,s,,,2682,2694,4
sfa,all,,RDSF,0.018,1
sfa,all,,conv,-10,Tblk
mpdata,hf,10,1,31.36,96.5,106.3,112,116,119.3
mpdata,hf,10,7,122.2

asel,s,,,2697
sfa,all,,RDSF,0.018,1
sfa,all,,conv,-11,Tblk
mpdata,hf,11,1,35.12,108,119,125.4,129.9,133.6
mpdata,hf,11,7,136.8

!* Side walls
asel,s,loc,x,-2*d1/7-tol,-2*d1/7+tol
asel,a,loc,y,-2*d2/7-tol,-2*d2/7+tol
asel,a,,,639
sfa,all,,RDSF,0.7,1
sfa,all,,conv,-5,Tblk
mpdata,hf,5,1,45.17,75.29,85.99,95.3,103.8,112.1
mpdata,hf,5,7,120.4

asel,s,loc,y,460e-06-tol,475e-06+tol
asel,a,loc,y,595e-06-tol,610e-06+tol
asel,r,loc,x,285e-06-tol,285e-06+tol
asel,r,loc,z,504.93e-06-tol,506.73e-06+tol
cm,anchors1,area

asel,s,loc,y,460e-06-tol,475e-06+tol
asel,a,loc,y,595e-06-tol,610e-06+tol
asel,r,loc,x,715e-06-tol,715e-06+tol
asel,r,loc,z,504.93e-06-tol,506.73e-06+tol
cm,anchors2,area

allsel

asel,s,loc,x,465e-06-tol,465e-06+tol
asel,a,loc,x,535e-06-tol,535e-06+tol
asel,u,loc,y,415e-06+tol,655e-06-tol
asel,a,loc,x,215e-06-tol,215e-06+tol
asel,a,loc,x,285e-06-tol,285e-06+tol
```

```
asel,a,loc,x,415e-06-tol,415e-06+tol
asel,a,loc,x,585e-06-tol,585e-06+tol
asel,a,loc,x,715e-06-tol,715e-06+tol
asel,a,loc,x,785e-06-tol,785e-06+tol
asel,a,loc,y,215e-06-tol,215e-06+tol
asel,a,loc,y,385e-06-tol,385e-06+tol
asel,a,loc,y,415e-06-tol,415e-06+tol
asel,a,loc,y,447.5e-06-tol,447.5e-06+tol
asel,a,loc,y,622.5e-06-tol,622.5e-06+tol
asel,a,loc,y,655e-06-tol,655e-06+tol
asel,a,loc,y,685e-06-tol,685e-06+tol
asel,a,loc,y,855e-06-tol,855e-06+tol
asel,r,loc,z,501.93e-06-tol,507.13e-06+tol
cmsel,u,anchors1,area
cmsel,u,anchors2,area
sfa,all,,RDSF,0.018,1
sfa,all,,conv,-7,Tblk
mpdata,hf,7,1,3524,4424,5191,5954,6684,7424
mpdata,hf,7,7,8167

allsel

vsel,s,,,607,611,1
aslv,s
asel,r,loc,z,504.93e-06+tol, 507.13e-06-tol
asel,u,,,2680,2692,4
asel,u,,,2641
asel,u,,,2657
asel,u,,,2671
asel,u,,,2677
sfa,all,,RDSF,0.018,1
sfa,all,,conv,-8,Tblk
mpdata,hf,8,1,8298,1.0349E+4,1.2153E+4,1.3951E+4,1.5672E+4,1.742E+4
mpdata,hf,8,7,1.9174E+4

allsel
!!----------------------------------------!!
```

The final set of commands defines the actual simulation execution in ANSYS Multiphysics and is reported below. The subsequent figures report the results of the simulation after its completion. Figure 4.22 shows the temperature distribution after the simulation (2 s of heating). No temperature changes are visible across the silicon wafer piece as the heating is localized in a small region around the RF-MEMS switch, which is not evident in the view of the complete structure. Differently, Figures 4.23 and 4.24 show restricted areas of the silicon piece corresponding to the RF-MEMS switch and the anchoring area above the activated polysilicon serpentine, respectively. A temperature increase of about 13 K is predicted with respect to the surrounding room temperature condition. Finally, the vertical upward deformation of the central membrane, consequent to the gold membrane induced heating, is displayed in Figure 4.25, it being around 0.7 µm.

4.4 Coupled Thermoelectromechanical Simulation of the RF-MEMS Ohmic Switch ... | 119

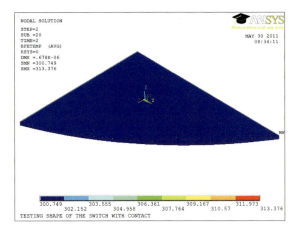

Figure 4.22 Temperature distribution across the whole 3D structure after the simulation is complete (2 s of heating). No changes are visible in the structure because the heating is localized around the RF-MEMS ohmic switch.

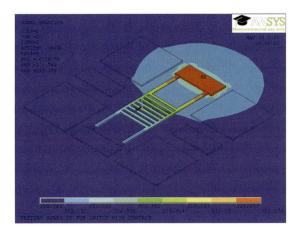

Figure 4.23 Close-up of the RF-MEMS switch, highlighting the temperature increase of about 13 K due to the activation of the polysilicon heater.

```
!!----------------------------------------!!
/solu
antype,trans
tref,300
tunif,300
nlgeom,on
outres,all,all
rescontrol,,none
trnopt,full            ! full transient dynamic analysis
TIMINT,OFF             ! Time integration effects off for static solution
SSTIF,ON
TIME,.0001             ! Small time interval
NSUBST,1               ! Two sub-steps
```

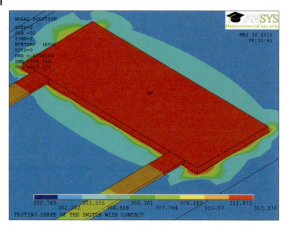

Figure 4.24 Close-up of the gold anchoring area above the activated polysilicon heater.

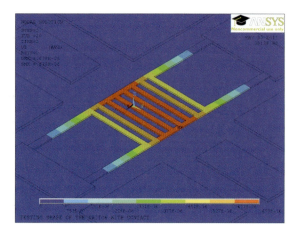

Figure 4.25 Visualization of the vertical upward deformation of the gold membrane due to the thermal expansion induced by the activation of the microheater.

```
KBC,1                        ! Stepped loads
allsel

LSWRITE

esel,s,mat,,1
nsle,s,all
nsel,r,loc,x,180e-06-tol,  205e-06+tol
nsel,r,loc,y,437.5e-06-tol, 462.5e-06+tol
cp,1,VOLT,all
n_gr_1=ndnext(0)
!d,all,UX,0,,,,UY,UZ
d,n_gr_1,VOLT,0

esel,s,mat,,1
```

```
nsle,s,all
nsel,r,loc,x,180e-06-tol, 205e-06+tol
nsel,r,loc,y,607.5e-06-tol, 632.5e-06+tol
cp,2,VOLT,all
n_vlt_1=ndnext(0)
!d,all,UX,0,,,,UY,UZ
d,n_vlt_1,VOLT,Vlt
nsel,all

allsel

nsel,all

allsel

TIMINT,ON ! Time-integration effects on for transient solution
time,2
nsubst,20,20,20
autots,on
KBC,1 ! Stepped loads
nlgeom,on
outres,all,all

LSWRITE ! Write 2nd load step
SAVE
LSSOLVE,1,2 ! Initiate multiple load step solution
FINISH
!!----------------------------------------!!
```

4.5
RF Simulation (S-parameters) of the RF-MEMS Variable Capacitor (Dev B1) in ANSYS HFSS™

The focus of this section is the simulation and prediction of the electromagnetic (i.e., RF) properties of MEMS devices through the use of a commercial FEM simulation tool. The software tool adopted is ANSYS HFSS (high-frequency structure simulator), previously developed by Ansoft Corporation, it being a 3D full-wave electromagnetic field simulator [157]. The RF-MEMS variable capacitor sample referred to as Dev B1 in Chapter 2 is employed here as a case study in order to briefly show and discuss the main steps to set the S-parameter simulation of an RF-MEMS structure in HFSS. Figure 4.26 shows the GUI of the HFSS environment, where the design space is integrated by lateral bars reporting the operations performed, such as the instantiation of blocks for the definition of the structure to be analyzed, or the selected materials.

HFSS features several functionalities to design a 2D or 3D structure to be simulated, such as the generation of geometrical shapes or 3D blocks and the execution of Boolean operations to merge or subtract objects. However, since in this case the 3D model of Dev B1 is available in SAT format, it is imported directly into the HF-

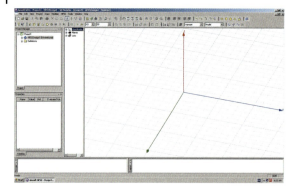

Figure 4.26 Graphical user interface of the ANSYS HFSS environment as it is when the program is first launched.

SS environment. Figure 4.27 shows the Dev B1 3D model after the execution of the import operation in HFSS, while Figure 4.28 reports a close-up of the RF-MEMS microdevice.

After the import of the 3D structure, the material properties have to be assigned to all the layers constituting the device shown in Figure 4.27. The HFSS simula-

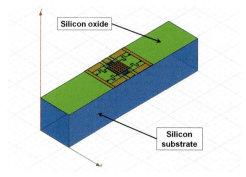

Figure 4.27 Three-dimensional schematic of Dev B1 imported into HFSS as a file in SAT format.

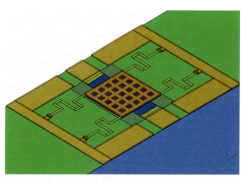

Figure 4.28 Close-up of the RF-MEMS variable capacitor microdevice in HFSS.

4.5 RF Simulation (S-parameters) of the RF-MEMS Variable Capacitor (Dev B1) in ANSYS HFSS™

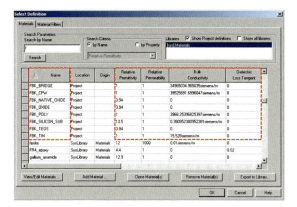

Figure 4.29 HFSS mask for the selection, definition, and modification of the materials to be assigned to the designed structure and of their properties.

tor offers a rather wide built-in library of materials (both conductive and insulating). However, the user is also allowed to include additional materials with custom-defined properties, and the latter is the option chosen regarding the example discussed in this section. Figure 4.29 shows the material properties mask of HFSS through which the materials can be selected, added, and modified.

The blocks marked with dashed lines report the names of the user-defined layers for the FBK surface micromachining technology [50], while on the right of the mask the inserted values regarding the dielectric constant (for insulating materials) and the conductivity (for conductors) are highlighted. More details concerning the names used to identify the layers and their properties can be found in Chapter 3.

To complete the design, a block of air is defined above the RF-MEMS device and the whole silicon piece on which it lies (see Figure 4.30). The air block is necessary in order to have a consistent distribution of the electromagnetic field above the MEMS structure. However, in order to avoid any issue related to the intersection of different blocks, a Boolean operation (i.e., subtraction) has to be performed

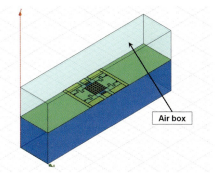

Figure 4.30 HFSS 3D schematic of the RF-MEMS variable capacitor under test after an air block has been defined above the whole silicon piece.

between the air block and all the other parts above the silicon. Details of such an operation are not given here for the sake of brevity.

Among the few different types of electromagnetic analyses HFSS can perform, the so-called `Driven Modal` option is chosen for this example as it features the definition of electromagnetic excitations, that is, sources of RF power to determine the S-parameter response of the analyzed RF-MEMS variable capacitor. For clarification purposes, another possible HFSS analysis option is `Eigenmode`, where the characteristic resonant frequencies of the analyzed structure are observed, and user-defined electromagnetic excitations cannot be applied.

The subsequent step is related to the definition of the electromagnetic excitations, for which, in this case study, there must be two as the RF-MEMS variable capacitor is a two-port device (with one input and one output termination). HFSS allows the definition of different types of excitation, depending on the structure to be analyzed. In this particular case, the type of excitation chosen is the so-called `Wave Port`, which is a 2D surface through which an RF signal is driven in or out of the structure [158]. The electromagnetic field is solved on the surface of the ports according to the Maxwell equations [59], and in the particular case of an RF-MEMS device configured as a coplanar waveguide, the ports have to be placed in such a way to bring the electromagnetic excitation on the cross section of the transmission line. Thinking of what is the real situation of a fabricated RF-MEMS device to be experimentally characterized helps understand the physical meaning of the excitation ports and the particular way they are placed in the structure. The on-wafer characterization of RF-MEMS devices is performed by placing the sample on a probe station and then connecting the device under test to a vector network analyzer by means of microprobes contacting the metal pads of the RF-MEMS (ground–signal–ground) in both input and output [159] in order to build the S-parameter matrix. The two excitations defined in the HFSS structure basically reproduce the microprobes meant for measurements as they represent the reference planes at which the S parameters are observed. Before the definition of the actual excitation, two surfaces have to be defined in the 3D model, corresponding to the input and output ports of the RF-MEMS device, as reported in Figure 4.31.

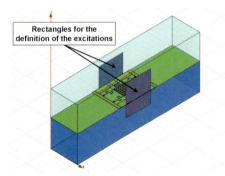

Figure 4.31 HFSS 3D model where two rectangles for the definition of the electromagnetic excitations have been added.

4.5 RF Simulation (S-parameters) of the RF-MEMS Variable Capacitor (Dev B1) in ANSYS HFSS™

The surfaces to define the Wave Port excitations have to be shaped according to certain criteria depending on several factors, such as the dimensions of the structure to be analyzed. As a general rule, the port surface should be large enough to capture all the significant contribution of the electromagnetic field, but not too large in order to avoid the inclusion of spurious (i.e., unwanted) transmission modes. Additional information on the guidelines for a proper definition of the excitations can be found in [160]. Once the two rectangles have been included in the structure, the next step is to assign to them the Wave Port excitation, which consists in choosing a proper vector representing the calibration line along which the electric field is patterned on the port. The calibration line is used in HFSS for all the calculations related to the electromagnetic properties of the structure, such as the characteristic impedance. Figure 4.32 shows the integration line defined for one of the two ports, where the vector connects the central RF conductor to one of the two ground planes, while Figure 4.33 shows a close-up of the same integration line.

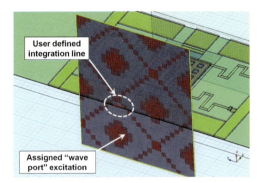

Figure 4.32 Visualization of one of the two surfaces meant for the definition of the port after the Wave Port excitation condition has been assigned to it. The integration line connecting the central RF line to the ground plane is visible.

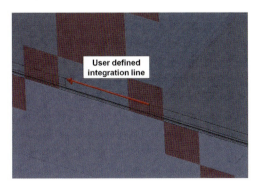

Figure 4.33 Close-up of the integration line defined for one of the two Wave Port excitations exciting the RF-MEMS variable capacitor under test.

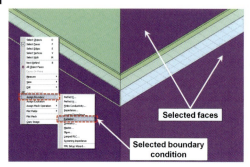

Figure 4.34 Assignment of the `Radiation` boundary condition to the selected vertical faces.

When a `Wave Port` excitation is assigned to a surface in HFSS, its four edges are automatically forced to RF ground. It is then clear that in a coplanar waveguide, like in this case, both vertical edges of the port surface have to cross the two RF ground planes of the transmission line.

The next step consists in the definition of proper boundary conditions for the faces of the whole block to be simulated. The assignment of proper boundary conditions is necessary in order to reproduce in the simulator a situation as close as possible to the real case. When testing real RF-MEMS devices on a probe station, we have an *infinite* space around the samples, while in the simulation we had to define a very limited volume around the device to be analyzed. It would be pointless to massively extend the silicon block around the RF-MEMS device (see Figure 4.30) and the air block around and surrounding the device to be analyzed as the computational load would dramatically increase, leading to unmanageable simulation times. It is more appropriate to define suitable boundary conditions that impose a continuity of the electromagnetic field on the physical boundaries of the 3D model defined in HFSS so that the solution is not performed within a small box, but on an infinite space around and above the RF-MEMS device, that is, a condition consistent with the real case. HFSS allows the definition of different types of boundary conditions, and the most appropriate to reproduce the condition just described is called `Radiation`. What the user has to do is to select all the external faces of the 3D model, and assign to them the `Radiation` condition, as reported in Figure 4.34.

The operations on the 3D structure are now complete, and the next steps refer to setting the actual simulation to be performed on the RF-MEMS variable capacitor structure. For this purpose, the `Solution Setup` window has to be opened, as reported in Figure 4.35.

The `Solution Setup` window provides options for the definition of the 3D structure's mesh characteristics. The most relevant ones are highlighted in Figure 4.35. First of all, the field `Solution Frequency` allows the user to define the frequency at which the initial mesh is generated and the fields are computed at the ports. Typically, in broadband analyses like in this case, it should correspond to the upper frequency limit at which the S parameters have to be simulated. In this example, the frequency range of interest is from 100 MHz to 15 GHz, so the field `Solution Frequency` is set to 15 GHz. The initial mesh is calculated according to iterative

4.5 RF Simulation (S-parameters) of the RF-MEMS Variable Capacitor (Dev B1) in ANSYS HFSS™

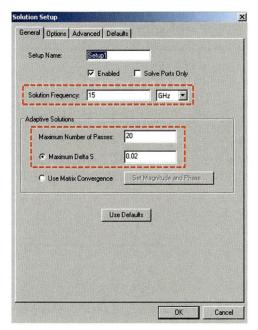

Figure 4.35 HFSS Solution Setup window for the definition of the mesh characteristics.

steps in order to refine the error of the calculated S parameters from one step to the next one. The field Maximum Number of Passes (see Figure 4.35) defines how many steps have to be performed in the refinement of the mesh (in this case 20 steps), while the field Maximum Delta S identifies the maximum error of the S parameters that has to be targeted (in this case 2% with respect to the previous step). If the error target is reached before 20 steps have been performed in the mesh recalculation, remaining steps are skipped and the actual simulation starts. On the other hand, if the error target is not reached within 20 steps, the simulation starts in any case with the initial mesh error computed at step number 20.

The last step before launching the actual computation is the definition of the simulation features, and Figure 4.36 shows the masks that the user has to access in order to modify such settings. HFSS offers three different ways to perform the simulation, namely, Discrete, Fast Discrete, and Interpolating. The last option is the one chosen for this example (see the highlighted rectangle in Figure 4.36), and it calculates the solution at given frequencies in the frequency range (that still has to be defined) and then interpolates the solution over the whole span. In the Frequency Setup mask (also highlighted in Figure 4.36), the type of frequency steps (linear steps in this case) is defined, as are the start and stop frequencies (100 MHz and 15 GHz, respectively) and the frequency step (100 MHz).

Now the actual solution is ready to be performed. The 3D geometry will be checked first, possibly producing errors or warnings if there are problems in the structure (e.g., overlapped volume), as well as if boundary conditions are missing, material properties are not assigned to one or more blocks, and so on. After the

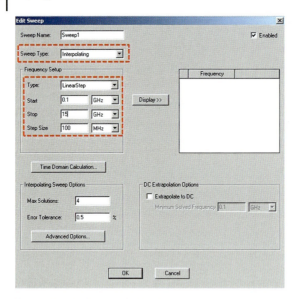

Figure 4.36 HFSS masks for the definition of the solution frequency and the frequency range to be simulated.

check, the initial mesh calculation is performed by HFSS following the criteria described above, and subsequently the simulation is performed. Depending on the complexity of the structure, the type of analysis, and the computational power of the computer, the simulation can take from several seconds to a few hours to be properly executed.

After the simulation is completed, many postprocessing operations can be performed in HFSS. In this case, we are interested in looking at the behavior of the S parameters over the analyzed frequency range. The data can be plotted in several ways, for example, on a rectangular plot as well as on a Smith chart, and using different magnitudes (e.g., decibels, magnitude and phase, and real and imaginary part). In this case we choose to display S11 (reflection parameter) and S21 (transmission parameter) in decibels on a rectangular plot. The sequence of commands reported in Figure 4.37 allows one to choose the rectangular plot visualization option.

If the option just described is selected, the window reported in Figure 4.38 appears, where Category, Quantity, and Function are specified, they being in this case S parameter, S11, and decibels, respectively.

The S11 characteristic simulated in HFSS is reported in Figure 4.39. If the same operation is repeated but S21 is selected and one chooses to create a new plot, the S21 simulated characteristic is similarly displayed as Figure 4.40 shows. As can be easily understood from the plots in Figures 4.39 and 4.40, the RF-MEMS shunt variable capacitor has been simulated in the rest position (MEMS actuator in the off state and switch in the closed configuration). In this case, the shunt capacitance to ground implemented by the device realizes the minimum value, and the RF

4.5 RF Simulation (S-parameters) of the RF-MEMS Variable Capacitor (Dev B1) in ANSYS HFSS™

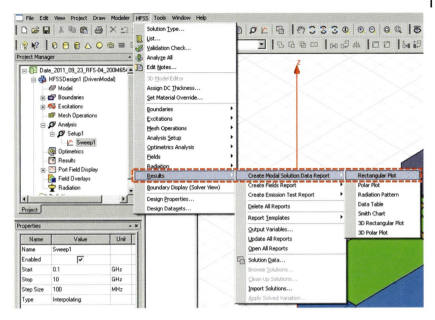

Figure 4.37 Sequence of commands in HFSS in order to display the simulated results on a rectangular plot.

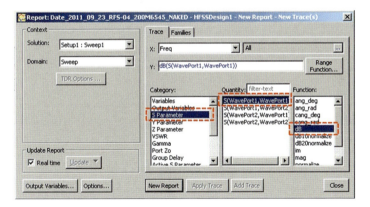

Figure 4.38 Window for the definition of Category, Quantity, and Function to be plotted (in this case S parameter, S11, and decibels, respectively).

signal in the input is passed to the output. In Chapter 7, it will be discussed how to simulate an RF-MEMS device in HFSS taking into account both the off state and the on state of the device.

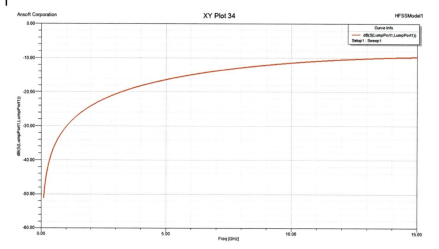

Figure 4.39 HFSS plot reporting the simulated S11 parameter (reflection) of the RF-MEMS variable capacitor discussed as a case study.

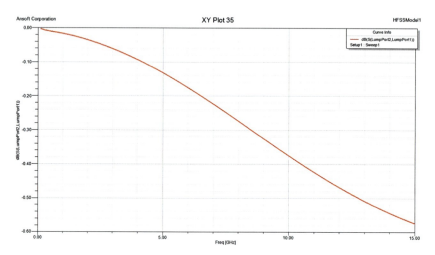

Figure 4.40 HFSS plot reporting the simulated S21 parameter (transmission) of the RF-MEMS variable capacitor discussed as a case study.

4.6
Conclusions

This chapter considered the prediction of the behavior of RF-MEMS devices obtained by means of the use of commercial simulation tools. Given the multiphysical characteristics of RF-MEMS devices, the use of different tools is proposed in order to address both the prediction of the mixed electromechanical behavior and the study of the electromagnetic (RF) characteristics of a certain device. The case studies shown in the chapter are referred to some of the RF-MEMS devices chosen earlier in the book as examples to be developed throughout this work. The approach followed to present how the software tools are used was as methodological as possible, in such a way to make the steps discussed applicable to the use of tools other than the ones chosen in this case. The content of this chapter is complementary to the information that will be provided in the following chapters concerning different simulation and prediction approaches also applicable to RF-MEMS devices.

Acknowledgments The author would like to thank Dr. Alena Repchankova (formerly with MEMS Research Unit at Fondazione Bruno Kessler, Italy) for her precious support in defining, implementing, and performing the mixed-domain simulations performed in ANSYS Multiphysics described throughout the chapter. The author would also like to thank Dr. Paola Farinelli (DIEI Department, University of Perugia, Italy and RF Microtech Srl, Perugia, Italy) for providing access to the HFSS electromagnetic simulation software tool used for the RF FEM analyses discussed in this chapter.

5
On-Purpose Simulation Tools

Abstract: This chapter focuses on the development of an on-purpose simulation tool for the prediction of the behavior of RF-MEMS devices based on the compact modeling approach. The trend based on the decomposition of the geometry of a radio frequency (RF) microelectromechanical system (MEMS) into simpler subblocks is discussed first. Subsequently, the mathematical models describing the mechanical and electrical characteristics of MEMS subelements, as well as the transduction between them, are presented in detail, stressing their suitability to be implemented in a programming language. Finally, the use of the models presented in a commercial circuit simulator is reported, with the validation of their prediction capabilities being performed by comparison against experimental measurements. The chapter is completed by reporting some examples of simulations of complete RF-MEMS devices performed with the compact models presented.

The main focus of this chapter concerning the description of the models concerns the MEMS rigid plate electrostatic transducer and the flexible straight beam. However, for those readers who are interested in significantly greater detail, there are two appendices at the end of the book. In particular, Appendix A contains details of the mechanical, electrostatic, and dynamic characteristics of the rigid plate model. On the other hand, Appendix B focuses on the same features regarding the flexible straight beam.

5.1
Introduction

Simulation of radio frequency (RF) microelectromechanical systems (MEMS) is a very wide field, and it can be faced having in mind different targets, depending on which one of the several approaches that are available is most suitable with respect to all the others. What is treated in this work deals with two different simulation approaches that are both diffused and practiced within the scientific community. On one hand, there is the approach based on finite element method (FEM) software tools, which was treated in detail in Chapter 4. On the other hand, a different approach is based on compact modeling based on commercial or in-house-developed software tools. With respect to the FEM approach, compact modeling addresses a lower level of accuracy, as it is based on simplified and approximate mathemat-

Practical Guide to RF-MEMS, First Edition. Jacopo Iannacci.
© 2013 WILEY-VCH Verlag GmbH & Co. KGaA. Published 2013 by WILEY-VCH Verlag GmbH & Co. KGaA.

ical models describing the physical behavior of certain devices and components. Moreover, the range of validity of such models is ensured only within well-defined condition ranges and under given hypotheses. However, a significant advantage of compact models is that they enable very fast analysis compared with FEM-based tools. Furthermore, as the computational load required by compact models is in general quite low, they enable the behavioral simulation of components and blocks with an elevated level of complexity, still maintaining reduced simulation times. In the following pages a MEMS compact model library developed by the author in a suitable programming language will be introduced. The library features elementary models that combined together at the schematic level enable the fast assembly and simulation of complete RF-MEMS devices and components. After the mathematical theories upon which the compact models are based have been introduced, the preliminary validation of the models will be discussed by comparison with FEM simulations. Finally, the cosimulation of RF-MEMS and CMOS standard components in the same schematic, enabled by the compact model library, will also be introduced, reporting a hybrid voltage-controlled oscillator (VCO).

5.2
MEMS Compact Model Library

The software library including MEMS elementary components was developed by means of the Verilog-A [101] programming language in the Cadence simulation environment. Two relevant electromechanical models are available in the library, namely, a suspended rigid plate, that is, an air gap, and a flexible straight beam realizing deformable suspensions. Their proper interconnection leads to the composition of complete MEMS devices, for example, a switch based on a central rigid plate and four suspending beams connected to its corners. The library also includes anchors for the definition of mechanical constraints and stimuli, such as force and displacement. Quantities belonging to different physical domains, that is, electrical and mechanical, are easily handled by the simulator since it treats them according to the Kirchhoff laws as *through* quantities, like electrical currents and mechanical forces, and *across* quantities, like voltages and mechanical displacements [161].

5.2.1
Suspended Rigid Plate Electromechanical Transducer

The rigid plate is connected to other elements through its four vertexes and all the forces and torques applied to these points are transferred to the center of mass. Figure 5.1 shows an example concerning the application of a force to the southeast (SE) node.

The initial nondisplaced plate position is visible in Figure 5.1 (plate not filled with color) and the coordinates of each vertex in the plate local reference system are reported. After the force F_{SE} has been applied, the plate rotates at an angle θ_Z around the vertical Z axis. With use of the rotation matrix [162], which defines

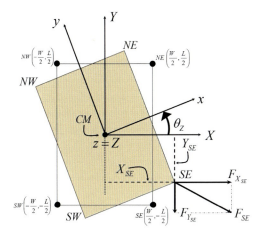

Figure 5.1 Schematic of the rigid plate. The coordinates of the four vertexes in the rest (initial) position, namely, northeast (NE), southeast (SE), southwest (SW), and northwest (NW), are reported. Moreover, the rotation of plate after a force has been applied to the SE node is shown.

the relations between a fixed and a movable reference system, the two arms X_{SE} and Y_{SE} are calculated and the torque around the Z axis in the center of mass is expressed as follows:

$$T_{\theta Z_{CM}} = -I_Z \frac{d^2 \theta_Z}{dt^2} + |Y_{SE}| F_{\Delta X_{SE}} - |X_{SE}| F_{\Delta Y_{SE}} \tag{5.1}$$

where I_Z is the moment of inertia of the plate with respect to the Z axis, while $F_{\Delta X_{SE}}$ and $F_{\Delta Y_{SE}}$ are the effective forces applied to the SE node along the X and Y axes, defined as:

$$F_{\Delta X_{SE}} = F_{X_{SE}} - F_{X_{CM}} \tag{5.2}$$

with

$$F_{X_{CM}} = -M \frac{d^2}{dt^2}(P_{X_{CM}}) \tag{5.3}$$

and

$$F_{\Delta Y_{SE}} = F_{Y_{SE}} - F_{Y_{CM}} \tag{5.4}$$

with

$$F_{Y_{CM}} = -M \frac{d^2}{dt^2}(P_{Y_{CM}}) \tag{5.5}$$

M being the mass of the plate, while $P_{X_{CM}}$ and $P_{Y_{CM}}$ are the coordinates of the center of mass in the XY plane [163]. Equations (5.3) and (5.5) express the inertial behavior of the center of mass of the plate. $F_{X_{CM}}$ and $F_{Y_{CM}}$ represent *through* variables associated with the nodes defined in the Verilog-A code, while $P_{X_{CM}}$ and $P_{Y_{CM}}$ are *across* variables.

On the basis of similar assumptions the model is extended to six degrees of freedom (DOFs), in which displacements and rotations along and around the X, Y, and Z axes are allowed. The mechanical model also accounts for the contact force when the plate collapses onto the underlying substrate. For this purpose, a fictitious elastic constant with a large value (e.g., 1×10^9 N/m) defines a vertical reactive force that assumes nonzero values when the plate touches the underlying electrode. The on/off states of the reactive force are determined by the instantaneous vertical displacement and are selected by proper flag variables. In particular, in the implementation of the plate with four DOFs (plate always parallel to the substrate) the reactive force due to the contact is applied directly to the center of mass. In contrast, in the extension to six DOFs the plate can assume any position in space and, consequently, it can be tilted with respect to the underlying surface. In this case, the contact force is selectively applied only to the vertex/vertexes (NE, SE, SW, NW) that is/are actually collapsing onto the substrate, resulting in a force and a torque applied to the center of mass. In fact, given the nondeformability of the plate, only three cases of contact are possible, namely, only one touching vertex, two adjacent touching vertexes, and all four vertexes touching.

The electrostatic model determines the transduction between the electrical and mechanical physical domains. It determines the capacitance and attractive force between the suspended plate and the lower electrode when a voltage is applied. The model is based on a generalization of the well-known formulas for the capacitance and electrostatic force of a parallel plate capacitor with lateral dimensions W and L:

$$C = \varepsilon_{AIR} \frac{WL}{g} \tag{5.6}$$

$$F = \frac{1}{2} \varepsilon_{AIR} V^2 \frac{WL}{g^2} \tag{5.7}$$

where ε_{AIR} is the permittivity of air, g is the gap between the two plates, and V is the applied voltage. In the rigid plate model with six DOFs, (5.6) and (5.7) are generalized to a closed double integral, taking into account that each point on the surface of the plate has a different distance from the substrate when the plate is not parallel to it [163]. This makes the electrostatic and mechanical six DOFs models consistent. Figure 5.2 shows schematically how, given a generic point P identified by the x_p, y_p, and z_p coordinates in the xyz local reference system, the absolute distance from the substrate can be calculated in the XYZ main reference system.

In more detail, such a distance is given by the vertical coordinate of the point P in the XYZ reference system $Z_{P_{INST}}$, expressed as a function of its local coordinates and of the rotation angles θ_X, θ_Y, and θ_Z, plus the initial air gap Z_{AIR} and the instantaneous vertical displacement of the center of mass of the plate $Z_{CM_{INST}}$. This leads to the following double integral expressions for the total capacitance and electrostatic attraction force, corresponding to (5.6) and (5.7) in the condition with

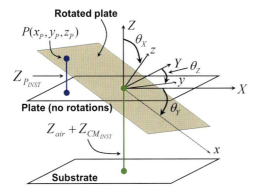

Figure 5.2 Schematic of the plate with six degrees of freedom (DOFs) in a configuration with all three axes of the xyz local reference system rotated with respect to the XYZ main reference system (angles θ_X, θ_Y, and θ_Z). The projection on the substrate of a generic point P lying on the surface of the plate is shown.

six DOFs:

$$C = \frac{\varepsilon_{\text{AIR}}}{\sigma_{\text{Arc}}} \int_{-\frac{W}{2}}^{\frac{W}{2}} \int_{-\frac{L}{2}}^{\frac{L}{2}} \frac{dx\,dy}{(Z_{\text{AIR}} + Z_{\text{CM}_{\text{INST}}} + Z_{P_{\text{INST}}}(x_p, y_p, z_p))} \tag{5.8}$$

$$F = \frac{1}{2} \frac{\varepsilon_{\text{AIR}} V^2}{\sigma_{\text{Arc}}^2} \int_{-\frac{W}{2}}^{\frac{W}{2}} \int_{-\frac{L}{2}}^{\frac{L}{2}} \frac{dx\,dy}{(Z_{\text{AIR}} + Z_{\text{CM}_{\text{INST}}} + Z_{P_{\text{INST}}}(x_p, y_p, z_p))^2} \tag{5.9}$$

The parameter σ_{Arc} accounts for the curvature of the electric field lines when the plate is not parallel to the substrate and is determined analytically [164]. It is also included in all the following formulas related to electrostatic effects throughout this book. The influence of the presence of an oxide layer deposited onto the underlying electrode on the capacitance and electrostatic force is also accounted for [163]. Its contribution is negligible when the plate is not actuated, but becomes significant when the suspended membrane pulls in onto the lower electrode. Furthermore, the plate electrostatic model also accounts for the presence of holes on the surface (required during the processing for the proper release of the suspended membranes). By use of the linearity of (5.6) and (5.7) with respect to the area, the contribution to the capacitance and force due to the holes is determined and is subtracted from (5.6) and (5.7) in the configuration with four DOFs and from (5.8) and (5.9) in the model with six DOFs, calculated in both cases for the plate without holes. Figure 5.3 shows a schematic of the plate with a matrix of m by n rectangular holes.

The expressions for the distances S_x and S_y between adjacent holes are

$$S_x = \frac{W - \Delta_{xL} - \Delta_{xR} - W_H m}{m - 1} \tag{5.10}$$

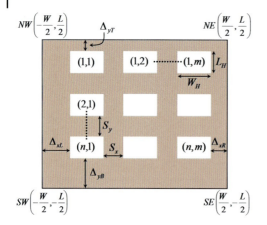

Figure 5.3 Schematic of the plate surface with a matrix of rectangular holes distributed on it. Holes are necessary in the fabrication process in order to remove the sacrificial layer defining the air gap. Holes are indexed and all the distances between adjacent holes as well as between the holes and the edges of the plate are reported.

$$S_y = \frac{L - \Delta_{yT} - \Delta_{yB} - L_H n}{n - 1} \tag{5.11}$$

where the geometrical parameters are defined in Figure 5.3. The capacitance and electrostatic force expressions then become

$$C = \frac{\varepsilon_{AIR}}{\sigma_{Arc}} \left[\int_{-\frac{W}{2}}^{\frac{W}{2}} \int_{-\frac{L}{2}}^{\frac{L}{2}} \frac{dx\,dy}{(Z_{AIR} + Z_{CM_{INST}} + Z_{P_{INST}}(x_p, y_p, z_p))} \right]$$

$$- \frac{\varepsilon_{AIR}}{\sigma_{Arc}} \left[\sum_{i=0}^{m-1} \sum_{j=0}^{n-1} \int_{a(i)}^{b(i)} \int_{c(j)}^{d(j)} \frac{dx\,dy}{(Z_{AIR} + Z_{CM_{INST}} + Z_{P_{INST}}(x_p, y_p, z_p))} \right]
\tag{5.12}$$

$$F = \frac{1}{2} \frac{\varepsilon_{AIR} V^2}{\sigma_{Arc}^2} \left[\int_{-\frac{W}{2}}^{\frac{W}{2}} \int_{-\frac{L}{2}}^{\frac{L}{2}} \frac{dx\,dy}{(Z_{AIR} + Z_{CM_{INST}} + Z_{P_{INST}}(x_p, y_p, z_p))^2} \right]$$

$$- \frac{1}{2} \frac{\varepsilon_{AIR} V^2}{\sigma_{Arc}^2} \left[\sum_{i=0}^{m-1} \sum_{j=0}^{n-1} \int_{a(i)}^{b(i)} \int_{c(j)}^{d(j)} \frac{dx\,dy}{(Z_{AIR} + Z_{CM_{INST}} + Z_{P_{INST}}(x_p, y_p, z_p))^2} \right]
\tag{5.13}$$

where the integration extrema $a(i)$, $b(i)$, $c(j)$, and $d(j)$ are expressed as

$$a(i) = -\frac{W}{2} + \Delta_{xL} + i(W_H + S_x) \tag{5.14}$$

$$b(i) = -\frac{W}{2} + \Delta_{xL} + i(W_H + S_x) + W_H \tag{5.15}$$

$$c(j) = \frac{L}{2} - \Delta_{yT} - L_H - j(L_H + S_y) \tag{5.16}$$

$$d(j) = \frac{L}{2} - \Delta_{yT} - j(L_H + S_y) \tag{5.17}$$

In order to complete the electrostatic model, an empirical description of the fringing field in the vicinity of the plate and hole edges is implemented in Verilog-A. Two contributions are accounted for: the first is due to distortion of the electric field lines in a small area near the plate and hole edges, while the second acts only when the plate is tilted, and accounts for a small portion of the vertical plate and hole areas facing the underlying electrodes in the tilted plate case. Both of the aforementioned fringing contributions are expressed by integral closed formulas, similar to the ones shown above for the main capacitance and electrostatic force, and their implementation in Verilog-A is straightforward [164]. The validation of the electrostatic model against FEM simulations, performed in COMSOL Multiphysics® for different dimensions of the plate and holes, as well as for different nonparallel positions of the plate with respect to the substrate, has been performed [163]. In all cases studied, the error of the Verilog-A model compared with FEM results is on the order of small percentage. Finally, the implementation of the viscous damping model completes the dynamic behavior of the rigid plate. The squeeze-film damping theory for vertical displacements is applied [165]. The effect of holes is accounted for by using the model developed by Bao et al. [166]. Such a model is based on the assumption that the fluid surrounding the moving object (e.g., air) is incompressible. On the other hand, a Couette-type flow model for the in-plane (along XY) displacements is implemented [167], assuming a vertical linear velocity profile for the fluid underneath the moving structure. The effect of the surface vicinity for both viscous damping contributions is also considered. Indeed, when the gap between the plate and the substrate is small enough to be comparable with the mean free path of air molecules, the assumption of incompressible fluid is no longer valid [23]. This effect is accounted for by replacing the constant value of the fluid viscosity μ with an expression depending on the vertical distance between the two surfaces. The expression proposed by Veijola et al. [168] for this effective viscosity value is adopted

$$\mu_{\text{eff}}(z) = \frac{\mu}{1 + 9.638 Kn^{1.159}} \tag{5.18}$$

In (5.18) Kn is the Knudsen number, defined as the ratio of the mean free path of gas molecules λ to the vertical distance of the plate from the substrate z:

$$Kn = \frac{\lambda}{z} \tag{5.19}$$

For those readers who are interested in studying the rigid plate model in a significantly deeper level of detail, see Appendix A at the end of the book.

5.2.2
Flexible Beam

The mechanical model for the flexible straight beam relies on the theory of structural mechanics based on matrices. The flexible beam is connected to other elements by means of two nodes placed at its ends (nodes A and B in Figure 5.4), characterized by six DOFs (i.e., 12 DOFs for the complete beam).

The relations between the forces (torques) applied at the two ends and the corresponding deformations (linear and angular) are defined by the 12×12 stiffness matrix (k) [169]. The implemented k matrix does not account for nonlinear effects because typically the beam undergoes small deformations (in the range of a few microns along the vertical axis for a beam length of at least 100 μm). These displacements are still within the linear stress–strain curve region of typical materials used for flexible beams (e.g., gold, copper, silver). With this approach, the inertial behavior is defined by the 12×12 mass matrix (m) and finally the viscous damping effect is described through the damping matrix (b) [170]. The viscous damping effect model relies on the squeeze-film damping theory for the vertical displacements and the Couette-type laminar flow for the in-plane displacements like in the rigid plate model described in the previous subsection. This leads to the following expression modeling the complete static and dynamic behavior of the flexible beam:

$$F = ku + b\frac{du}{dt} + m\frac{d^2u}{dt^2} \qquad (5.20)$$

where F is a 12×1 vector with all the forces and torques associated with each DOF of the beam, while u is a 12×1 vector with the displacements (linear and angular) for all 12 DOFs. The electrostatic model for the flexible beam is based on an approach similar to that implemented for the rigid plate. The beam length is split into n subsections, and for each of them the surface is approximated as

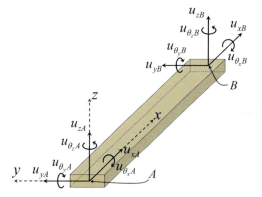

Figure 5.4 Schematic of a flexible beam with 12 DOFs. The model includes six DOFs per beam end (A and B): three linear displacements along the X, Y, and Z axes and three rotational DOFs around the same axes. Two local systems for ends A and B are necessary because, the beam being a flexible structure, their deformations can be different.

a small rigid plate with a certain orientation in space. The total capacitance and electrostatic force is the sum of the contributions given by each subelement. The force is then transferred to the two beam ends and weighted by means of Hermitian shape functions (HSFs) [169]. The expression for the total capacitance is

$$C_{AIR} = L \frac{\varepsilon_{AIR}}{\sigma_{Arc}} \sum_{i=0}^{n-1} \int_{\frac{i}{n}}^{\frac{i+1}{n}} \int_{-\frac{W}{2}}^{\frac{W}{2}} \frac{d\xi\, dy}{Z_P(\xi, y, \theta_{x_i}, \theta_{y_i})} \quad (5.21)$$

In (5.21) Z_P is the vertical coordinate of the generic point on the ith small plate, and depends on the ξ and y local coordinates as well as on the rotation angles θ_{x_i} and θ_{y_i} of the ith subplate. The ξ coordinate is normalized by the beam length L. The HSFs for the DOF of the vertical beam at ends A and B are

$$\psi_{zA}(\xi) = 1 - 3\xi^2 + 2\xi^3 \quad (5.22)$$

and

$$\psi_{zB}(\xi) = 3\xi^2 - 2\xi^3 \quad (5.23)$$

It is now possible to express the total electrostatic attraction force to be applied to beam ends A and B, respectively, as

$$F_{zA} = \frac{L}{2}\frac{\varepsilon_{AIR}}{\sigma_{Arc}^2} \sum_{i=0}^{n-1} \left\{ \psi_{zA}\left[\frac{1}{n}\left(i+\frac{1}{2}\right)\right] \int_{\frac{i}{n}}^{\frac{i+1}{n}} \int_{-\frac{W}{2}}^{\frac{W}{2}} \frac{V^2 d\xi\, dy}{Z_P(\xi, y, \theta_{x_i}, \theta_{y_i})^2} \right\} \quad (5.24)$$

$$F_{zB} = \frac{L}{2}\frac{\varepsilon_{AIR}}{\sigma_{Arc}^2} \sum_{i=0}^{n-1} \left\{ \psi_{zB}\left[\frac{1}{n}\left(i+\frac{1}{2}\right)\right] \int_{\frac{i}{n}}^{\frac{i+1}{n}} \int_{-\frac{W}{2}}^{\frac{W}{2}} \frac{V^2 d\xi\, dy}{Z_P(\xi, y, \theta_{x_i}, \theta_{y_i})^2} \right\} \quad (5.25)$$

The accuracy in predicting the deformations is verified by comparing the results against FEM simulations in COMSOL Multiphysics, in both the static and the dynamic regime [163]. Concerning the latter case, the results of the eigenfrequency calculation in COMSOL Multiphysics are compared against the AC small signal analysis in Spectre® for a cantilever structure, that is, a suspended beam with one end anchored and the other end free. Figure 5.5a shows the COMSOL Multiphysics schematic of the analyzed cantilever with a typical deformed shape associated with a resonant mode.

Figure 5.5b shows the Spectre schematic for the same structure. It is composed of the straight beam with an anchor connected to its left end and a stimulus source (force) at the right end. In this case, only the resonant modes on the XY plane are accounted for and consequently the force source in the Spectre schematic is set to provide a stimulus with equal components along the X and Y axes and no vertical component.

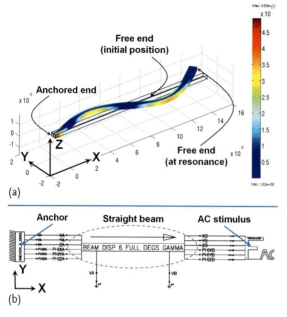

Figure 5.5 (a) COMSOL Multiphysics schematic of the cantilever with the typical deformed shape related to the second-higher mode on the X Y plane. The color scale represents the von Mises stress distribution. (b) Spectre schematic of the same cantilever implemented with the compact models discussed in the chapter.

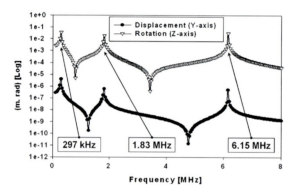

Figure 5.6 Semilogarithmic plot of the displacement of the free end of the beam along the Y axis and the rotation around the Z axis versus frequency (Spectre simulation). The peaks corresponding to the resonant frequencies are visible. The simulation is performed by using just one beam element, as reported in Figure 5.5b.

The plot in Figure 5.6 reports on a semilogarithmic scale the Y displacement and the rotation around the vertical axis of the free end of the beam produced by the Spectre AC simulation up to 8 MHz. The frequencies are visible and highlighted in the graph. In this case the Spectre schematic features a single beam element (see Figure 5.5b).

Table 5.1 Comparison of the first three resonant mode frequencies of the cantilever on the XY and XZ planes simulated in Spectre and COMSOL Multiphysics. The compact model predictions are in good agreement with the finite element method simulations for all the cases analyzed.

Mode	Spectre	COMSOL Multiphysics	Error
Fundamental (XY)	289.1 kHz	289.7 kHz	+0.20%
First higher (XY)	1.821 MHz	1.808 MHz	+0.72%
Second higher (XY)	5.117 MHz	5.027 MHz	+1.79%
Fundamental (XZ)	173.5 kHz	173.9 kHz	−0.22%
First higher (XZ)	1.092 MHz	1.088 MHz	+0.37%
Second higher (XZ)	3.069 MHz	3.039 MHz	+0.10%

However, it has been noted that such a schematic configuration in Spectre is not able to accurately predict the resonant frequencies of the modes above the fundamental mode. This is because in the beam model the displaced shape of the structure is based on the HSFs, which are not suitable to reproduce the deformed profile of the beam corresponding to resonant modes higher than the fundamental mode [169]. This issue can be effectively overcome by splitting the whole beam structure in the schematic into multiple shorter beams connected in series. This solution, discussed in detail in [163], introduces additional DOFs along the beam length, indeed enabling one to reproduce the typical deformed shape of the structure in relation to resonant modes of higher order than the fundamental mode. As an empirical rule, it can be stated that, starting from a beam of length L, and splitting it into n beams of length L/n, the Spectre simulation can accurately predict the resonant frequencies up to the nth mode, where $n = 1$ represents the fundamental mode of the structure. However, this approach has some limitations, as the mechanical model describing the elastic behavior of the beam is valid as long as one dimension of the structure (typically the length) is significantly larger than the other two dimensions (i.e., the section). Consequently, when the length of each elemental subbeam decreases (because n is increasing), the mechanical model is less and less accurate in predicting the elastic behavior of the structure.

In the case discussed in this section (see Figure 5.5), the beam was divided into three subbeams in the Spectre schematic in order to predict up to the higher second-higher resonant mode. Table 5.1 reports the mode frequencies simulated with COMSOL Multiphysics and Spectre up to 8 MHz, confirming the good accuracy achieved by the compact models in predicting the dynamic behavior of the analyzed cantilever also for the XZ modes. In all the cases analyzed, the Spectre error is less than 2%.

Such results also confirm the effectiveness of the technique described above in order to achieve better accuracy in the prediction of resonant frequencies higher than the fundamental one. For this purpose, the resonant frequencies predicted in Spectre with the schematic in Figure 5.5b (and depicted in the plot in Figure 5.6)

Table 5.2 Comparison of the first three resonant mode frequencies of the cantilever on the XY plane simulated in Spectre with a single beam schematic (of length L), and with a structure fragmented into three subelements (of length $L/3$). The error of the prediction of the frequencies obtained with the first model with respect to the latter one is reported in the rightmost column.

Mode	One beam of length L	Three beams of length $L/3$	Error
Fundamental (XY)	297.0 kHz	289.1 kHz	+2.73%
First higher (XY)	1.830 MHz	1.821 MHz	+0.05%
Second higher (XY)	6.150 MHz	5.117 MHz	+20.2%

and the ones computed for a beam split into three subelements (each one-third of the length of the whole beam) are summarized in Table 5.2.

The error of the prediction of the frequencies obtained with the single beam model with respect to the model featuring three subbeams connected in series is reported in Table 5.2, and is particularly evident for the second-higher resonant frequency observed.

For those readers who are interested in studying the flexible straight beam model in a significantly deeper level of detail, see Appendix B at the end of the book.

5.2.3
Simulation Validation of a MEMS Toggle Switch

The MEMS toggle switch is a particular implementation of a microrelay with a more complex actuation mechanism compared with the traditional one based on a single suspended rigid plate [171]. A schematic top view and a schematic cross section of a MEMS toggle switch are reported in Figures 5.7 and 5.8, respectively.

The toggle switch is based on a central suspended rigid plate which implements the switch or varactor function. This is connected by means of flexible beams to two

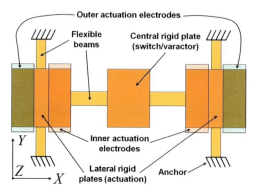

Figure 5.7 Schematic top view of a microelectromechanical system (MEMS) toggle switch. The central switch is connected to two lateral actuation structures.

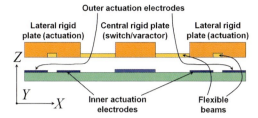

Figure 5.8 Schematic side view of the cross section of a MEMS toggle switch. The lateral actuation structures are kept suspended by two short flexible beams.

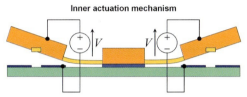

Figure 5.9 Schematic of the inner actuation mechanism of the MEMS toggle switch.

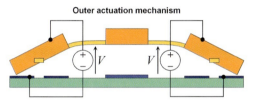

Figure 5.10 Schematic of the outer actuation mechanism of the MEMS toggle switch.

lateral actuation rigid plates. The latter are kept suspended by two lateral short flexible beams anchored at the other ends. A pair of electrodes is deployed under each actuation rigid plate, forming a couple of inner and outer controlling electrodes (see Figure 5.8).

When a biasing voltage V is applied between the inner electrodes and the two actuation plates, the attraction force acts on their area closer to the central switch/varactor. This causes the rotation of the two controlling plates as shown in Figure 5.9, and the consequent lowering of the central plate toward the underlying electrode.

On the other hand, when the voltage is applied to the outer electrodes, the actuation structures rotate around the Y axis in the other direction, as Figure 5.10 shows. This time the central plate moves upward, increasing the gap with respect to the initial position. When the toggle structure is used as a varactor, the last configuration allows one to widen the capacitance range with respect to the standard implementation based on a single suspended rigid plate with springs connected to its four corners [172]. The suspending beams must be short enough in order to have a small deformation along the Z axis with respect to the torque around the Y axis [173].

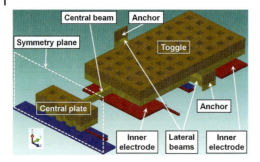

Figure 5.11 Schematic view of the toggle switch implemented in CoventorWare. The symmetry condition is used so that just half of the structure is implemented.

A toggle structure is implemented in CoventorWare®, and its 3D FEM model is reported in Figure 5.11. Due to the symmetry of the structure, just one half of it structure is simulated in order to reduce both the computational complexity and the simulation time. Symmetry boundary conditions for both the mechanical and the electrostatic simulations are applied in CoventorWare to the plane shown in Figure 5.11. The Cadence Virtuoso® schematic of the half-toggle structure implemented in Spectre, with the compact models discussed in this chapter, is shown in Figure 5.12. Mechanical constraints due to the symmetry condition are applied to the central plate by means of anchors. Due to the symmetry, the central plate can only move in the vertical direction (Z axis), without any in-plane displacement (in the XY plane) and without any rotation around the three axes (i.e., the central plate is always parallel to the substrate). The dimensions of the toggle structure along the X, Y, and Z axes are 220, 420, and 4.8 µm, respectively. It has five holes (20×20 µm^2) along the X axis and 10 holes along the Y axis. The lateral suspending beams are 60 µm long, while their width and thickness are 20 and 1.8 µm, respectively. The beam connecting the toggle to the central signal plate is 150 µm long. Its width is 10 µm and the thickness is 1.8 µm. The central signal plate's X, Y, and Z dimensions are 180, 140, and 4.8 µm, respectively (4×3 holes of area 20×20 µm^2). Finally, the length of the inner and outer electrodes is 100 µm along the Y axis, while they are as wide as the toggle structure (i.e., 220 µm) and the vertical air gap is 3.23 µm. The suspended structure is made of gold and the design of the toggle switch is based on Fondazione Bruno Kessler technology [174]).

Static DC simulations of the structure shown in Figure 5.12 were performed in Spectre and compared with the results produced by CoventorWare. First of all, a biasing voltage was applied to the inner electrode (see Figure 5.9) to detect the pull-in of the toggle switch. The biasing voltage was swept from 0 to 12 V in steps of 100 mV in Spectre. The comparison of the static pull-in characteristic simulated with the compact models and the CoventorWare prediction is shown in Figure 5.13.

The vertical displacement is referred to the edge of the toggle switch connected to the straight beam constrained to the central plate (see Figure 5.11). Spectre predicts the pull-in at 10.6 V, which is in very good accordance with the CoventorWare pull-in voltage of 10.7 V (corresponding to an error of −0.9% with respect

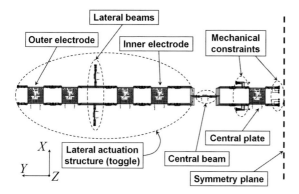

Figure 5.12 Cadence Virtuoso schematic of the toggle switch. By use of the symmetry of the structure, just half of the toggle switch is implemented. The Cadence Virtuoso schematic corresponds to the structure in Figure 5.11 (refer to the indicated reference systems to understand the analogy between the CoventorWare and the Cadence Virtuoso schematics).

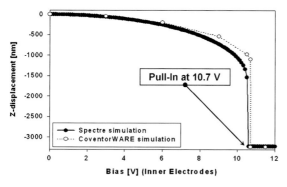

Figure 5.13 Pull-in characteristic of the toggle switch predicted in Spectre, compared with the CoventorWare simulation.

to CoventorWare). The simulation time saving introduced by compact models in Spectre with respect to FEM-based software is considerable. The simulation just mentioned took 30 s to complete with the compact models in Verilog-A. On the other hand, CoventorWare determined the pull-in in 10 steps, and the whole simulation needed about 30 min to be accomplished. The simulation platform on which the Spectre simulations were run was a Spark workstation with one 700 MHz CPU and 1 GB RAM. On the other hand, CoventorWare ran on a desktop PC with two 2.8 GHz Intel Pentium® IV processors and 2 GB RAM.

Finally, the vertical displacement of the same edge as in Figure 5.13 is shown in Figure 5.14 when the controlling voltage is applied to the outer electrodes.

The considerations just discussed are still valid. In particular, also in this case the error in the Spectre pull-in prediction is −0.9% with respect to the CoventorWare pull-in voltage. The CoventorWare schematic views of the deformed toggle switch when a biasing voltage is applied to the inner and outer electrodes are shown in Figures 5.15 and 5.16, respectively.

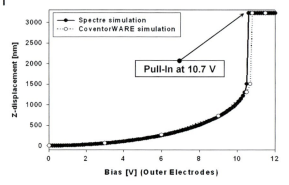

Figure 5.14 Pull-in characteristic of the toggle switch predicted in Spectre, compared with the CoventorWare simulation. The controlling voltage is applied to the outer electrodes.

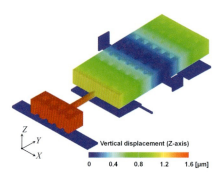

Figure 5.15 Schematic view in CoventorWare of the deformed toggle switch when the controlling voltage is applied to the inner electrodes. The color scale represents the vertical displacement.

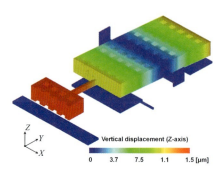

Figure 5.16 Schematic view in CoventorWare of the deformed toggle switch when the controlling voltage is applied to the outer electrodes. The color scale represents the vertical displacement.

5.3
A Hybrid RF-MEMS/CMOS VCO

This section reports the hybrid implementation of a VCO [175] in order to show the capabilities of the compact modeling approach in the simulation of functional hybrid multiphysical and multitechnology blocks. The active sustaining circuitry is designed with standard CMOS technology and is implemented with the design kit provided by AMS (0.35 μm HBT BiCMOS S35 technology) [176]. Whereas, the varactors of the LC tank are implemented with MEMS technology with the compact models previously shown throughout this chapter. The Cadence Virtuoso schematic of the complete VCO is shown in Figure 5.17. The two symbols representing the tunable capacitors are realized with a suspended rigid plate with four straight beams connected to its corners [177]. Each of them corresponds to the Cadence Virtuoso schematic in Figure 5.18. On the other hand, the two inductors in the LC tank in Figure 5.17 are realized with standard technology and are available in the design kit provided by AMS. Two RF-MEMS varactors are included in the symmetric LC-tank scheme, also providing the decoupling of the controlling DC voltage from the oscillator RF outputs. For the same reason, a capacitor (10 pF) is placed between the controlling voltage generator and the bias voltage supply ($V_{DD} = 3.3$ V). Concerning the structure in Figure 5.18, both the width and the length of the central

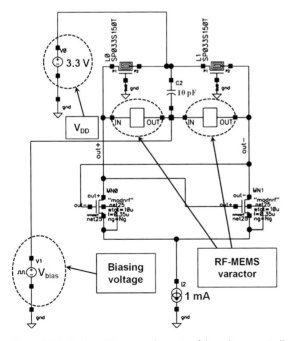

Figure 5.17 Cadence Virtuoso schematic of the voltage-controlled oscillator (VCO) with the CMOS active sustaining circuitry realized with AMS technology and the radio frequency (RF) MEMS LC tank.

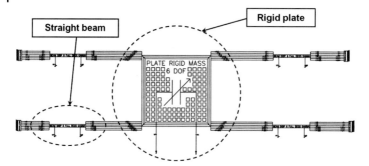

Figure 5.18 Cadence Virtuoso schematic corresponding to each of the two RF-MEMS varactors in the schematic of the VCO in Figure 5.17.

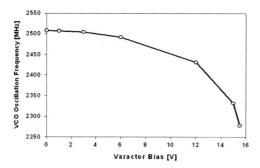

Figure 5.19 VCO oscillation frequency versus biasing voltage applied to the RF-MEMS varactors (tuning characteristic).

plate are 500 μm, while the thickness is 4.8 μm. On its surface, 12 holes are distributed along both the X axis and the Y axis (dimensions 20 × 20 μm²). The four straight beams are 75 μm long, while their width and thickness are 10 and 1.8 μm, respectively. The gap between the suspended plate and the fixed electrode is 3 μm. Depending on the controlling voltage applied to the common node between the RF-MEMS varactors, their capacitance changes and consequently the oscillation frequency of the entire VCO. Transient analyses were performed in Spectre for different controlling voltages lower than the pull-in of the structure in Figure 5.18 (i.e., 15.6 V). The VCO tuning characteristic (frequency versus biasing voltage) is shown in Figure 5.19. The capacitance of each RF-MEMS varactor and the corresponding VCO oscillation frequency are reported in Table 5.3. The VCO output voltage versus time for the transient simulation with zero biasing voltage is shown in Figure 5.20. It can be seen that after a few periods the oscillation becomes stable. The RF-MEMS/CMOS VCO implementation described represents a significant example of the utilization of the mixed-domain simulation environment (electrical and mechanical) discussed in this chapter.

Table 5.3 Dependence of the oscillation frequency of the voltage-controlled oscillator (VCO) on the biasing voltage applied to the radio frequency microelectromechanical system varactors (below the pull-in) of the LC tank.

Biasing voltage (V)	Capacitance (fF)	VCO frequency (GHz)
0	597	2.508
1	598	2.507
3	601	2.504
6	611	2.492
12	671	2.431
15	775	2.332
15.5	838	2.278

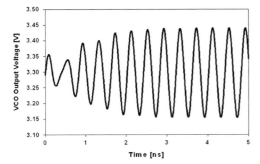

Figure 5.20 VCO output voltage versus time simulated in Spectre (transient simulation) when no biasing voltage is applied to the RF-MEMS varactors.

5.4 Excerpts of Verilog-A Code Implemented for MEMS Models

The scope of this section is to provide the reader with some excerpts of Verilog-A code implemented for some of the most important models included in the library, as well as to describe their salient physical features and characteristics. The code shown in the following pages is divided into blocks depending on the particular characteristic described (e.g., implementation of the capacitance or the structural elasticity), and is annotated with comments to improve understanding. Moreover, the Spectre schematic symbol of each element discussed is also reported, highlighting the correspondence between the input/output nodes at the schematic level and those defined in the code written in Verilog-A. The approach followed in reporting the MEMS library models is based on incremental complexity. For this reason, the flexible beam is reported after the boundary condition elements, which are also included in the library. The rigid plate transducer is not discussed in this section, as the structure of the Verilog-A code is similar to that for the flexible beam model.

Before the MEMS models are described more in detail, the typical structure of a complete routine in Verilog-A is briefly discussed. Regardless of the specific structure implemented, the sequence and arrangement of the code blocks are always the same, and are described as follows:

- Definition of the module name and of its nodes:
  module   name_of_the_module(node_1, node_2, node_3,..., node_N);
- Statement of the node types (input, output, or both):
  input   node_1, node_2, node_3,..., node_N;
  output   node_1, node_2, node_3,..., node_N;
  inout   node_1, node_2, node_3,..., node_N;
- Statement of the discipline/s associated with the nodes:
  discipline name   node_1, node_2, node_3,..., node_N;
- Statement of the internal nodes (optional):
  discipline name   int_node_1, int_node_2, int_node_3,..., int_node_N;
- Statement of the parameters and their default value:
  parameter   *parameter type, parameter name = default value;*
- Statement of variables and constants:
  variable type   *variable name 1;*
  variable type   *variable name 2;*
  ...
  variable type   *variable name N;*
- Analog code block (routine core):
  analog begin
  Initial step block:
  @(initial step) begin
  ...
  end
  ...
  instructions...
  Final step block:
  @(final step) begin
  ...
  end
  endmodule

5.4.1
Anchor Point

The anchor point is a boundary condition element enabling one to fix to zero the linear displacement and angular rotation of all the nodes wired to its terminals. The Cadence schematic reported in Figure 5.21 shows the symbol for an anchor point with six DOFs, namely, the linear displacements along the X, Y, and Z axes, and the rotations around the same three axes.

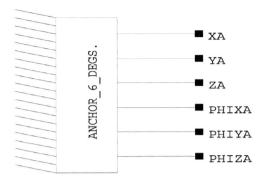

Figure 5.21 Cadence schematic symbol for the anchor point with six DOFs. The nodes for the linear displacements are XA, YA, and ZA, while the ones for the rotations are PHIXA, PHIYA, and PHIZA.

As it is easy to understand, the main code for the anchor point is rather simple, as all the displacements (linear and rotational) of the nodes must be forced to zero. The Verilog-A code for the anchor point is given below.

```
!!----------------------------------------!!
// Implementation of an anchor point, forcing all the displacements to zero
// (6 degrees of freedom)

// Inclusion of the file with the definition of all the physical features
// of the physical domain/s involved (e.g. mechanical, electrical nodes, etc.)
'include "../../discipline/discipline.h"

// Main module, defining the input/output nodes of the structure
// The nodes must correspond with the ones defined in the device symbol
module anchor_6_degs_verilog(PHIXA, PHIYA, PHIZA, XA, YA, ZA);

// Rotational displacements, torques are flow signals
inout PHIXA;
mechanical_angular PHIXA;
inout PHIYA;
mechanical_angular PHIYA;
inout PHIZA;
mechanical_angular PHIZA;

// Orthogonal displacements, forces are flow signals.
inout XA;
mechanical_linear XA;
inout YA;
mechanical_linear YA;
inout ZA;
mechanical_linear ZA;

// Beginning of the 'analog' section.
// All the instructions included in this section are computed per each
// iteration occurring during the simulation.
```

```
// Differently, the instructions not included in the 'analog' section
// are computed once per simulation, and are related e.g. to the
// variable definitions and initialization to certain values.
analog begin

    // Forcing to zero all the across variables of the nodes.
    // In the mechanical domain, the across variable is the
    // displacement (linear or angular), while the through
    // variable is the force or torque.
    disp(XA) <+ 0;
    disp(YA) <+ 0;
    disp(ZA) <+ 0;
    ang(PHIYA) <+ 0;
    ang(PHIZA) <+ 0;
    ang(PHIXA) <+ 0;

end
endmodule
!!----------------------------------------!!
```

5.4.2
Force Source

Another simple device implemented in the MEMS model library is the force stimulus source. It is necessary in order to impose a mechanical force and/or a torque on one or more nodes of a certain structure to observe its mechanical behavior. The Cadence schematic reported in Figure 5.22 shows the symbol for the force stimulus source with six DOFs, namely, the linear displacements along the X, Y, and Z axes, and the rotations around the same three axes.

The model features four user-defined parameters. The first one is the force modulus (expressed in newtons). Two additional parameters define the orientation angles around the y and z axes, depending upon which the force modulus is decomposed into its contributions along the x, y, and z axes. The fourth parameter defines the torque imposed around the x axis. The Verilog-A code for the force stimulus source is given below.

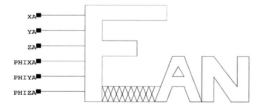

Figure 5.22 Cadence schematic symbol for the stimulus source (force) with six DOFs. The nodes for the linear displacements are XA, YA, and ZA, while the ones for the rotations are PHIXA, PHIYA, and PHIZA.

5.4 Excerpts of Verilog-A Code Implemented for MEMS Models

```
!!----------------------------------------!!
// Implementation of a stimulus (force) applied to a node with a certain rotation
// angle around the y and z axis, and torque around the x axis.
// The torque around the x axis is decomposed in three components around the
// x, y and z axis as a function of the rotation angles around the
// y and z axis, being the latter ones given as initial parameters.
// 6 degrees of freedom.

`include "../../discipline/discipline.h"

module force_ang_6_full_degs_veriloga(PHIXA, PHIYA, PHIZA, XA, YA, ZA);

// Rotational displacements, torque are flow signals.
inout PHIXA;
mechanical_angular PHIXA;
inout PHIYA;
mechanical_angular PHIYA;
inout PHIZA;
mechanical_angular PHIZA;

// Orthogonal displacements, forces are flow signals.
inout XA;
mechanical_linear XA;
inout YA;
mechanical_linear YA;
inout ZA;
mechanical_linear ZA;

// ----------------------------------- //
// --------- DEFINITION OF THE MODEL PARAMETERS --------- //
// ----------------------------------- //
// Applied force magnitude.
parameter real force = 1e-6;

// Rotation angle around the y axis.
parameter real angle_y = 0;

// Rotation angle around the z axis.
parameter real angle_z = 0;

// Torque around the x axis.
parameter real torque_x = 0;
!!----------------------------------------!!
```

The four model parameters defined in the last portion of the Verilog-A code just reported are available to the user when the symbol corresponding to the force stimulus is instanced in a Cadence schematic. Through the object properties mask, reported in Figure 5.23, the user can access the model parameters (highlighted in Figure 5.23) and set their value. If nothing is specified by the user, the variables have the default value, as reported in the Verilog-A code above.

The part for the force source is given below, and comprises the statement of variables needed for the calculation of the force and torque decomposition, as well

5 On-Purpose Simulation Tools

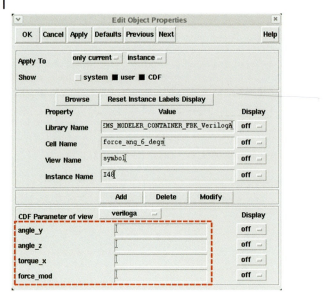

Figure 5.23 Object properties mask (Cadence) for the force stimulus model, through which the user can modify the values of the model parameters.

as the analog part of the code, where the computed values are assigned to the model input/output nodes.

```
!!----------------------------------------!!
// Angles in radians.
real rad_y, rad_z;

// Trigonometric function helping the calculations.
real cos_phi_y, sin_phi_y, cos_phi_z, sin_phi_z;

// Applied forces, in the chip reference system.
real Fachipxa, Fachipya, Fachipza;

// Applied torques, in the chip reference system.
real Tqchipxa, Tqchipya, Tqchipza;
real pi;

analog begin

@(initial_step) begin

pi = 3.141592654;

// Transformation of an angle in radians.
rad_y = angle_y * pi / 180;
rad_z = angle_z * pi / 180;

// Pre-calculation of the rotation coefficients.
```

```
cos_phi_y = cos(rad_y);
sin_phi_y = sin(rad_y);
cos_phi_z = cos(rad_z);
sin_phi_z = sin(rad_z);

// Forces calculation in the chip reference system.
Fachipxa = +force * cos_phi_y * cos_phi_z;
Fachipya = +force * cos_phi_y * sin_phi_z;
Fachipza = -force * sin_phi_y;

// Torques calculation in the chip reference system.
Tqchipxa = +torque_x * cos_phi_y * cos_phi_z;
Tqchipya = +torque_x * cos_phi_y * sin_phi_z;
Tqchipza = -torque_x * sin_phi_y;

end

// Assignment of the calculated values to the 'through' variables of the
// input/output nodes.
frc(XA) <+ Fachipxa;
frc(YA) <+ Fachipya;
frc(ZA) <+ Fachipza;
trq(PHIXA) <+ Tqchipxa;
trq(PHIYA) <+ Tqchipya;
trq(PHIZA) <+ Tqchipza;

end
endmodule
!!----------------------------------------!!
```

5.4.3
Flexible Beam

The flexible beam model is more complex, as it features multiple DOFs and encompasses the theory of elasticity, the electrostatic transduction, and so on. The Cadence schematic for the flexible beam model is reported in Figure 5.24, and it comprises two ends, each with six DOFs (three linear displacements and three rotations).

Figure 5.24 Cadence schematic symbol for the flexible straight beam with 12 DOFs. The beam features two ends, namely, A and B, each characterized by six DOFs (XA, YA, ZA, PHIXA, PHIYA, and PHIZA for end A, and XB, YB, ZB, PHIXB, PHIYB, and PHIZB for end B). The beam symbol also features two electrical nodes (to apply the controlling voltage and to read the capacitance), namely, VA and VB.

The Verilog-A code for the whole flexible beam model is not reported for the sake of brevity. In contrast, some relevant portions of the beam software implementation are presented and discussed in order to explain its features and their realization. The first code excerpt follows, and refers to the definition of the beam nodes (mechanical and electrical).

```
!!----------------------------------------!!
module beam\_6dof\_linear (PHIXA, PHIXB, PHIYA, PHIYB, PHIZA, PHIZB,
  VA, VB, XA, XB, YA, YB, ZA, ZB);

// Rotational displacements around the x, y and z axis.
inout PHIXA, PHIXB, PHIYA, PHIZA, PHIYB, PHIZB;
mechanical_angular PHIXA, PHIXB, PHIYA, PHIZA, PHIYB, PHIZB;

// Electrical signals.
inout VA, VB;
electrical VA, VB;

// Linear displacements along the x, y and z axis.
inout XA, XB, YA, YB, ZA, ZB;
mechanical_linear XA, XB, YA, YB, ZA, ZB;
!!----------------------------------------!!
```

The next Verilog-A code block comprises the definition and initialization of the model variables (accessible to the user via an object properties mask similar to the one reported in Figure 5.23), as well as the statement of the internal nodes. The latter are nodes similar to the ones listed in Figure 5.24, but are not associated with any pin of the symbol (i.e., not externally available). Such nodes are necessary to perform operations such as the time derivative of the displacement (to determine the velocity of a node) and the time derivative of the velocity (to determine the acceleration of a node).

```
!!----------------------------------------!!
// ********************* //
// Geometrical parameters //
// ********************* //

// Beam width [meters].
parameter real w = 3.0e-6 from(0:inf);

// Beam length [meters].
parameter real L = 10.0e-6 from(0:inf);

// Beam thickness [meters].
parameter real t = 1.0e-6 from(0:inf);

// Orientation angle of the beam around the y axis [degrees].
parameter real angle_y = 0.0;

// Orientation angle of the beam around the z axis [degrees].
parameter real angle_z = 0.0;
```

5.4 Excerpts of Verilog-A Code Implemented for MEMS Models

```
// Other parameters
parameter integer n_step = 10;
parameter integer lin_flag = 0;
parameter real gap = 1.0e-6;
parameter integer out_flag = 0;
parameter real delta_x = 1.0e-7;
parameter real delta_y = 1.0e-7;
parameter real fict_zero = 1.0e-6;
parameter real el_sx = +0.0e-6;
parameter real el_dx = +0.0e-6;
parameter integer split\_el = 0;

// Internal nodes for computing velocities.
mechanical_linear_velocity VXA, VXB, VYA, VYB, VZA, VZB;
mechanical_angular_velocity VPHIXA, VPHIXB, VPHIYA, VPHIYB, VPHIZA, VPHIZB;

// Internal nodes for computing accelerations.
mechanical_linear_acceleration AXA, AXB, AYA, AYB, AZA, AZB;
mechanical_angular_acceleration APHIXA, APHIXB, APHIYA, APHIYB, APHIZA, APHIZB;

// Intrinsic capacitor nodes
electrical VP_INT, VN_INT;
!!----------------------------------------!!
```

The following code block shows part of the statements for all the variables and is necessary to support all the calculations required from the beam model chosen.

```
!!----------------------------------------!!
// ****************** //
// Physical quantities //
// ****************** //

real Jx; // Second moment of inertia (x axis).
real Iy; // Moment of inertia (y axis).
real Iz; // Moment of inertia (z axis).

real rad_y; // Angle (in radians) around the y axis.
real rad_z; // Angle (in radians) around the z axis.

real Bx, By, Bz; // Damping coefficients.

real A; // Cross-sectional area.
real r; // Resistance thought the beam;
real Mult; // Intermediate constant (needed for calculations).
real G; // Shear modulus.

// Coefficients for the spring constants (Stiffness Matrix).
real k11, k22, k33, k44, k55, k66, k62, k53, k11_5, k12_6;

// Shear coefficients
real shr_y, shr_z;
```

```
// Coefficients for the concentrated mass (Mass Matrix).
real m11, m22, m44, m55, m53, m82, m95, m10_4, m11_5, m71;

// Rotation Matrix Elements (Direct Transformation)
real a_11_dir, a_12_dir, a_13_dir;
real a_21_dir, a_22_dir, a_23_dir;
real a_31_dir, a_32_dir, a_33_dir;
!!----------------------------------------!!
```

The following block reports some of the calculations needed in order to implement the beam model in Verilog-A. Among them, the calculation of the coefficients of the stiffness and mass matrices is particularly important to determine the elastic and inertial behavior of each DOF of the beam.

```
!!----------------------------------------!!
// Transformation of angles in radians.
rad_y = angle_y * pi/180.0;
rad_z = angle_z * pi/180.0;

// Moment of inertia of the beam (x axis).
Jx = (2.0 * pow((w * t) , 3)) / (7.0 * (pow(w , 2) + pow(t , 2)));

// Moment of inertia of the beam (y axis).
Iy = ((w * pow(t , 3)) / 12.0);

// Moment of inertia of the beam (z axis).
Iz = ((t * pow(w , 3)) / 12.0);

// Electrical resistivity of the beam.
r = rho * (L / w);

// Cross-sectional area of the beam.
A = w * t;

// Intermediate value for the lumped masses calculation.
Mult = dens * A * L;

// Shear modulus.
G=E/(2.0*(1.0 + Poi_nu));

shr_y = (12.0 * E * Iy) /(G * w * t * pow(L,2));
shr_z = (12.0 * E * Iz) /(G * w * t * pow(L,2));

// Stiffness matrix coefficients calculation
k11 = +(A * E) / L;
k22 = +(12.0 * E * Iz) / (pow(L,3) * (1.0 + shr_y));
k33 = +(12.0 * E * Iy) / (pow(L,3) * (1.0 + shr_z));
k44 = +(G * Jx) / L;
k55 = +((4.0 + shr_z) * E * Iy) / (L*(1.0 + shr_z));
k66 = +((4.0 + shr_y) * E * Iz) / (L*(1.0 + shr_y));
k53 = -(6.0 * E * Iy) / (pow(L, 2) * (1.0 + shr_z));
k62 = +(6.0 * E * Iz) / (pow(L, 2) * (1.0 + shr_y));
k11_5 = +((2.0 - shr_z) * E * Iy) / (L * (1.0 + shr_z));
k12_6 = +((2.0 - shr_y) * E * Iz) / (L * (1.0 + shr_y));
```

```
// Mass matrix coefficients calculation
m11 = +Mult / 3.0;
m71 = +Mult / 6.0;
m22 = +Mult * (13.0 / 35.0);
m44 = +Mult * ((Iy + Iz) / (3.0 * A));
m55 = +Mult * (pow(L , 2) / 105.0);
m53 = -Mult * ((11 * L) / 210.0);
m82  = -Mult * (9.0 / 70.0);
m95 = -Mult * ((13 * L) / 420.0);
m10_4 = +Mult * ((Iy + Iz) / (6.0 * A));
m11_5 = -Mult * (pow(L , 2) / 140.0);

// Trigonometric calculations
sin_phi_xa = sin(ist_ang_xa);
cos_phi_xa = cos(ist_ang_xa);
sin_phi_ya = sin(ist_ang_ya);
cos_phi_ya = cos(ist_ang_ya);
sin_phi_za = sin(ist_ang_za);
cos_phi_za = cos(ist_ang_za);
!!----------------------------------------!!
```

The next Verilog-A commands are needed to calculate the velocities and accelerations to be assigned to the corresponding internal nodes, and to transform all the displacements of the beam's DOFs from the chip reference system to the beam local reference system.

```
!!----------------------------------------!!
// Velocity nodes calculations and assignment.
vel(VXA) <+ ddt(disp(XA));
vel(VXB) <+ ddt(disp(XB));
vel(VYA) <+ ddt(disp(YA));
vel(VYB) <+ ddt(disp(YB));
vel(VZA) <+ ddt(disp(ZA));
vel(VZB) <+ ddt(disp(ZB));
avel(VPHIXA) <+ ddt(ang(PHIXA));
avel(VPHIXB) <+ ddt(ang(PHIXB));
avel(VPHIYA) <+ ddt(ang(PHIYA));
avel(VPHIYB) <+ ddt(ang(PHIYB));
avel(VPHIZA) <+ ddt(ang(PHIZA));
avel(VPHIZB) <+ ddt(ang(PHIZB));

// Acceleration nodes calculations and assignment.
acc(AXA) <+ ddt(vel(VXA));
acc(AXB) <+ ddt(vel(VXB));
acc(AYA) <+ ddt(vel(VYA));
acc(AYB) <+ ddt(vel(VYB));
acc(AZA) <+ ddt(vel(VZA));
acc(AZB) <+ ddt(vel(VZB));
aacc(APHIXA) <+ ddt(avel(VPHIXA));
aacc(APHIXB) <+ ddt(avel(VPHIXB));
aacc(APHIYA) <+ ddt(avel(VPHIYA));
aacc(APHIYB) <+ ddt(avel(VPHIYB));
aacc(APHIZA) <+ ddt(avel(VPHIZA));
```

```
aacc(APHIZB) <+ ddt(avel(VPHIZB));

// Transformation of the displacements from the chip to the beam
// reference system.
l_xa = a_11_inv*disp(XA) + a_21_inv*disp(YA) + a_31_inv*disp(ZA);
l_xb = a_11_inv*disp(XB) + a_21_inv*disp(YB) + a_31_inv*disp(ZB);
l_ya = a_12_inv*disp(XA) + a_22_inv*disp(YA) + a_32_inv*disp(ZA);
l_yb = a_12_inv*disp(XB) + a_22_inv*disp(YB) + a_32_inv*disp(ZB);
l_za = a_13_inv*disp(XA) + a_23_inv*disp(YA) + a_33_inv*disp(ZA);
l_zb = a_13_inv*disp(XB) + a_23_inv*disp(YB) + a_33_inv*disp(ZB);

// Transformation of the rotation angles from the chip to the beam
// reference system.
l_phi_xa = a_11_inv*ang(PHIXA) + a_21_inv*ang(PHIYA) + a_31_inv*ang(PHIZA);
l_phi_xb = a_11_inv*ang(PHIXB) + a_21_inv*ang(PHIYB) + a_31_inv*ang(PHIZB);
l_phi_ya = a_12_inv*ang(PHIXA) + a_22_inv*ang(PHIYA) + a_32_inv*ang(PHIZA);
l_phi_yb = a_12_inv*ang(PHIXB) + a_22_inv*ang(PHIYB) + a_32_inv*ang(PHIZB);
l_phi_za = a_13_inv*ang(PHIXA) + a_23_inv*ang(PHIYA) + a_33_inv*ang(PHIZA);
l_phi_zb = a_13_inv*ang(PHIXB) + a_23_inv*ang(PHIYB) + a_33_inv*ang(PHIZB);
!!----------------------------------------!!
```

Subsequently, the forces and torques for each beam DOF are calculated in the beam local reference system.

```
!!----------------------------------------!!
// Computation of forces and torques (due to the elastic behavior)
// in the beam reference system.
Fr_ya = -k22 * dy + k62 * (l_phi_za + l_phi_zb);
Fr_yb = +k22 * dy - k62 * (l_phi_za + l_phi_zb);
Fr_za = -k33 * dz + k53 * (l_phi_ya + l_phi_yb);
Fr_zb = +k33 * dz - k53 * (l_phi_ya + l_phi_yb);
Tq_xa = -k44 * dphix;
Tq_xb = +k44 * dphix;
Tq_ya = -k53 * dz + k55 * l_phi_ya + k11_5 * l_phi_yb;
Tq_yb = -k53 * dz + k55 * l_phi_yb + k11_5 * l_phi_ya;
Tq_za = -k62 * dy + k66 * l_phi_za + k12_6 * l_phi_zb;
Tq_zb = -k62 * dy + k66 * l_phi_zb + k12_6 * l_phi_za;

// Computation of forces and torques (due to the inertial behavior)
// in the beam reference system.
Fil_xa = +m11 * a_xa + m71 * a_xb;
Fil_xb = +m71 * a_xa + m11 * a_xb;
Fil_ya = +m22 * a_ya - m53 * a_phi_za + m82 * a_yb + m95 * a_phi_zb;
Fil_yb = +m82 * a_ya - m95 * a_phi_za + m22 * a_yb + m53 * a_phi_zb;
Fil_za = +m22 * a_za + m53 * a_phi_ya + m82 * a_zb - m95 * a_phi_yb;
Fil_zb = +m82 * a_za + m95 * a_phi_ya + m22 * a_zb - m53 * a_phi_yb;
Tlq_xa = +m44 * a_phi_xa + m10_4 * a_phi_xb;
Tlq_xb = +m10_4 * a_phi_xa + m44 * a_phi_xb;
Tlq_ya = +m53 * a_za + m55 * a_phi_ya + m95 * a_zb + m11_5 * a_phi_yb;
Tlq_yb = -m95 * a_za + m11_5 * a_phi_ya - m53 * a_zb + m55 * a_phi_yb;
Tlq_za = -m53 * a_ya + m55 * a_phi_za - m95 * a_yb + m11_5 * a_phi_zb;
Tlq_zb = +m95 * a_ya + m11_5 * a_phi_za + m53 * a_yb + m55 * a_phi_zb;
!!----------------------------------------!!
```

5.4 Excerpts of Verilog-A Code Implemented for MEMS Models

The next block reports the calculations to determine the attraction electrostatic force when a bias voltage is applied between the suspended beam and the fixed underlying electrode. The calculations are split into a few conditioned code parts depending on the instantaneous value of the orientation angles. Such code duplication is necessary in order to avoid numerical problems (e.g., division by zero) when one or more orientation angles are very close to zero.

```
!!----------------------------------------!!
// ELECTROSTATIC FORCE CALCULATION WHEN PHI_Y = 0
   if((abs(theta_y_ij) < fict_zero))
   begin
   // Total electrostatic force between the plate and the substrate
   Fz_loc_tot_end_a = abs((+1.0-3.0*pow(x_ist_sw_se_mid,2)
                   + 2.0*pow(x_ist_sw_se_mid,3))*
                   (0.5*(eps0*delta_x*w*pow(vlt,2)))/pow(z_mid,2));

   // Total electrostatic torque between the plate and the substrate
   My_loc_tot_end_a = +0.0;
      end

// ELECTROSTATIC FORCE CALCULATION WHEN PHI_X = 0
   else
   begin
   // Total electrostatic force between the plate and the substrate
   Fz_loc_tot_end_a = abs((+1.0-3.0*pow(x_ist_sw_se_mid,2)
                   + 2.0*pow(x_ist_sw_se_mid,3))*
                   (((pow(vlt,2)*eps0)*w)/ab_31)*
                (1.0/(2.0*z_mid-ab_31*delta_x)
                 - 1.0/(2.0*z_mid+ab_31*delta_x)));

   // Total electrostatic torque between the plate and the substrate
   My_loc_tot_end_a = +((-x_ist_sw_se_mid+2.0*pow(x_ist_sw_se_mid,2)
                   -pow(x_ist_sw_se_mid,3)))*(0.5*pow(vlt,2)*
                   eps0*abs(+ abs(z_mid + a_31 + (w / 2.0)) *
   abs(+ ln(abs(2.0*z_mid + 2.0*a_31 + delta_x + w) /
            abs(2.0*z_mid + 2.0*a_31 + w))
      - ln(abs(2.0*z_mid + 2.0*a_31 + w) /
           abs(2.0*z_mid + 2.0*a_31 - delta_x + w))) +
           + abs(z_mid + a_31 - (w / 2.0)) *
   abs(+ ln(abs(2.0*z_mid + 2.0*a_31 - w) /
      abs(2.0*z_mid + 2.0*a_31 + delta_x - w))
      - ln(abs(2.0*z_mid + 2.0*a_31 - delta_x - w) /
               abs(2.0*z_mid + 2.0*a_31 - w)))));
   end

// ELECTROSTATIC FORCE CALCULATION WHEN PHI_Y = 0
   if((abs(theta_y_ij) < fict_zero))
   begin
   // Total electrostatic force between the plate and the substrate
   Fz_loc_tot_end_b = abs((+3.0*pow(x_ist_sw_se_mid,2)
                   - 2.0*pow(x_ist_sw_se_mid,3))*
                   (0.5*(eps0*delta_x*w*pow(vlt,2)))/pow(z_mid,2));
```

```
        // Total electrostatic torque between the plate and the substrate
        My_loc_tot_end_b = +0.0;
                end

        // ELECTROSTATIC FORCE CALCULATION WHEN PHI_X = 0
        else
        begin
        // Total electrostatic force between the plate and the substrate
        Fz_loc_tot_end_b = abs((+3.0*pow(x_ist_sw_se_mid,2)
                        - 2.0*pow(x_ist_sw_se_mid,3))*
                        (((pow(vlt,2)*eps0)*w)/ab_31)*
                   (1.0/(2.0*z_mid-ab_31*delta_x)
                   -1.0/(2.0*z_mid+ab_31*delta_x)));

        // Total electrostatic force between the plate and the substrate
        My_loc_tot_end_b = +((pow(x_ist_sw_se_mid,2)
                        -pow(x_ist_sw_se_mid,3)))*
                        (0.5*pow(vlt,2)*eps0*
                        abs(+ abs(z_mid + a_31 + (w / 2.0)) *
    abs(+ ln(abs(2.0*z_mid + 2.0*a_31 + delta_x + w) /
            abs(2.0*z_mid + 2.0*a_31 + w))
         - ln(abs(2.0*z_mid + 2.0*a_31 + w) /
            abs(2.0*z_mid + 2.0*a_31 - delta_x + w))) +
                        + abs(z_mid + a_31 - (w / 2.0)) *
    abs(+ ln(abs(2.0*z_mid + 2.0*a_31 - w) /
        abs(2.0*z_mid + 2.0*a_31 + delta_x - w))
         - ln(abs(2.0*z_mid + 2.0*a_31 - delta_x - w) /
                    abs(2.0*z_mid + 2.0*a_31 - w)))));
        end
!!----------------------------------------!!
```

The last Verilog-A code, reported below, shows the transformation of all the sums of forces (electrostatic, elastic, etc.) from the beam reference system back to the chip reference system, and the final assignment of such values to the beam nodes. The transformation back to the chip reference system is necessary because, in order to ensure consistency, the reference system of the whole schematic (e.g., the beams connected to the rigid plate, realizing a MEMS switch) must be unique.

```
!!----------------------------------------!!
// Transformation of the calculated forces, from the beam local
// reference system back to the chip reference system.
Fx_chip_tot_end_a = + Fx_chip_tot_end_a + a_13*Fz_loc_tot_end_a;
Fx_chip_tot_end_b = + Fx_chip_tot_end_b + a_13*Fz_loc_tot_end_b;
Fy_chip_tot_end_a = + Fy_chip_tot_end_a + a_23*Fz_loc_tot_end_a;
Fy_chip_tot_end_b = + Fy_chip_tot_end_b + a_23*Fz_loc_tot_end_b;
Fz_chip_tot_end_a = + Fz_chip_tot_end_a + ab_33*Fz_loc_tot_end_a;
Fz_chip_tot_end_b = + Fz_chip_tot_end_b + ab_33*Fz_loc_tot_end_b;

// Transformation of the calculated torques, from the beam local
// reference system back to the chip reference system.
Mx_chip_tot_end_a = + Mx_chip_tot_end_a + a_11*Mx_loc_tot_end_a
```

```
    + a_12*My_loc_tot_end_a;
Mx_chip_tot_end_b = + Mx_chip_tot_end_b + a_11*Mx_loc_tot_end_b
    + a_12*My_loc_tot_end_b;
My_chip_tot_end_a = + My_chip_tot_end_a + a_21*Mx_loc_tot_end_a
    + a_22*My_loc_tot_end_a;
My_chip_tot_end_b = + My_chip_tot_end_b + a_21*Mx_loc_tot_end_b
    + a_22*My_loc_tot_end_b;
Mz_chip_tot_end_a = + Mz_chip_tot_end_a + a_31*Mx_loc_tot_end_a
    + a_32*My_loc_tot_end_a;
Mz_chip_tot_end_b = + Mz_chip_tot_end_b + a_31*Mx_loc_tot_end_b
    + a_32*My_loc_tot_end_b;

// Forces assignment to the beam nodes.
frc(XA) <+ +((- Fchipxa + Fchip_xa_NL + Fx_chip_tot_end_a));
frc(XB) <+ +((- Fchipxb + Fchip_xb_NL + Fx_chip_tot_end_b));
frc(YA) <+ -((+ Fchipya + Fchip_ya_NL + Fy_chip_tot_end_a));
frc(YB) <+ -((+ Fchipyb + Fchip_yb_NL + Fy_chip_tot_end_b));
frc(ZA) <+ -((+ Fchipza + Fchip_za_NL - Fz_touch_end_a + Fz_chip_tot_end_a));
frc(ZB) <+ -((+ Fchipzb + Fchip_zb_NL - Fz_touch_end_b + Fz_chip_tot_end_b));

// Torques assignment to the beam nodes.
trq(PHIXA) <+ -((+ Tqchip_xa + Tqchip_xa_NL + Mx_chip_tot_end_a));
trq(PHIXB) <+ -((+ Tqchip_xb + Tqchip_xb_NL + Mx_chip_tot_end_b));
trq(PHIYA) <+ -((+ Tqchip_ya + Tqchip_ya_NL + My_chip_tot_end_a));
trq(PHIYB) <+ -((+ Tqchip_yb + Tqchip_yb_NL + My_chip_tot_end_b));
trq(PHIZA) <+ -((+ Tqchip_za + Tqchip_za_NL + Mz_chip_tot_end_a));
trq(PHIZB) <+ -((+ Tqchip_zb + Tqchip_zb_NL + Mz_chip_tot_end_b));
!!--------------------------------------!!
```

5.5 Conclusions

This chapter reported the development of a software model library for the simulation and prediction of the behavior of RF-MEMS devices in a commercial circuit simulator. The methodological approach of compact modeling was discussed first, and then the mathematical models capturing the salient aspects of the multiphysical behavior of MEMS were reported. Such models were then implemented in a programming language and simulation cases of a few RF-MEMS structures were reported for both validation purposes and as explicative examples of the use of such a tool.

Acknowledgment The author would like to acknowledge the Advanced Research Center on Electronic Systems for Information and Communication Technologies E. De Castro (ARCES) of the University of Bologna (Italy) for the support provided in 2006 and 2007 concerning the use of the CoventorWare software tool.

6
Packaging and Integration

Abstract: The critical aspects of packaging and integration of radio frequency (RF) microelectromechanical system (MEMS) devices are treated in this chapter, they being of significant relevance concerning both the protection and encapsulation of the typically very fragile RF-MEMS devices and their interfacing with circuits and blocks realized with other technologies (i.e., standard semiconductor technologies). Among the different approaches available for the fabrication of a protective package for RF-MEMS devices, one solution is chosen and discussed in detail concerning technology aspects of the cap fabrication and its application to RF-MEMS devices. The chapter is completed by reporting experimental evidence of fabricated packages and their alignment and bonding to actual RF-MEMS devices. Another two relevant aspects connected with the packaging of RF-MEMS devices are also reported in this chapter. The first one concerns the influence of the cap on the electromagnetic (RF) characteristics and performance of the encapsulated RF-MEMS devices, which are unavoidably affected by the application of a physical element (i.e., the package) on them. For this purpose, experimental measurements of the RF performance of a few test structures with and without an applied package are compared. The second one refers to the proper simulation of an in-package RF-MEMS device, showing how the cap design can be optimized in order to reduce the impact on the RF performance. Finally, aspects concerning the integration of in-package RF-MEMS devices within RF circuits and systems are introduced.

6.1
Introduction

The focus of this chapter concentrates on the demonstration of a packaging solution for the encapsulation of radio frequency (RF) microelectromechanical system (MEMS) devices. Chapter 1 gave an overview of the mainly diffused approaches and technology solutions for the fabrication of a package and for the subsequent alignment with and bonding to an RF-MEMS device wafer or wafer die. Given the information provided in the section on the state of the art, an example of encapsulation of RF-MEMS devices performed by means of the wafer-level packaging (WLP) approach will be described in detail. The use of the same WLP solution in order to enable and make easier the integration of components based on RF-MEMS technology with circuitry based on standard semiconductor (CMOS) technology is also presented, showing the key role played by the packaging in obtaining functional blocks based on the hybridization of different noncompatible technologies.

6.2
A WLP Solution for RF-MEMS Devices and Networks

The solution for the encapsulation of RF-MEMS devices and networks that will be discussed in detail in this chapter is based on the WLP approach. The material employed for the capping part is silicon and the redistribution of the electrical signals from the MEMS/RF-MEMS device wafer to the outside world is provided by through-wafer vias etched in the capping substrate and subsequently filled with copper. Recesses are etched in the bottom side of the package in order to accommodate the RF-MEMS devices, while the wafer-to-wafer bonding (i.e., the package and the device wafer interfacing) can be performed by reflow of solder bumps or via the use of electrically conductive adhesives (ECAs) [178, 179].

Before going into the details of the package fabrication process, we will focus on the WLP solution. Starting from a device wafer onto which the RF-MEMS devices and networks are arranged within dies distributed on its surface, a second wafer, that is, the package, is necessary. Vertical interconnects have to be opened through its vertical thickness and distributed in relation to the signal pads on the dies of the device substrate. A schematic view of both the RF-MEMS device and the capping wafer is shown in Figure 6.1. After the capping part has been aligned with the device wafer, the two substrates must be brought into contact.

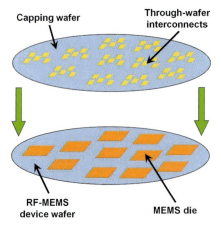

Figure 6.1 In the wafer-level packaging approach, a capping wafer, with a scheme of vertical through-wafer interconnects compatible with the distribution of electrical signal pads on the radio frequency (RF) microelectromechanical system (MEMS) device substrate, is necessary.

6.2 A WLP Solution for RF-MEMS Devices and Networks | 169

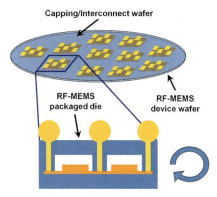

Figure 6.2 After the alignment of the capping part with the MEMS/RF-MEMS substrate, the wafer-to-wafer bonding is performed. Moreover, after the subsequent singulation, the packaged RF-MEMS dies are available for standard surface-mount technologies (SMT) (see Figure 6.3).

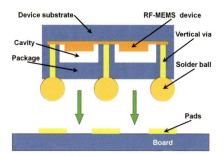

Figure 6.3 The final step is flipping the packaged MEMS/RF-MEMS die over the board (implementing a high-level function, such as an RF transceiver) and performing the bonding step by reflow of solder balls.

Once the wafer-to-wafer bonding has been performed, the capped RF-MEMS wafer is ready for the subsequent singulation (i.e., separation or cutting into dies/chips) as Figure 6.2 reports. After this step, the packaged RF-MEMS dies are ready for the flip-chip on-board mounting by means of reflow of solder balls, as shown in Figure 6.3.

6.2.1
Package Fabrication Process

Let us now discuss the details of the capping part fabrication process. The process is realized with the facilities available at the DIMES Technology Center (Technical University of Delft, the Netherlands) and employs silicon wafers. The process flow steps are as follows:

Step 1 The silicon wafer is first polished and made ready for the subsequent steps (see Figure 6.4). Its initial thickness is around 525 µm.

170 | 6 Packaging and Integration

Figure 6.4 Step 1: The silicon wafer is polished and prepared for the processing.

Step 2 Recesses are etched from the bottom side of the wafer by means of the Bosch deep reactive ion etching (DRIE) process [180] (see Figure 6.5). The recess depth typically ranges from a few tens of microns up to 120 – 150 μm.

Figure 6.5 Step 2: Recesses are etched in the bottom side of the wafer by using the Bosch deep reactive ion etching (DRIE) process.

Step 3 The Bosch DRIE process is used once again to etch vertical vias from the top side of the wafer as Figure 6.6 shows. Their diameter ranges from about 10–15 μm to 100–120 μm. It is not required to etch vias through the whole wafer thickness as the wafer is then ground (i.e., thinned) from the back side (next step).

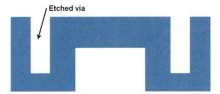

Figure 6.6 Step 3: Vertical vias are etched from the top side of the wafer by using the Bosch DRIE process.

Step 4 The wafer with etched recesses and vertical vias is thinned from the bottom side by means of grinding performed with a lapping machine [181, 182] (see Figure 6.7). Its thickness is reduced as long as the vias are opened at the bottom, and corresponds to a final wafer height of about 350 μm. Optionally, it is possible to further thin the wafer by bringing its thickness down to about 230–250 μm.

Step 5 A seed layer is sputtered [183] above the top face of the wafer and onto the sidewalls of the vias (see Figure 6.8), after that an insulating silicon oxide layer (not shown in Figure 6.8) is grown on the two sides of the

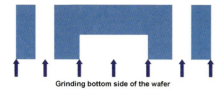

Figure 6.7 Step 4: The silicon wafer is thinned by grinding it from the back side. The bottom vias openings are released.

wafer and sidewalls of the vias. The seed layer is necessary to perform the subsequent electrodeposition of copper.

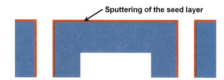

Figure 6.8 Step 5: The seed layer is sputtered on the top side of the wafer and onto the sidewalls of the vias. Before this step, a silicon oxide layer (not shown here) is grown on the wafer and the surface of the vias.

Step 6 A photoresist film is patterned on the top side of the wafer to define the features of the copper pads which are going to be electrodeposited, and on the bottom side to define where the bumps will be deposited (see Figure 6.9).

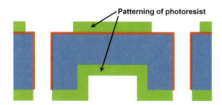

Figure 6.9 Step 6: A photoresist film is deposited on both sides of the wafer in order to pattern the top copper pads and bottom bumps, which will be subsequently electrodeposited.

Step 7 Top-side copper pads are electrodeposited on the silicon wafer as shown in Figure 6.10. Such pads will ensure the electrical interconnection with the RF-MEMS devices once the vias are filled with copper. The top-side pads can be used both for direct measurements of capped MEMS devices (i.e., probes can be placed on them) and for hosting electrodeposited solder balls for the final on-board mount (see Figure 6.3).

Step 8 Vertical etched vias are filled bottom-up with copper as Figure 6.11 shows. In order to perform this step, the wafer has to be flipped upside-down and temporary chromium metallizations (not shown in Figure 6.11) are placed on the vertical wafer faces to ensure electrical continuity with the seed layer.

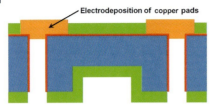

Figure 6.10 Step 7: The top-side copper bumps are electrodeposited by using the seed layer.

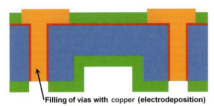

Figure 6.11 Step 8: The vertical vias are filled bottom-up with electrodeposited copper.

Step 9 Bumps are electrodeposited on the bottom capping side of the vias (see Figure 6.12). Depending on the solution chosen to perform the wafer-to-wafer bonding, the bumps will be made of copper or a special alloy. More details about the bonding step will be given later in this chapter.

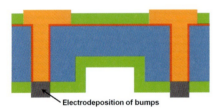

Figure 6.12 Step 9: Bumps are electrodeposited at the bottom capping side of the vias. Depending on the wafer-to-wafer bonding solution selected, the latter will be made of copper or a special alloy.

Step 10 As a final step the photoresist is removed from the top and bottom sides of the wafer as is the seed layer lying underneath it on the top side (see Figure 6.13).

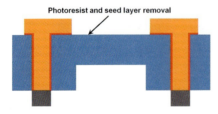

Figure 6.13 Step 10: Photoresist residuals are removed from both wafer faces as is the seed layer lying underneath the photoresist on the top capping side.

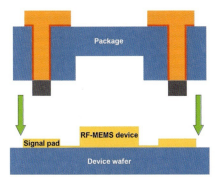

Figure 6.14 The package is bonded to the RF-MEMS device wafer after the proper alignment between the two substrates has been achieved.

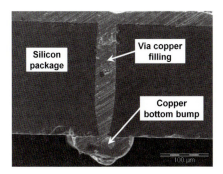

Figure 6.15 Scanning electron microscopy (SEM) microphotograph of a capping test wafer which is cut along the cross section of a vertical via in order to make the profile visible. The via exhibits a high aspect ratio and a very uniform copper filling.

The capping part is then ready to be aligned with the RF-MEMS device wafer for the wafer-to-wafer bonding step. A schematic of such a final step is shown in Figure 6.14.

A scanning electron microscopy (SEM) microphotograph of the cross section of a test capping wafer is shown in Figure 6.15. The test wafer was cut to make the via profile visible. The very high aspect ratio of the via achieved with the Bosch DRIE process and the uniform copper filling of it are noticeable. The narrowing of the via in the lower part (close to the bump) is an artifact due to the nonparallel direction followed during the sawing of the wafer to produce the vertical cross section of the via.

6.2.2
Wafer-to-Wafer Bonding Solutions

It has been briefly mentioned before that the wafer-to-wafer bonding step, chosen for this example, can be performed in two different ways. The first one is based on the reflow of solder bumps, while the second one employs the use of ECAs. More details about both solutions will be given below.

Reflow of Solder Bumps

In the first solution the bumps deposited on the bottom capping face (see Figure 6.12) are melted (i.e., reflowed) once they are brought into contact with the signal pads of the RF-MEMS device, as shown in Figure 6.16. The reflow is achieved by heating the two wafers until the melting temperature of the bumps is reached [184].

The bumps then start to spread out, increasing the contact surface of the package-to-device wafers as depicted in Figure 6.17. The subsequent cooling of the bumps ensures both package-to-device wafer adhesion and proper electrical interconnection. Since the melting temperature of the more commonly used metals (e.g., copper, gold) is generally too high to be tolerated both by the package and by the device wafers, the material for the bumps must be chosen in an attempt to minimize the thermal budget required for their reflow. This is the reason why suitable alloys composed of different materials to be deposited as bumps are investigated. Indeed, depending on the metals chosen for the creation of the alloy and on their relative percentages, one can considerably reduce the melting temperature of the bumps, making the bonding step tolerable for the entire capped wafer. This method is well known in the literature as eutectic bonding [185]. In the process discussed in this example, the alloy is made of gold and tin. Their volume percentages are 80 and 20%, respectively [186]. This allows one to perform the bump reflow at a temperature of about 300 °C.

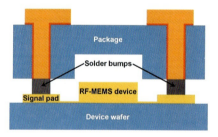

Figure 6.16 Scheme of the package aligned and in contact with the RF-MEMS device wafer *before* the reflow of the solder bumps is performed.

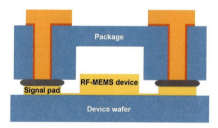

Figure 6.17 Scheme of the package aligned and in contact with the RF-MEMS device wafer *after* the reflow of the solder bumps is performed.

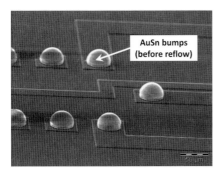

Figure 6.18 SEM microphotograph of the AuSn bumps on a dummy wafer *before* the reflow step.

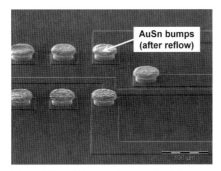

Figure 6.19 SEM microphotograph of the AuSn bumps on a dummy wafer *after* the reflow step.

SEM pictures taken before and after performing the bump reflow on a dummy wafer (and removing the cap from the target wafer after the soldering is complete) are shown in Figures 6.18 and 6.19, respectively. The initial shape of the bumps is like half a sphere (see Figure 6.18), while after the reflow and cooling, their upper part spreads around, creating a flat and circular area (see Figure 6.19), which significantly improves the contact surface when the package is bonded to an actual RF-MEMS device wafer to be protected, ensuring both mechanical adhesion and electrical connectivity.

Electrically Conductive Adhesives
The other wafer-to-wafer bonding solution relies on the use of special adhesive materials, generally referred to as electrically conductive adhesives (ECA), as they are able to ensure both the adhesion and the electrical interconnection of the two surfaces brought into physical contact. ECAs are constituted of a polymer-based matrix that basically behaves as an electrically insulating glue, and are available in the form of paste or films. Within the matrix a certain amount of conductive particles is distributed [187]. The conductive particles are made of various metals, for instance, gold, copper, or aluminum. Such particles can be small spheres (a few

microns in diameter), flakes, or even bigger spheres made of insulating material coated with a conductive material (13–15 μm in diameter) [188].

Depending on the dimensions of the conductive particles and on their volume percentage with respect to the insulating glue within which they are distributed, one usually divides ECAs into two categories, namely, anisotropic conductive adhesives (ACAs) and isotropic conductive adhesives (ICAs). If the volume of conductive particles is about 5–10% of the entire ECA volume, we are dealing with an ACA material [189]. This means that when the polymer with particles inside is normally uncompressed, it behaves as a dielectric material, whereas when the ACA is compressed, the conductive particles get closer to each other, forming an electrical path through the polymer. A scheme of an ACA in the uncompressed and compressed conditions is shown in Figure 6.20. The fact that the insulating material can locally become conductive results in the quality of being anisotropic.

On the other hand, when the volume of conductive particles is higher (25–35%), the material is classified as an ICA. This means that the distribution of particles is large enough to ensure the electrical conductivity independently of the condition of compression or noncompression of the polymer matrix. Referring to the technology solution described in previous pages, the use of an ACA or an ICA involves different solutions for its patterning. Concerning the first one, due to its anisotropic conductivity property, it can be patterned (e.g., rolled) over the entire bottom side of the package. When the wafer-to-wafer bonding is performed, the ACA is compressed only where there are copper bumps on the RF-MEMS signal pads (see Figure 6.14), which is where the electrical interconnection is needed. On the other hand, an ICA material should be patterned only on the copper bumps (e.g., distributed with a needle or screen printed), otherwise it would cause the shorting of

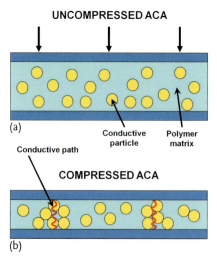

Figure 6.20 (a) When the anisotropic conductive adhesive (ACA) is uncompressed, the particles are far from each other and the material is electrically insulating. (b) When the ACA is compressed, the particles get closer and form an electrically conductive path.

all the electric signals. In conclusion, ACA/ICA materials require a lower temperature (180–200°C) to be cured with respect to the bump reflow solution and also exhibit good adhesion strength [190].

The ECA materials have pros and cons compared with the reflow of solder bumps. Briefly, ECAs are easier to pattern and require a lower thermal budget to be cured when compared with metal bumps. On the other hand, they typically introduce larger contact resistance and do not ensure hermeticity of the sealed cavities.

6.3
Encapsulation of RF-MEMS Devices

In the packaging schemes presented so far, the only contact area among the capping-to-device wafers is represented by the bumps on the RF-MEMS signal pads. Of course this does not represent a complete encapsulation solution as the vertical gap between the two wafers still remains open, and the MEMS devices are not protected at all against external factors such as moisture and dust particles. A possible solution to circumvent this issue is discussed here, and it is based on the deployment of a peripheral ring surrounding the entire RF-MEMS die and, additionally, surrounding also each single device, as the schematic top view in Figure 6.21 shows.

If more RF-MEMS devices are electrically connected to each other on the device wafer, the protective ring can be patterned in such a way to leave the necessary room for the interconnects. The surrounding ring covers the whole vertical gap between the two wafers (i.e., the package and the device wafer), ensuring the encapsulation of the RF-MEMS devices. It can be realized by employing different materials, for instance, SU-8 [191], benzocyclobutene [192], or ACA/ICA materials. All of these solutions ensure the protection of the RF-MEMS devices from direct exposure to harmful factors although, on the other hand, they are not able to maintain a her-

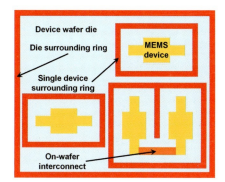

Figure 6.21 Schematic top view of a die on the RF-MEMS device wafer surrounded by a protective peripheral ring. Additionally, a protective ring can also be patterned around each MEMS device.

metic sealing condition (vacuum encapsulation). Nonetheless, if the sealing ring is made of the same alloy used for the solder bumps and if it is reflowed together with them, it would be possible to achieve a stable hermetic sealing of the packaged cavities [193]. In the applications of interest linked to the examples reported in this chapter, the availability of a vacuum packaging solution is not critical. Indeed, such examples refer to RF-MEMS switches and varactors, whose performance is not considerably enhanced when operated in a vacuum. However, if the same packaging solution is applied, for instance, to mechanical resonators realized with MEMS technology, the benefits in terms of higher quality factor, and the consequent lower losses and higher frequency selectivity, enabled by the vacuum condition would be prominent [122].

6.3.1
The Issue of Wafer-to-Wafer Alignment

The success of the application of the package to the device substrate is critically dependent on the quality of the wafer-to-wafer alignment achieved before the bonding step (i.e., adhesion) is performed. If the alignment is coarse, it might happen that vertical vias and bumps are not centered on the RF-MEMS signal pads. This can lead to a poor quality of the package-to-device wafer electrical interfacing (i.e., increased contact resistance) as well as, in the severest cases, to a totally missed interconnection with the RF-MEMS devices as their pads remain floating. Figure 6.22 shows a few possible cases of good electrical interconnection, while Figure 6.23 highlights the cases of poor and totally missed electrical interconnection.

Figure 6.22 Top view of a package electrical interconnect (via plus bump) superimposed on a MEMS signal pad. Examples of good alignment in which the proper electrical connection between the device wafer and the package is achieved.

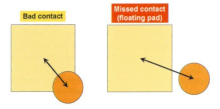

Figure 6.23 Top view of a package electrical interconnect (via plus bump) superimposed on a MEMS signal pad. Examples of bad alignment in which the proper electrical connection between the device wafer and the package is coarse or totally missing.

The latter cases would compromise the utilization of the capped RF-MEMS devices. The example to ensure a precise alignment proposed in this chapter is rather simple. It is based on direct observation under the microscope of the RF-MEMS device wafer through a window opened in the capping part. This allows one to align the two wafers by observing reference structures on the device wafer, which have been chosen as alignment marks prior to the window opening. First of all, the window is preformed by etching (Bosch DRIE process) a narrow through-wafer peripheral groove. Small tips (i.e., anchor points) are left nonetched at the four corners in order to keep the inner silicon part in place [194]. Once the wafer has been brought out of the DRIE machine, the windows are released by ultrasonic cleaning, as the anchors break because of the vibrations. A SEM microphotograph of a cavity after the narrow groove etching is shown in Figure 6.24. The same cavity after the release by ultrasonic cleaning is reported in Figure 6.25.

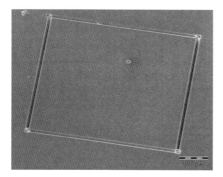

Figure 6.24 SEM microphotograph of a preformed window realized by etching (Bosch DRIE process) a narrow peripheral groove. The four anchors at the corners are visible.

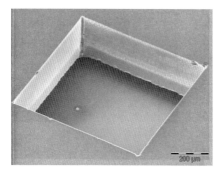

Figure 6.25 SEM microphotograph of the window after the release by breaking the anchor points is achieved via ultrasonic cleaning.

6.3.2
Hybrid Packaging Solutions for RF-MEMS Devices

The packaging solution proposed in this chapter enables the integration of capped RF-MEMS devices with the circuitry realized with standard (CMOS) technology (i.e., hybrid packaging). Basically, two different approaches are possible in order to use the capping part to make easier the interfacing of the two parts (RF-MEMS and CMOS). The first one is shown in Figure 6.26, where the CMOS chip is placed upside-down within a through-wafer cavity in the capping substrate with the same technique as described in Section 6.3.1.

A suitable matrix of interconnects must be deployed directly on the device wafer in order to properly contact the CMOS chip signal pads. This can be patterned at the same time as the fabrication of the RF-MEMS devices, or can be added later in a metal deposition step (postprocessing). The electrical interfacing of the MEMS to the CMOS part is provided by vertical vias and signal overpasses (i.e., coplanar waveguides, CPWs). In this way, the cavities where the RF-MEMS devices are accommodated are not affected by the opening of the cavity hosting the CMOS chip. Hence, if a bonding solution ensuring the vacuum sealing is employed, it is possible to keep such a hermetic condition even after the integration with the CMOS part. After the solder balls have been deposited on the electrical pads of the hybrid RF-MEMS/CMOS capped die, the latter is ready to be mounted upside-down on the board (see Figure 6.3). If the electrical signals are unevenly distributed over the surface of the capped die, it is possible to add some dummy solder balls (see Figure 6.26) to enhance the mechanical stability and strength of the flip-chip mount on the board.

Another possible solution to integrate the RF-MEMS with the CMOS part is shown in Figure 6.27. In this case the CMOS chip is placed upward (and glued) in a shallow etched cavity in the package. The electrical interfacing is done by means of wire bonds. It is unnecessary to deploy in advance any interconnect matrix scheme on the device wafer, and the CMOS chip can be freely placed anywhere in the package die. The area management can also be improved since it is possible to three-dimensionally stack the CMOS chip on top of the area of the RF-MEMS interconnects by means of waveguides, matching networks, and so on.

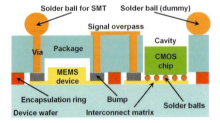

Figure 6.26 Hybrid packaging solution based on the placement of the CMOS chip upside-down within a cavity etched through the package.

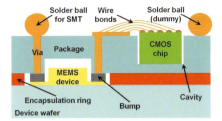

Figure 6.27 Hybrid packaging solution based on placement of the CMOS chip within a shallow cavity. The electrical interconnection of the two blocks is ensured by wire bonding.

6.4
Fabrication Run of Packaged Test Structures

In order to test the viability of the packaging solution discussed so far, a capping substrate to be applied to a wafer of 50 Ω CPWs and shorts was fabricated at the DIMES Technology Center. This choice was made mainly because the influence of the capping substrate is more obvious and can easily be interpreted when applied to structures with a predictable and well-known frequency response. Indeed, in this preliminary stage of the package process development it would not be useful to focus directly on its influence on the RF performance of specific MEMS devices. Moreover, there is also a practical reason supporting this choice. In the package process development, it is preferable to use simple test structures without the need to sacrifice actual RF-MEMS devices, which also represents a low-cost solution. The 50 Ω CPWs and shorts are arranged in a die which is repeated several times on the silicon wafer. The CPWs and shorts are made of evaporated gold on a silicon substrate. A microphotograph of an uncapped gold CPW is shown in Figure 6.28, where it is possible to notice the considerable roughness of the silicon surface.

This first packaging test run is not performed at wafer-to-wafer level, which means that not an entire capping wafer but a single die is bonded to the device test wafer. The reason is that at this preliminary stage confidence in processing has

Figure 6.28 Microphotograph of an uncapped evaporated gold coplanar waveguide (CPW). The roughness of the silicon surface is clearly noticeable. The probes for direct on-wafer S-parameter measurement are also visible.

182 | *6 Packaging and Integration*

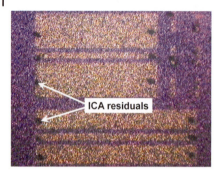

Figure 6.29 Microphotograph of a few test structures after debonding of the package. The location of the black stains (residuals of isotropic conductive adhesive, ICA) indicates a very good chip-to-wafer alignment.

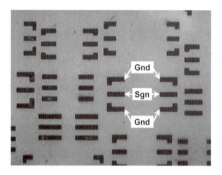

Figure 6.30 Top-side copper pads (on the capping chip) for some of the underlying test structures. "Gnd"' represents "ground" and "Sgn" represents "signal."

to mature and it is easier to perform several trials at chip level instead of at wafer-to-wafer level. The chip-to-wafer bonding is performed by means of an ICA (see Section 6.2.2) dispensed with a needle onto the capping back-side vias. The particular type of ICA used is CE 3103 WLV provided by Emerson & Cuming [186, 195]. In Figure 6.29, a few test structures are shown after the debonding of the capping chip. The visible black stains are the ICA residuals, which indicate a very good alignment of the package with the underlying structures. Figure 6.30 shows the top-side pads (see Figure 6.10) over the capping chip for some of the packaged test structures.

Experimental data have been collected both for the uncapped and the capped test structures mentioned above. The values of the S parameters of an uncapped transmission line are compared with those of the same device after the capping part has been applied in Figure 6.31 (reflection parameter – S11) and Figure 6.32 (transmission parameter – S21).

The CPW length is 1350 µm, while the widths of the ground and signal lines are 300 and 116 µm, respectively. The gap between the signal and ground lines is 65 µm, while the thickness of the evaporated gold layer is around 500 nm. The

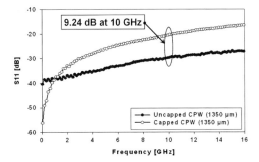

Figure 6.31 Experimental comparison of the S11 parameter for an uncapped and a capped CPW (line length 1350 μm).

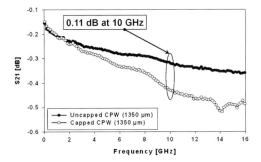

Figure 6.32 Experimental comparison of the S21 parameter for an uncapped and a capped CPW (line length 1350 μm).

thickness of the package is around 280 μm, the length of the top-side pads is 200 μm, while the via diameter is 50 μm. The bottom recesses are not etched in the package as CPWs and shorts do not require a distance clearance larger than the bumps themselves at this preliminary stage.

The reported S-parameter plots show acceptable offsets introduced by the package. For instance, the increase of the reflection parameter due to the capping is 9.24 dB at 10 GHz, while the decrease of the transmission parameter is 0.11 dB at the same frequency. This suggests that a not too large influence of the proposed packaging method on the RF performance of capped MEMS devices can be expected. The analyzed frequency range is limited to 16 GHz because of the equipment available for the experimental testing.

Other comparisons of S parameters related to an uncapped and a capped CPW are shown in Figure 6.33 (reflection parameter – S11) and Figure 6.34 (transmission parameter – S21). In this case, the transmission line is 1200 μm long and the widths of the signal and ground lines are 100 and 210 μm, respectively. The gap is 60 μm and the via diameter is 50 μm. The length of the top-side pads is 200 μm. The S-parameter plots for the uncapped and capped lines show an offset of 13.48 dB at 8 GHz for the reflection parameter and 0.21 dB at 8 GHz for the transmission parameter.

184 | *6 Packaging and Integration*

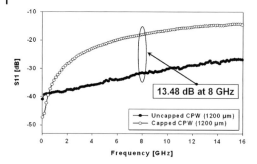

Figure 6.33 Experimental comparison of the S11 parameter for an uncapped and a capped CPW (line length 1200 μm).

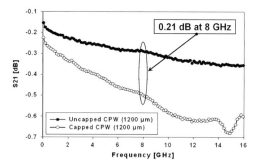

Figure 6.34 Experimental comparison of the S21 parameter for an uncapped and a capped CPW (line length 1200 μm).

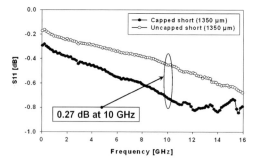

Figure 6.35 Experimental comparison of the S11 parameter for an uncapped and a capped short.

Finally, experimental data related to an uncapped and a capped short are reported in Figure 6.35 (S11, rectangular plot) and Figure 6.36 (S11, Smith chart).

The length of the short is 1350 μm and the widths of signal and ground lines are 116 and 300 μm, respectively. The gap is 65 μm, and, concerning the package, the diameter of the vertical vias is 50 μm and the length of the top-side pads is 200 μm. The offset of the reflection parameter, referred to the capped short, compared with

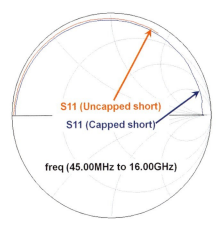

Figure 6.36 Experimental comparison (Smith chart) of the S11 parameter for an uncapped and a capped short.

that of the uncapped structure is rather small (see Figure 6.35) and, for instance, it is 0.27 dB at 10 GHz. In the Smith chart for the S11 parameter in Figure 6.36, one can see an extra rotation for the reflection parameter of the capped short with respect to that of the uncapped short. Indeed, the electrical signal has to travel along a longer path due to the vertical vias and the top-side pads.

In conclusion, the experimental data described above prove the viability of the packaging solution chosen from the technology point of view. Moreover, data concerning losses and mismatch, introduced by the effect of the capping part on the RF performances of test structures, are rather promising. From the point of view of the yield, in all the chip-to-wafer tests a large part of the capped devices was measurable (about 50%), which suggests a rather high level of confidence in technology issues such as alignment and bonding. Also the achieved repeatability of experiments is rather good, as multiple measurements of the same capped devices on different dies bonded in different experiments gave very similar results. However, accurate electromagnetic optimization is a further necessary step in order to develop a successful packaging method for RF-MEMS devices, and this will be discussed in the following pages.

6.5
Electromagnetic Characterization of the Package

According to what has been discussed so far in this chapter about packaging of RF-MEMS, the capping part has to meet several requirements. First of all, its main task is to provide appropriate protection to vulnerable devices from factors that represent a jeopardy, such as moisture, dust particles, shocks, and vibration. Moreover, it has also been stated that the package substrate must facilitate both integration with different functional blocks within the packaged chip itself (i.e., hybrid packaging) and the final mount by surface-mount technology on a board (e.g., flip-chip) to enable the operability of an entire system. Additionally, with reference to certain types of devices for which the performance and characteristics strictly depend on the operation in a vacuum (e.g., MEMS resonators), the establishment of the hermeticity condition is also desirable regarding the package. Beside all these considerations, a further requirement has to be satisfied by the protective substrate when the latter is conceived for devices operating in the RF range [196]. Depending on the type of strategy for redistribution of signals offered by the packaging solution (e.g., through vias, underpass, V-grooves), unwanted parasitics (capacitive and inductive couplings) will affect the RF performance of the capped devices. Thereafter, the protective substrate itself interferes with the electromagnetic field of the devices because of its closeness to them, introducing additional losses and mismatch. Experimental data of the first packaging test run shown above demonstrate that the technology solution presented in this chapter does not have a large impact on the RF characteristics of the capped RF-MEMS test structures. However, a further step will be taken in this direction throughout the current section, showing extensive electromagnetic optimization of the package in order to reduce losses and mismatch affecting the RF behavior of packaged MEMS devices as much as possible.

6.5.1
Validation of the S-parameter Simulations of Packaged Test Structures

The validation of the S-parameter (3D finite element method, FEM) simulations (in ANSYS HFSS™ [157]) in order to predict the electromagnetic behavior of actual capped structures is performed by comparing experimental data of capped CPWs shown in the previous section with the simulated results of the same structures. The use of the HFSS FEM simulation tool was discussed in detail in Section 4.5. Comparisons of the reflection parameter (S11) and the transmission parameter (S21) for the capped CPW, shown in Figures 6.31 and 6.32, with the simulated results are reported in Figures 6.37 and 6.38. The offset between the simulated and experimental data is rather small in the frequency range analyzed, concerning both the S11 parameter and the S21 parameter.

For example, for the simulated curve the difference for the S11 parameter is 5.68 dB at 6 GHz, while for the S21 parameter it is 0.27 dB at 6 GHz. The same type of comparison for another capped CPW geometry shown in Figures 6.33 and 6.34 is reported in Figures 6.39 and 6.40.

In conclusion, the comparisons just reported validate the 3D FEM-based electromagnetic simulation approach chosen for the prediction of the S-parameter behavior of packaged test structure. This is a necessary step before using the FEM tool for the optimization of the available degrees of freedom (DOFs) in the package design, aiming to minimize the impact of the package on the RF performance and characteristics of the capped devices.

6.5 Electromagnetic Characterization of the Package | 187

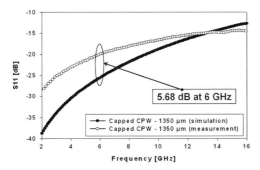

Figure 6.37 Comparison of the experimental S11 parameter (shown in Figure 6.31) with the S11 parameter predicted by the 3D finite element method (FEM) simulation for a CPW of length 1350 μm.

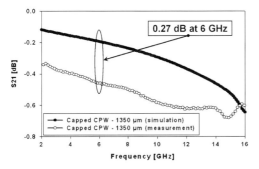

Figure 6.38 Comparison of the experimental S21 parameter (shown in Figure 6.32) with the S21 parameter predicted by the 3D FEM simulation for a CPW of length 1350 μm.

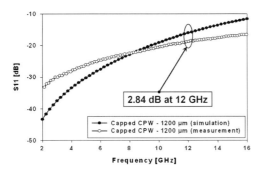

Figure 6.39 Comparison of the experimental S11 parameter (shown in Figure 6.33) with the S11 parameter predicted by the 3D FEM simulation for a CPW of length 1200 μm.

6.5.2
Parameterized S-parameter Simulation of Packaged Test Structures

In order to make easier the electromagnetic optimization of the capping part, a fully parameterized model of a packaged transmission line was implemented in

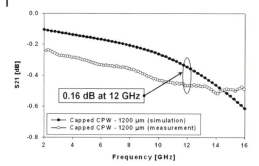

Figure 6.40 Comparison of the experimental S21 parameter (shown in Figure 6.34) with the S21 parameter predicted by the 3D FEM simulation for a CPW of length 1200 μm.

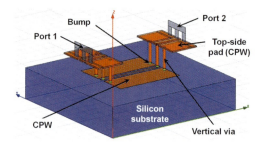

Figure 6.41 Schematic view in ANSYS HFSS of a capped CPW. The capping substrate was made invisible to get a plain view of the underlying CPW as well as the vertical through-wafer vias.

HFSS [186]. In such a model, all the features of the package fabrication process (i.e., DOFs) which can be modified are included, for example, via diameter, height of the capping substrate, and recess depth. Such features are defined as parameters so that automated (parametric) simulations in which they are modified within ranges defined by the user can be launched. The observation of the results allows one to recognize the influence of each variation on the reflection and transmission parameters (S11 and S21, respectively) for each technology DOF, leading to the definition of guidelines to be followed in order to achieve the optimum design from the point of view of RF losses and the characteristic impedance-mismatch reduction. A schematic view of a capped CPW in HFSS is shown in Figure 6.41. In this case the excitation of the structure is defined in HFSS as Lumped Port rather than Wave Port (for details refer to Section 4.5).

An example of a two DOFs optimization, performed with the 3D FEM simulator, by using the parameterized model just discussed, is shown in Figures 6.42 and 6.43. The transmission line is 2000 μm long. The widths of the signal and ground lines are 150 and 900 μm, respectively, while the gap is 30 μm. The height of the capping part is 350 μm and the frequency of this analysis is set to 15 GHz. The two technology DOFs taken into account are the via diameter and the resistivity of the silicon substrate employed for the fabrication of the capping part. Both of these DOFs are reported on the XY plane in the previous 3D plots, while on the

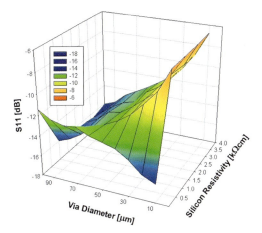

Figure 6.42 Reflection parameter (S11) for the two degrees of freedom (DOFs) optimization (via diameter and package substrate resistivity) performed at 15 GHz with the parameterized model in HFSS.

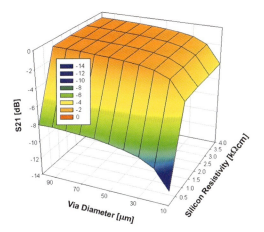

Figure 6.43 Transmission parameter (S21) for the two DOFs optimization (via diameter and package substrate resistivity) performed at 15 GHz with the parameterized model in HFSS.

vertical axis the reflection and transmission parameters (S11 and S21, respectively) are reported. The via diameter ranges between 10 and 100 µm, which is compatible with the proposed process. Concerning the second DOF, wafers with various resistivities are on stock at DIMES Technology Center. It is possible to use, for example, low-resistivity silicon (LRS) wafers, with resistivity of about 15 Ω cm, or four high-resistivity silicon (HRS) wafers with resistivity of 1, 2, 3, and 4 kΩ cm. The 3D plot in Figure 6.42 shows that lower values of the reflection parameter (S11) are achieved for smaller values of the via diameter with the LRS (15 Ω cm). For instance, the S11 parameter is 16.1 and 13.6 dB at 15 GHz for via diameters of 10 and 40 µm, respectively. This (somehow) unexpected result is, in fact, a false optimum as will be shown in the analysis of the S21 parameter. More appropriately, low values for the S11 parameter are also observable when the via diameter chosen is large (over 60 μm) and as HRS substrate is employed (1–4 kΩ cm). Indeed, for the 2 kΩ cm substrate, the reflection parameter ranges between 13.4 and 16.5 dB at 15 GHz for an increase in via diameter from 60 to 100 μm.

The same analysis is shown for the transmission parameter S21 in the 3D plot in Figure 6.43. It is obvious that the poorest values are obtained for narrow vias etched in an LRS substrate. In general, the 15 Ω cm silicon must be avoided since when it is used the S21 parameter ranges between 12.34 and 8.10 dB at 15 GHz for a via diameter from 10 to 100 μm, respectively. This consideration complies with the presence of substrate losses due to eddy currents surrounding the vias. In contrast, high values of the transmission parameter (S21) are achieved for an HRS substrate (1–4 kΩ cm) in combination with large vias (above 60 μm). For instance, it falls in the interval from 1 to 0.64 dB at 15 GHz for a via diameter ranging between 50 and 100 μm when a 2 kΩ cm silicon substrate is employed. Eventually, the optimization of both the reflection parameter (S11) and the transmission parameter (S21) with the 3D FEM electromagnetic simulator shows that the best choice is represented by vias with a diameter of at least 60 μm. Concerning the type of substrate, while the use of the LRS is not an option, the choice of the proper high-resistivity substrate brings up a possible trade-off between performances and costs. In this respect, since the benefits achieved in terms of S parameters with very high ohmic substrates (3–4 kΩ cm) are not significantly greater than those achieved with 1 and 2 kΩ cm silicon wafers, the best choice would probably fall between one of the latter two, the use of the 2 kΩ cm silicon wafer being the most reasonable choice.

After extensive simulations including all the technology DOFs had been run, the ones that exhibit the greatest influence on the RF behavior of various capped CPW geometries were identified. These are listed in Table 6.1 together with the trend for each of them regarding additional losses and mismatch introduced by the capping part as a consequence of their increase.

For instance, the increase of the recess depth as well as of via diameter leads to a reduction of parasitic effects associated with the package, while increasing the height of capping part leads to an undesired increase of losses and mismatch. This information allows one to define appropriate ranges within which the values of the

Table 6.1 Summary of the most important technology degrees of freedom available in the package fabrication and the influence on the radio frequency performance of microelectromechanical systems (MEMS) associated with their increase.

Capping silicon substrate resistivity	↗	Losses and mismatch	↘
Capping silicon substrate height	↗	Losses and mismatch	↗
Recess depth	↗	Losses and mismatch	↘
Via diameter	↗	Losses and mismatch	↘
Ground–signal–ground via lateral distance	↗	Losses and mismatch	↘
Height of bumps	↗	Losses and mismatch	↘

package DOFs should fall in order to ensure that there are only small parasitics affecting the electromagnetic behavior of capped RF-MEMS devices. Of course, in the definition of the guidelines for the optimum design of the capping part, issues related to technology are also to be accounted for. For example, the package thinning leads to better RF performance, but a trade-off emerges with the mechanical strength of the cap itself, which decreases with its height. When this issue is accounted for, the optimum package height is found to fall between 250 and 300 µm, which is a consequence of the specific technology platform employed. Concerning the substrate resistivity, the choice satisfying the trade-off that involves reduction of the parasitics and the costs is the 2 kΩ cm HRS one. Moreover, the recess depth should be large but not too large, according to the mechanical strength reduction of the cap just mentioned. When a 250–300 µm thick package is employed, a reasonable recess depth should be around 100 µm. Finally, the diameters of the vias should be as large as possible (not smaller than 60–70 µm) in order to reduce their resistance, and the horizontal spacing between the signal and ground vias has to be large as well (more than 250 µm) in order to reduce the capacitive coupling between them. The height of the bumps must be large (not smaller than 20–40 µm) in order to keep the largest gap possible between the device and the capping wafers. Finally, noncritical DOFs concerning the package-related parasitic effects are also identified, such as the oxide layer's thickness on the via sidewalls, which can then be freely chosen in the process without particular restrictions. All these considerations are based on the results of parametric simulations performed over the frequency range from 2 to 16 GHz.

6.6
Influence of Uncompressed ACA on the RF Performance of Capped MEMS Devices

As previously discussed in Section 6.2.2, one option to achieve adhesion relies on the use of ACAs. The characteristic of such a material is to behave as an electrically nonconductive material when it is not compressed. Nonetheless, isolated conductive particles are somehow distributed within the ACA and this might represent a source of additional unwanted capacitive couplings at high frequency. For this purpose, FEM electromagnetic simulations are performed to carry out a preliminary study and assess whether the presence of conductive particles can significantly affect the RF behavior of capped MEMS structures. Very simple test structures are chosen for this analysis, that is, 50 Ω CPWs and open structures. Such structures are simulated by introducing a layer of insulating material with a certain number of conductive spheres (made of gold) distributed within it. The percentage of conductive material with respect to the ECA volume has been assumed to be in the range of 5–10% [178] (see the considerations before in this chapter concerning ECA materials). The results of simulations for the structures including the ACA layer are compared with the ones of the same structures without the material. Typical cases that can occur during the wafer-to-wafer bonding concerning the placement of the layer of uncompressed ACA are taken into account. The first structure analyzed is

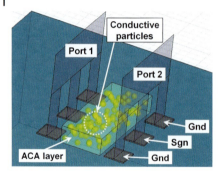

Figure 6.44 HFSS 3D schematic view of the open structure with a layer of uncompressed ACA in between the two ports.

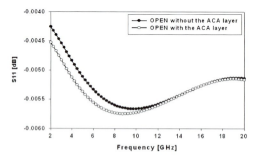

Figure 6.45 S11 parameter plot for the open structure with a layer of uncompressed ACA between the two ports compared with the one for the same structure without the ACA layer.

an open structure [197]. An HFSS 3D schematic view of it with a layer of uncompressed ACA in between the two ports is shown in Figure 6.44. Simulations were performed from 2 to 20 GHz.

The comparison of the S-parameter plots for the structure in Figure 6.44 with those for the same structure without a layer of uncompressed ACA is reported in Figure 6.45 (reflection parameter – S11) and Figure 6.46 (transmission parameter –

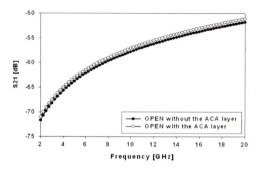

Figure 6.46 S21 parameter plot for the open structure with a layer of uncompressed ACA between the two ports compared with the one for the same structure without the ACA layer.

6.6 Influence of Uncompressed ACA on the RF Performance of Capped MEMS Devices

S21). It can be seen that both the S11 parameter curve and S21 parameter curve for the open structure with an ACA layer are very close to those for the open structure without a layer of uncompressed ACA. Consequently, a rather small influence due to the presence of the layer of uncompressed ACA is to be expected when it is placed, for example, between adjacent RF-MEMS structures in a real device wafer, as the cross talk between them is not significantly increased.

Another realistic case that might occur in a device wafer is when an RF-MEMS structure is surrounded by a layer of uncompressed ACA (i.e., sealing ring). For this purpose, an ACA layer is patterned around a CPW; an HFSS 3D view of the test case is shown in Figure 6.47. Also in this case the RF behavior of the CPW does not seem to be significantly affected by the presence of the layer of uncompressed ACA, as is noticeable from the S-parameter plots in Figures 6.48 and 6.49. Moreover, the case in which a layer of uncompressed ACA is suspended above a CPW is also taken into account [197], confirming the same scarce influence on the RF characteristics of the test structures. The example is not reported here for the sake of brevity.

In conclusion, according to this preliminary study, the presence of uncompressed ACA does not seem to introduce significant additional parasitic effects on the RF behavior of capped structures in any of the cases analyzed.

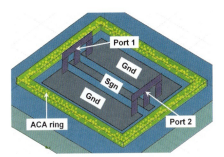

Figure 6.47 HFSS 3D schematic view of a CPW surrounded by a layer of uncompressed ACA.

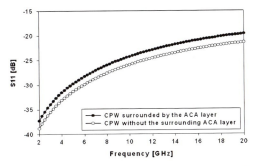

Figure 6.48 S11 parameter plot for a CPW surrounded by a layer of uncompressed ACA compared with the one for a CPW without a surrounding ACA layer.

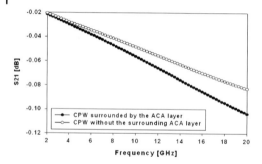

Figure 6.49 S21 parameter plot for a CPW surrounded by a layer of uncompressed ACA compared to the one for a CPW without a surrounding ACA layer.

6.7
Conclusions

The focus of this chapter was the packaging and integration of RF-MEMS devices. Technology aspects related to the realization of a proper protective cap for microsystem-based devices were reported, choosing one among the different fabrication approaches available as a case study. The fabrication was presented step-by-step and the issues of alignment with and adhesion of the package to the RF-MEMS device wafer/chip were discussed. Moreover, the influence of the package on the RF performance and characteristics of the encapsulated RF-MEMS devices was also stressed, illustrated by the comparison of measurements related to test structures with and without a package, as well as simulation methods in order to optimize the package design were reported, aiming at the reduction of its influence on the RF performance of a certain device. Finally, some indications about how to perform the integration of in-package RF-MEMS devices within RF circuits and blocks were provided as well.

Acknowledgments The author would like to acknowledge the DIMES Research Center, Technical University of Delft, the Netherlands, for providing access to the technology resources and supporting the development of the packaging solution for RF-MEMS devices discussed in this chapter. Moreover, the author would like to acknowledge the DIMES Research Center for providing access to the use of the ANSYS (formerly Ansoft) HFSS electromagnetic simulation tool.

7
Postfabrication Modeling and Simulations

Abstract: This chapter can be considered as a practical summary of most of the information, details, and methods discussed in the previous chapters. The behavioral simulation of several of the radio frequency microelectromechanical system case studies introduced at the beginning of the book is performed in this chapter according to different approaches, concerning both the electromechanical and the electromagnetic behavior of microsystems. Differently from the simulation approaches previously described throughout the book, in this chapter the modeling of nonidealities arising from technology aspects is comprehensively considered, starting from the explanation of their physical bases, and then providing their interpretation both qualitatively and quantitatively, having in mind their inclusion in the simulation phases.

7.1
Introduction

Up to now, significant examples concerning the most popular simulation approaches and techniques for the prediction and optimization of the behavior of radio frequency (RF) microelectromechanical system (MEMS) devices have been reported and discussed. In particular, the finite element method (FEM) approach, based on commercial tools, was discussed in Chapter 4, while the development of on-purpose compact models for the fast and parameterized simulation of RF-MEMS devices and complex networks was reported in Chapter 5. However, the examples and procedures presented in previous chapters refer to nominal values and parameters of the given technology platform employed for the manufacturing of RF-MEMS devices. In this chapter, a significant step forward is going to be taken, as the nonidealities and the characteristic aspects of the technology will be observed by referring to the results of experimental testing and characterization of fabricated specimens. Afterward, the nominal values will be tuned in the simulations in order to match the simulated results to the experimental results. Such a refinement of the simulated results should not appear as a mere attempt to fix and adjust the difference between the preliminary simulation and the experimental results. On the other hand, it should be interpreted as a sensible step in order to tailor the simulations to the actual possibilities and characteristics of the technology platform, which, regardless of its level of development and control, will never match perfectly the expected nominal values in terms of material properties, layers thicknesses, in-plane dimensions, surface roughness, electrical and electromagnetic properties, and so on. Only after performing the tailoring step of the simulations according to the actual and effective characteristics of a given technology the simulation tools adopted will actually become a critical resource for the correct optimization of new RF-MEMS layouts, having in mind the specifications with which they are expected to comply. For these reasons, the tailoring of simulations must be performed not by changing parameters, seeking better matching with results, but by fully understanding the physical bases of behavioral deviations of real devices with respect to nominal values.

7.2
Electromechanical Simulation of an RF-MEMS Varactor (Dev B2) with Compact Models

The MEMS device analyzed in this section is the parallel plate variable capacitor with the upper (suspended) plate movable with respect to the lower (fixed) electrode, already discussed in Chapter 2, and referred to as Dev B2. The simulation tool selected for this example is the compact model library developed in-house and discussed in Chapter 5. Figure 7.1 reports the 3D profile of the device obtained experimentally by means of a profilometer based on optical interferometry (Wyko NT1100 DMEMS from Veeco; www.veeco.com (accessed 27 April 2013)).

The capacitance is in shunt-to-ground configuration and the ground–signal–ground (GSG) lines implement a short coplanar waveguide (CPW) structure in input and output. The capacitance presents its minimum value when no DC bias is applied between the two plates. In this case, the whole device implements a closed switch for RF signals. Otherwise, when the plate collapses onto the underlying ox-

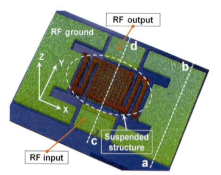

Figure 7.1 Measured 3D static profile of a microelectromechanical system (MEMS) variable capacitor fabricated with Fondazione Bruno Kessler radio frequency (RF) MEMS technology. The color scale refers to the vertical (Z axis) height. The central suspended capacitor plate and its two folded suspensions are visible, as are the ground–signal–ground (GSG) input/output coplanar waveguide (CPW)-like access lines.

7.2 Electromechanical Simulation of an RF-MEMS Varactor (Dev B2) with Compact Models

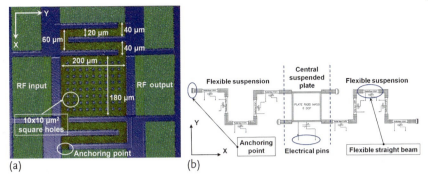

Figure 7.2 (a) Experimental top view of the variable capacitor reported in Figure 7.1. All the most important XY in-plane geometrical features are indicated. (b) Spectre schematic of the MEMS structure implemented with the basic models available in the MEMS software library. Symbols are placed in order to maintain a visual link with the appearance of the real device (e.g., rotated beam symbols composing the folded suspensions).

ide layer, the maximum capacitance is reached and the MEMS device behaves as a capacitive short to ground for RF signals (i.e., open switch). More details about the geometrical features of the RF-MEMS device analyzed are available in the top view shown in Figure 7.2a, also measured with the profilometer. The Cadence schematic, corresponding to the geometry shown in Figure 7.2a, assembled with the elementary MEMS components of the software library, is depicted in Figure 7.2b.

Different symbols are highlighted concerning the flexible beams connected together in order to form the serpentine-based suspensions and the central rigid capacitor plate. Moreover, the anchoring points are visible and they define the mechanical constraints for the ends of the folded beams. All the symbols are connected to each other in the same way as traditional electronic circuits. From a comparison of Figure 7.2a,b, it is easy to understand the relationship between the real MEMS device and the Spectre® schematic. It should be noted that, despite some of the straight beam symbols in the schematic being rotated by 90° to recall the actual folded-suspension configuration, this does not influence their mechanical behavior. In fact, in order to actually rotate the beam model, it is necessary to specify the rotation angle in a set of fields, accessible through the schematic window, where also other parameters can be chosen, such as the beam dimensions and the air gap [163]. First of all, the 3D static profile of the device under test (DUT) is measured in order to analyze the flatness of the suspended parts as well as possible variations of the thickness of the gold against the nominal values reported in the description of the technology (see Chapter 3). Looking again at Figure 7.1, two lines are visible, namely, ab, crossing the device along the X axis on the RF ground, and cd, also crossing the suspended part. The vertical profiles corresponding to the aforementioned lines are reported in Figure 7.3.

The line ab allows one to determine the mean thickness of the two gold metallizations, that is, 4.3 µm, which is in fact lower than the nominal value, 4.8 µm. Moreover, from the line cd, the mean value of the absolute thickness of the sus-

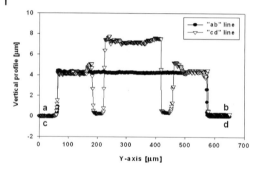

Figure 7.3 Vertical profiles of the MEMS variable capacitor reported in Figure 7.1 along the *ab* and *cd* cross sections obtained with the optical profilometer. The line *ab* allows the calculation of the mean thickness of the two electrodeposited gold layers, while from the line *cd*, the average air gap of the central plate is extracted.

pended plate is determined and, by subtracting from it the effective gold thickness, an air gap of about 2.8 µm is estimated. The suspended central plate is arched and exhibits a difference of about 300–360 nm between the center and the edges induced by the residual stress accumulated within gold during processing. This nonideality will mainly influence the pull-out voltage V_{PO}, which in ideal conditions depends on the thickness of the insulating layer t_{ox}:

$$V_{PO} = \sqrt{\frac{2k t_{ox}^2 g}{\varepsilon_{ox} W L}} \qquad (7.1)$$

where k is the elastic constant of the suspensions, g is the gap between the two faces, ε_{ox} is the oxide permittivity, and W and L are the dimensions of the faces [92]. The curvature of the real device, together with the roughness of the surfaces in contact, introduces a residual gap of air in the actuated state that, on one hand, increases the pull-out voltage and, on the other hand, reduces the on-state capacitance of the MEMS variable capacitor (as will be discussed later this chapter). Figure 7.4 reports a schematic view of what happens in a real device, where the bowing induced by the residual stress gradient within the suspended metal layer plus the surface roughness reduces the actual contact surface.

The presence of a residual air gap between the pulled-in membrane and the underlying oxide can be easily accounted for in the simulations by including a constant effective residual air gap, schematically shown in Figure 7.4. The pull-in/pull-out static characteristic is experimentally determined by cycling the DUT with a 20 Hz zero mean value triangular voltage ranging from −20 to +20 V. The bias frequency is low enough to disregard all the dynamic effects due to inertia and viscous damping. The measurement is performed with an optical profiling system based on interferometry equipped with a pulsed illuminator (i.e., stroboscopic light) synchronized via a General Purpose Interface Bus with a waveform generator providing the biasing stimulus for the observed device. The stroboscopic light source is pulsed at the same frequency, but from one measurement step to the next its phase with respect to the bias is changed. In this way, it is possible to take

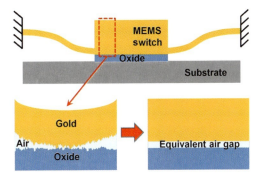

Figure 7.4 Schematic representation of the gold-to-oxide nonideal surface contact (top). The surfaces roughness together with the bowing of the suspended gold membrane (magnified in the sketch) leads to the presence of a residual air gap between the contact areas (bottom left). This effect can be accounted for in the simulations by including a constant equivalent residual air layer (bottom right).

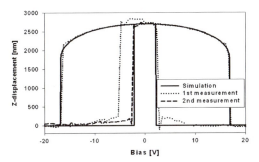

Figure 7.5 Experimental versus simulated pull-in/pull-out characteristic of the MEMS variable capacitor shown in Figure 7.1. The experimental vertical Z-displacement is measured with the profilometer, and corresponds to an average value over the plate area. The controlling signal is a 20 Hz, zero mean value, 40 V peak-to-peak triangular voltage. The first quarter of the second measurement is superimposed on the first one, showing how the pull-out voltage shifts as a result of the smoothing of the contact surface.

a snapshot of the moving suspended structure at a different point of its behavior at each phase change [30]. Eventually, the composition of such points allows one to reconstruct the displacement characteristic of the DUT. This technique is necessary because the acquisition time of the profilometer is too long to perform a real-time observation of a moving device. Concerning the schematic in Figure 7.2b, a DC static simulation is performed in Spectre by applying the same stimulus to bias the measured sample. A voltage source from a standard Cadence library of electronic components generates such a waveform. Figure 7.5 reports the measured and simulated pull-in/pull-out characteristics. First of all, the good match of the simulated and measured pull-in, occurring in both cases at an applied bias of 16.9 V, is visible on both the positive and the negative voltage axis.

The vertical displacement characteristic from the 0 V applied bias to the plate collapse is also accurately predicted by the Spectre simulation. A few considerations are necessary concerning the measured pull-out behavior of the DUT. In Figure 7.5, we see there is a discrepancy between the first pull-out (labeled as "1st measurement") occurring in the first quarter of the biasing signal (corresponding to the bias rising from -20 to 0 V), which occurs at -4.7 V and the second measured pull-out in the third bias quarter (from 20 to 0 V), which occurs at 2.33 V. As explained above, dynamic measurements with stroboscopic light require the DUT to undergo a large number of cycles. The first pull-out ("1st measurement") occurs in the first quarter of the biasing signal, that is, after a limited number of actuation cycles, while the second pull-out occurs in the third quarter of the triangular voltage, after the tunable capacitor has touched the underlying oxide a significant number of times. It is reasonable to assume that cycle after cycle, small plastic deformations occur on the two surfaces in contact [198], leading to progressive slight smoothing of them and, as a final result, to a reduction of the unwanted air layer between them, accompanied by a decrease of the pull-out voltage. On this basis, the second measured pull-out voltage is closer to ideality. In order to further verify this assumption, the pull-in/pull-out characteristic of the same MEMS sample was measured a second time. Only the first quarter of the measurement, that is, the first pull-out, is reported in Figure 7.5, and is labeled as "2nd measurement." In this case, the pull-out occurs at about -2.5 V, very close to both the second pull-out of the first measurement and the simulated result. Concerning the issues related to charge accumulation within the oxide layer [141], which is particularly significant when the switch is in the actuated state, we assume they are negligible in the particular functioning condition analyzed in this case. Indeed, by application of a biasing signal that is symmetric with respect to the zero axis, the positive and negative charges injected into the oxide are balanced, compensating the electromechanical performance drift due to the built-in bias they cause [199].

Further considerations are necessary concerning the nonideal behavior of the MEMS sample introduced by the bowing of the suspended gold membrane. Looking again at the measurement in Figure 7.5, one can see that in the actuated state, when the voltage is in the range of ± 12 to ± 20 V, the central plate is flattened onto the oxide, because of the large attractive force (it should be noted here that the profilometer measures the average vertical displacement over the plate area). However, for a lower bias modulus the plate, while still in the actuated position, begins to bend up near the edges, tending to assume the configuration reported in Figure 7.3. From Figure 7.5 a difference of about 200–240 nm between the position of the plate at a bias of ± 12 to ± 20 V and at a bias slightly higher than V_{PO} can be estimated. Since the measured value is an average over the whole plate area, it roughly corresponds to a maximum total bending from the edges to the center of the plate of about 400 nm, which is consistent with the arching reported in Figure 7.3. Given the above considerations about the smoothing of the contact surfaces and the bowing of the plate, in the Spectre simulations reported in this section we set a residual air gap in the actuated state of 120 nm, which leads to an accurate prediction of the pull-out voltage as shown in Figure 7.5.

A further validation step of the MEMS compact models was performed in dynamic regime. We again measured the vertical displacement of the variable capaci-

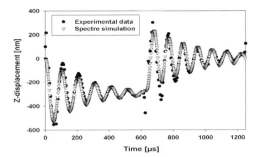

Figure 7.6 Measured versus simulated dynamic response of the MEMS variable capacitor biased with an 800 Hz square pulse voltage (0 V low value, 13 V high value, and 50% duty cycle). Compact models accurately predict the activation's and deactivation's damped oscillations.

tor with the profilometer in response to an applied biasing signal. However, in this case the biasing signal is a 0–13 V square wave with 800 Hz frequency and 50% duty cycle, so the MEMS dynamic behavior (due to inertia and viscous damping) is observable. The high voltage of the pulse is lower than the pull-in voltage. Concerning the simulation, a transient analysis was performed on the schematic in Figure 7.2b in order to include the dynamics of the compact models implemented. The measured versus simulated vertical displacement of the plate is reported in Figure 7.6.

The Spectre simulation predicts quite accurately the amplitude of the damped oscillations for both transitions, while some discrepancies are visible concerning the ringing frequency. In particular, in both transitions, a good superimposition of the first two/three peaks is visible, while in the last (more damped) part of the transient the measured oscillations are faster than the simulated ones. Since the elastic constant k of the simulated MEMS device fits well with the one for the measured sample, as confirmed by the accurate prediction of the pull-in voltage reported in Figure 7.5, the discrepancies in Figure 7.6 could be caused by an overestimation of the simulated suspended mass compared with the real one. However, other important aspects have to be taken into account. As opposed to the static pull-in/pull-out measurement, the square pulse applied in the measurement shown in Figure 7.6 is unipolar and therefore more prone to charge injection from the bottom (fixed) electrode into the silicon oxide. Moreover, in order to capture the dynamic characteristic of the DUT, the stroboscopic light phase change is slower than that used for the measurement in Figure 7.5, and consequently the square pulse is applied to the MEMS sample a greater number of times. The charge accumulated within the oxide when the square pulse is on causes a built-in potential that can be partially responsible for the difference between the simulated and measured transient behavior. In conclusion, the dynamic behavior is the most difficult one to predict, as it also involves the inertia and the viscous damping effects, which apparently are simulated with a good level of accuracy. The differences between the simulations and experimental data highlighted can be traced back to second-order effects not modeled and implemented in the MEMS software library discussed (see Chapter 5). Finally, it is worth remarking that the only fitting parameter used in the simulations presented is the equivalent residual air gap, due to the strong nonidealities of the arched MEMS device in the actuated state, which is rather difficult to model as it depends on several factors (e.g., applied bias, residual stress, and number of cycles the DUT has already undergone). All the other critical parameters presenting small differences compared with the nominal ones, for example, effective gold thickness and air gap, are extracted from experimental measurements performed with the optical profilometer and are not arbitrarily chosen in order to fit measurements. This approach reflects the philosophy underlying the tailoring of simulations to the actual technology parameters, that is, basing the extraction of effective values as much as possible on experimental results and physics-based considerations.

7.3
RF Modeling of an RF-MEMS Varactor (Dev B2) with a Lumped Element Network

In this section, the tailoring approach of simulations with respect to the actual behavior of fabricated RF-MEMS samples will be presented concerning the prediction of electromagnetic (RF) characteristics. The analysis is still based on the compact modeling approach, and the procedure for the extraction of a lumped element network, capturing nonidealities of the DUT, will be discussed in detail. Effort will be expended, as it was for the previously discussed electromechanical behavior, in order to keep the extracted network as close as possible to the physics lying behind the nonidealities. The effectiveness of the proposed approach will be corroborated by discussing RF-MEMS samples with a geometry slightly different from Dev B2, showing that such a variation influences just the element/s within the lumped network describing that particular feature [139]. By joining the material presented in Section 7.2 with what is discussed here, we create a multidomain electromechanical/electromagnetic simulation environment entirely based on compact models. Since the MEMS sample discussed in the previous section was damaged during further measurements, we focus here on one of its (identical) replicas located in a different area of the same wafer. Consequently, the electromechanical properties and 3D features might exhibit small differences. In order to achieve a consistent modeling of the DUT, we performed a complete electromechanical characterization of it before studying the electromagnetic properties. Briefly, the measured gold thickness in this case is 4.2 µm and the average air gap is 2.7 µm, leading to a pull-in voltage of about 14.5 V. Finally, given the bowing of the suspended plate of about 400–450 nm (larger if compared with Figure 7.5), a residual air gap of 200 nm was set in the Spectre simulation in order to fit the measured pull-out voltage of about 2.5 V. After the electromechanical characterization, S-parameter measurements (two ports) were performed on the same sample on a probe station with GSG probes by means of an HP 8719C vector network analyzer (VNA) [200] in the frequency range from 150 MHz to 13.5 GHz, while the controlling voltage was applied directly to the RF probes by means of two bias tees. The DUT was biased at different (constant) voltages. The VNA calibration performed was a SOLT

7.3 RF Modeling of an RF-MEMS Varactor (Dev B2) with a Lumped Element Network

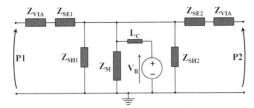

Figure 7.7 Schematic of the lumped element network that models the RF behavior of the intrinsic MEMS variable capacitor including all the surrounding parasitic effects. The voltage generator has a choke inductor L_C connected in series in order to decouple the DC and RF signals. Details of the impedance blocks composing the network are shown in Figure 7.8.

(i.e., short, open, load, thru) calibration [201] on a commercial impedance standard substrate, that is, the reference planes are brought to the GSG tips of the two probes. Consequently, the S parameters acquired include the behavior of the intrinsic variable capacitor (i.e., the MEMS suspended plate) as well as the contribution due to the input/output access CPWs (see Figure 7.1) and parasitic effects, that is, no de-embedding was performed. Given these assumptions, we used a well-known technique usually adopted in microwave transistor modeling [202] based on the extraction of lumped parasitic elements that are wrapped around the intrinsic device. Figure 7.7 shows the schematic of the intrinsic MEMS variable capacitor and the wrapping lumped element network for the surrounding parasitic effects.

In particular, the intrinsic MEMS impedance is labeled as Z_M, while Z_{SE1}, Z_{SH1}, Z_{SE2} and Z_{SH2} model the impedance of the access CPWs at ports P1 and P2, respectively. Furthermore, Z_{VIA} models the impedance due to the parasitic effects introduced by the gold to multimetal through vias (explained in detail later) and L_C is a choke inductor (1 mH) necessary in the Spectre simulations in order to decouple the DC bias from the RF signal. The lumped elements composing Z_M, Z_{SE1}, Z_{SE2}, Z_{SH1}, Z_{SH2}, and Z_{VIA} are shown in Figure 7.8.

The intrinsic MEMS variable capacitor is modeled as a shunt-to-ground capacitance (C_{MEMS}), in parallel with a resistor accounting for small dielectric losses (R_{MEMS}) and in series with an inductance (L_{MEMS}) accounting for the contribution of the two suspending folded beams reported in Figure 7.2 (see Figure 7.8a). The accessing CPWs are modeled according to a well-known lumped network scheme [59], shown in Figure 7.8b. It relies on a series RL section, accounting for the resistive losses within the metal and the line inductance, respectively, and a parallel RC shunt section to ground, modeling the losses within the substrate and the capacitive coupling between the signal and ground planes through the air and through the substrate. Finally, the network in Figure 7.8c accounts for the parasitic effects due to a technology issue linked to the opening of vias through the oxide (discussed in detail in [50]). In more detail, when an ohmic contact between the electrodeposited gold and the buried multimetal layer is required, a dry etching step is performed in order to remove all the oxide and to attack the top titanium layer forming the sandwiched multimetal layer (see Chapter 3, where the Fondazione Bruno Kessler (FBK) RF-MEMS technology is described). The aim in removing the titanium layer is to reach the underlying aluminum, which leads to a better ohmic

7 Postfabrication Modeling and Simulations

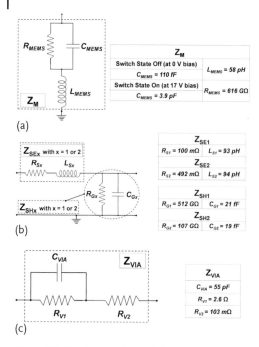

Figure 7.8 Details of the lumped elements composing the network in Figure 7.7 decomposed here into three subnetworks. (a) Intrinsic MEMS variable capacitor. A resistor in parallel with the variable capacitance models the losses, while a series-to-ground inductor accounts for the inductive contribution of the suspending folded beams. The table reports the extracted values of the three elements (also including the off-state and on-state capacitance). (b) Input/output lines. The topology is based on a well-known CPW lumped scheme. The table reports the values of the four elements composing the input (subscript 1) and output (subscript 2) CPWs. (c) Through-oxide via parasitic effects. This subnetwork models the nonidealities related to the opening of vertical vias through the oxide in order to establish an ohmic contact between the gold and the buried multimetal layer.

contact with gold. Because of an inappropriate end point of the dry etching process performed on the batch from which the variable capacitor discussed here came, the titanium layer was not completely removed.

During the wafer handling in a normal atmosphere in the subsequent processing, the residual titanium oxidizes in air, forming a very thin layer of titanium oxide. Optical inspection confirms the presence of titanium oxide. Figure 7.9 reports two microphotographs of a via (belonging to a test structure) concerning two previous fabricated batches (as no photographs of the optical inspections for the batch including the devices discussed here are available).

In one of these batches the opening of vias was done successfully, while in the others residual titanium oxide was present. The light-gray color in Figure 7.9a shows that aluminum is reached after opening of the via, while the brownish color of the via in Figure 7.9b confirms the presence of residual titanium oxide. Such a layer jeopardizes the establishment of a good ohmic contact with gold, introducing additional losses and a series parasitic large capacitance that mainly affects the RF

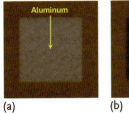 

Figure 7.9 Comparison between a via correctly opened (a) and one that has residual titanium oxide (b), both belonging to previously fabricated batches. The presence of an unwanted very thin titanium oxide layer is confirmed by a brown color that replaces the usual light-gray color of aluminum.

behavior of the variable capacitor in the low-frequency range (as will be discussed later in this section). In Figure 7.8c, we see this nonideality is modeled with a capacitance (C_{VIA}) in parallel with a resistor (R_{V1}) that models the losses in the low-frequency range, plus a series resistance (R_{V2}) accounting for the losses through the whole frequency range.

Once the whole topology of the lumped element network has been fixed, the specific values of all its components are tuned by using the optimization software tool available in the Agilent Advanced Design System (ADS) framework [203]. Suitable targets aiming at the reduction of the difference between measured and modeled S parameters are defined. The first optimization run is performed with the S parameters measured at 0 V. The optimized value of the intrinsic MEMS variable capacitor (110 fF) is compared with the analytical one (see (5.6)) in order to verify the consistency of the optimizer output. Other optimization runs are performed by replacing the target of the first run with the S parameters measured at an applied voltage of 1, 2, 3 V, and so on, up to 14 V, that is, right before the actuation of the plate. The consistency of the extracted lumped element values is monitored step-by-step [139]. To do this, the extracted intrinsic MEMS capacitance is cross-checked with the analytical value computed for each voltage from the vertical displacement known from the experimental measurements (see Figure 7.5). All the element values of the network in Figure 7.8, excluding C_{MEMS}, do not show any significant change with the applied voltage, as expected. Once all the lumped element values have been determined, they are kept fixed and only C_{MEMS} is allowed to change. A few other optimization steps are performed with the S parameters measured beyond the pull-in voltage, from 14 to 20 V. These further steps allow extraction of the MEMS capacitance in the actuated state, corresponding to a value of 3.9 pF. The C_{max}/C_{min} tuning range of the MEMS variable capacitor analyzed is 35.5, and the extracted values of all the other elements concerning the three subblocks of the whole network are reported in the tables included in Figure 7.8. Concerning the accessing CPW short lines, the only element that shows a significant difference between the two ports is the series resistance, equal to 100 mΩ at port 1 and 492 mΩ at port 2. This difference is likely due to the disparate contact resistance between the tips of RF probes and gold pads, which can be influenced by an asymmetric deterioration of the GSG probes. The two reactive elements (L and C) composing

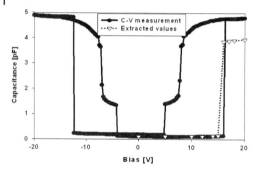

Figure 7.10 Comparison of the C–V measured characteristic of the MEMS structure in Figure 7.1 and its intrinsic capacitance values determined with the lumped element network in Figure 7.7 with the parameters extracted from experimental S parameters for different constant applied bias levels.

the input/output CPWs present very similar values, while, eventually, the losses to ground are so small that they can be disregarded. Concerning via parasitic effects, the series capacitance due to the titanium oxide is rather large (55 pF) and this explains the S-parameter behavior in the low-frequency range (reported and discussed later), while the resistance R_{V1} is mainly responsible for rather large losses. In order to gather further data validating the method presented, C–V (i.e., capacitance versus bias) measurements were performed on the same device. The bias was provided by an Agilent 4156C semiconductor parameter analyzer [204] and the MEMS capacitance was measured with an HP 4285C LCR meter [205]. The applied voltage was swept from 0 to $+20$ V, then back to -20 V and up to 0 V. Figure 7.10 reports the C–V measurement (solid line) compared with the C_{MEMS} values extracted with the optimization tool (dotted line) from the S parameters measured at different bias levels.

Good agreement of C_{MEMS} with the C–V measurements is evident when the capacitor is not actuated, while a difference of about 0.8 pF is detected in the actuated state, with the extracted on-state capacitance smaller than the measured one. Some remarks are therefore necessary. First, the voltage sweep in the C–V measurements is significantly slower than the one applied when using the profiling system described in Section 7.2 and more importantly starts from 0 V and not from -20 V. This means that the charge accumulated in the positive and negative parts of the bias waveform is not negligible and influences the pull-in and pull-out voltages. Indeed, due to the charge accumulated during the C–V measurements the pull-in occurs at $+16$ and -12 V, while in the electromechanical measurements it happens at ± 14.5 V. Concerning the S-parameter measurements, in order to allow the VNA to sweep the whole frequency range, the MEMS bias is kept for a rather long time (several seconds) at fixed values (1, 2, ..., 19, and 20 V). Consequently, the amount of charge accumulated in the actuated state is larger than in the C–V measurements and is not even compensated by applying negative voltages. This issue can easily lead to a reduction of the effective electrostatic force between the capacitor plates (i.e., voltage screening [140]) and, since the MEMS plate tends to be less flat as the attractive force decreases (see Section 7.2), to a small increase of the residual air gap between the two plates and, ultimately, to a reduction of the on-state capacitance. Such an effect is visible in the C–V measurements when the voltage decreases from +20 toward 0 V, as well as when it increases starting from −20 V. The bending up of the plate (when still actuated) due to the gold residual stress causes a capacitance reduction of about 1 pF between its value measured at ±20 V and at about ±7 V. Finally, the nonideal mechanical behavior of the DUT during the release is also visible in the C–V curve, where the pull-out presents a double step transition, suggesting that it is characterized by two abrupt upward plate movements, separated by a time interval where the bowing of the plate increases, causing a smoother capacitance reduction. In conclusion, the capacitance validation by means of C–V measurements allows verification of the correctness of the extracted value even if nonidealities of the device and differences in the measurement setup can make such a cross-check more qualitative than quantitative as happens in this case. After the complete lumped element network has been extracted, the C_{MEMS} element in Figure 7.7 is replaced in the Spectre schematic editor with the network based on the MEMS model library as discussed in Chapter 5 (see Figure 7.2b). In this case, the residual air gap has to be refitted, exhibiting values that are different from those extracted in Section 7.2. This aspect will be explained in more detail later in this section. S-parameter simulations are performed and the intrinsic MEMS capacitance is calculated by means of the compact models. This enables one to observe the change in S parameters dependent on the DC voltage applied to the suspended MEMS variable capacitor in Figure 7.2 (V_B in Figure 7.7). The measured and simulated S11 and S21 parameters at 0 V (nonactuated plate, closed switch) are reported in Figures 7.11 and 7.12, respectively.

The influence of the via parasitic effects are particularly visible up to 2 GHz, with a reflection parameter (S11) of about 25 dB and a transmission parameter (S21) worse than 0.2 dB. Since the analyzed MEMS device in the capacitance low-value state is supposed to behave as a standard CPW, that is, with very reduced loss and reflection at low frequencies, this discrepancy is due to the gold to mul-

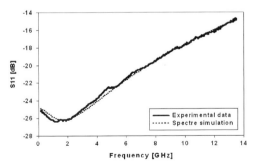

Figure 7.11 Measured and simulated S11 (reflection) parameter with 0 V applied bias (not actuated plate). The S-parameter simulation was performed on the schematic in Figure 7.7, where the intrinsic variable capacitor was implemented with the MEMS compact models as reported in Figure 7.2b.

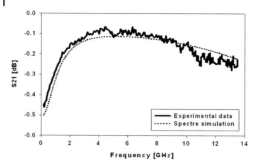

Figure 7.12 Measured and simulated S21 (transmission) parameter with 0 V applied bias (nonactuated plate). The S-parameter simulation was performed on the schematic in Figure 7.7, where the intrinsic variable capacitor was implemented with the MEMS compact models as shown in Figure 7.2b.

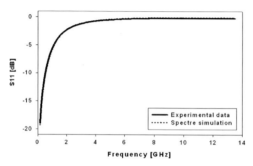

Figure 7.13 Measured and simulated S11 (reflection) parameter with 20 V applied bias (actuated plate). The S-parameter simulation was performed on the schematic in Figure 7.7, where the intrinsic variable capacitor was implemented with the MEMS compact models as reported in Figure 7.2b.

timetal transition parasitic effects discussed above. The simulated curves are in very good agreement with the measurements, confirming the correct prediction of the MEMS intrinsic capacitance described by the compact models developed. The same consideration is valid also when the MEMS device is pulled in. Figures 7.13 and 7.14 report the simulated and measured return loss (S11) and isolation (S21), respectively, when the applied DC bias is 17 V. In this case, the contribution of the via at low frequency is hidden by the large S11 parameter and the wide variation of S21. In order to prove the consistency and predictive capability of the modeling method presented, the RF behavior of two more switches with different geometrical sizes but topology similar to the one discussed above has been considered and will be presented.

Figure 7.15 reports the top view of the three devices observed with the optical profiling system. Sample (a) is the one discussed so far, while samples (b) and (c) have the same geometrical features except for the plate width, which is reduced to 160 and 100 μm, respectively. All the other dimensions are the same as reported in Figure 7.2a. Since only the plate width changes, the lumped element network

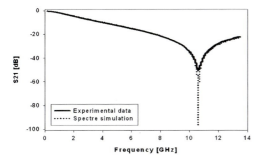

Figure 7.14 Measured and simulated S21 (isolation) parameter with 20 V applied bias (actuated plate). The S-parameter simulation was performed on the schematic of Figure 7.7, where the intrinsic variable capacitor was implemented with the MEMS compact models as reported in Figure 7.2b.

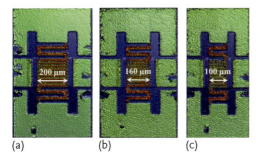

Figure 7.15 Top view of three MEMS variable capacitors having the same topology as the one in Figure 7.2a with different widths of the suspended plates (the one labeled as (a) is the same as the variable capacitor in Figure 7.2a). All the other geometrical features are the same for the three samples and the measured replicas are located in the same area of the wafer.

in Figures 7.7 and 7.8 is expected to model correctly the RF behavior also of samples (b) and (c) with a modification only of Z_M.

Starting from the measured S parameters for different bias levels, we used the ADS optimization tool once again to extract the network element values. The extracted values for the Z_M subnetwork are summarized in Table 7.1, while all the other elements have the same values previously reported. C–V measurements were also performed on sample (b) and are compared with the extracted C_{MEMS} values in Figure 7.16. As opposed to sample (a) (see Figure 7.10), the disagreement in the on state is only 0.3 pF. Concerning L_{MEMS}, a decrease is detected from sample (a) to sample (c), consistent with the decrease of the length of the folded suspension. For this purpose, analytical calculations were carried out in order to validate the extracted values. The inductance of a single folded suspension can be approximated as the result of two contributions: the first one given by the two parallel long arms, and the second one given by the sum of the three short beams connecting the suspended plate to the CPW (see Figure 7.2).

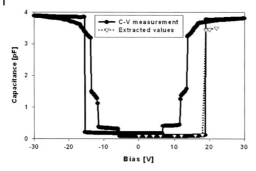

Figure 7.16 Comparison of the C–V measured characteristic of MEMS device (b) in Figure 7.15 and its intrinsic capacitance determined with the lumped element network shown in Figure 7.7 with the parameters extracted from experimental S parameters for different constant applied bias levels.

By use of well-known formulas, the self-inductance of the two parallel flexible arms was determined to be 81, 65, and 40 pH for a single suspension of samples (a)–(c), respectively. The second contribution, constant for three different topologies as the length of the short connecting beams does not change, was calculated with the formula for the inductance of a straight wire and is about 30 pH. The analytically calculated total inductance for a single suspension is then 111, 95, and 70 pH for samples (a)–(c), respectively. Comparing these values with the extracted ones (reported in Table 7.1) and considering that the latter refer to two folded suspensions in parallel, we find the difference is 5 pH for samples (a) and (b) and 10 pH for sample (c). This proves both the consistency of the extracted values and the achievement of a parameterized description of the changed geometrical features. The residual air gaps for the three samples are also reported in Table 7.1.

They were extracted using (5.6) expressed for two capacitances in series (air plus oxide) using the on-state capacitance values reported in the same table. The extracted air gaps are all in the range of a few tens of nanometers as expected. These values are definitely smaller than the effective residual air gap (120–200 nm) used in the electromechanical simulations in order to fit the pull-out voltage (see Section 7.2). This is because the S parameters used for extraction of the on-state capacitance are

Table 7.1 Values of the lumped elements describing the intrinsic microelectromechanical system (MEMS) variable capacitor (see Figure 7.8a) of the three samples in Figure 7.15. The effective residual air gap calculated analytically from the on-state capacitance is also reported.

Lumped element	Sample (a)	Sample (b)	Sample (c)
C_{MEMS} off	110 fF (at 0 V)	95 fF (at 0 V)	85 fF (at 0 V)
C_{MEMS} on	3.9 pF (at 17 V)	3.46 pF (at 22 V)	2.61 pF (at 23 V)
L_{MEMS}	58 pH	45 pH	41 pH
R_{MEMS}	233 GΩ	476 GΩ	908 GΩ
Residual air gap (on state)	40 nm	34 nm	24 nm

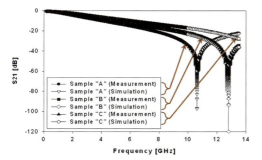

Figure 7.17 Comparison of the isolation (S21 parameter) for samples (a)–(c) (see Figure 7.15) measured and simulated with the network in Figure 7.7. The increase of the isolation peak frequency as the on-state capacitance is reduced from sample (a) to sample (c) is clearly visible.

measured for bias levels higher than the pull-in voltage (i.e., larger attractive force), while the pull-out takes place at lower voltages, when the bowing of the MEMS plate reaches the largest value. Moreover, despite the significant variability of the on-state capacitance due to the MEMS nonidealities, from Table 7.1, it is possible to notice a dependence of the residual air gap on the plate dimensions.

Figure 7.17 reports the measured isolation (S21) curves (i.e., on state) for the three samples in Figure 7.15 compared with those simulated with the network in Figure 7.7 using the values reported in Table 7.1. The plot proving that the approach based on a fixed framework lumped network with only the intrinsic MEMS device elements allowed to change leads to results supported by both physical and analytical considerations (such as the decrease of the inductance of folded suspensions going from sample (a) to sample (c)).

In conclusion, the approach described in this section concerning the electromagnetic modeling of MEMS devices allows the validity of the MEMS compact model library to be extended also to the assessment of the RF performances. The lumped element network extraction is not immediate, as it requires time to be carried out as well as a comprehensive set of experimental data, both electromechanical and electromagnetic, in order to cross-check and to validate the extracted values. However, once the extraction has been successfully performed, the MEMS designer is allowed to optimize the topology of the intrinsic MEMS device (e.g., plate dimensions, shape, width, and number of suspensions), checking in a very short time the effects of each change on the electromechanical and RF performance of the entire device. Once again, it should be remarked that also in this section about RF modeling, the only fitting parameter employed was the residual air gap in the on state as the other changing parameter, namely, the inductance of the suspensions, can be parameterized by applying well-known analytical formulas. To conclude this section, S-parameter measurements and simulations in Spectre for sample (a) are arranged in 3D plots reporting their behavior for the following applied control bias levels: 0, 5, 10, 15, and 16 V, that is, before the pull-in, and 17, 18, 19, and 20 V, that is, after the pull-in. The measured and simulated S11 parameter curves are

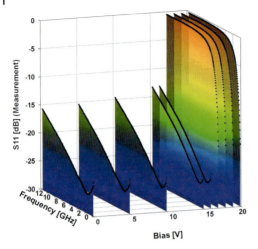

Figure 7.18 S11 parameter measured at several bias levels, ranging from 0 to 20 V, i.e., beyond the pull-in.

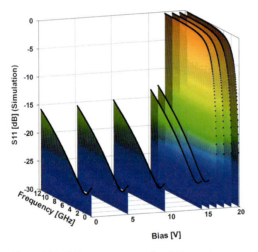

Figure 7.19 S11 parameter simulated (Spectre) at several bias levels, ranging from 0 to 20 V, i.e., beyond the pull-in.

reported in Figures 7.18 and 7.19, respectively, while the measured and simulated S21 parameter curves are shown in Figures 7.20 and 7.21, respectively.

7.4 Electromechanical Modeling of an RF-MEMS Series Ohmic Switch (Dev A) with Compact Models

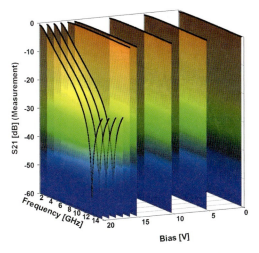

Figure 7.20 S21 parameter measured at several bias levels, ranging from 0 to 20 V, i.e., beyond the pull-in.

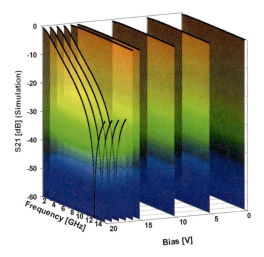

Figure 7.21 S21 parameter simulated (Spectre) at several bias levels, ranging from 0 to 20 V, i.e., beyond the pull-in.

7.4
Electromechanical Modeling of an RF-MEMS Series Ohmic Switch (Dev A) with Compact Models

In this section the modeling approach (i.e., based on compact models) discussed up to now is applied to a different RF-MEMS device, namely, the series ohmic switch discussed in Chapter 2 and referred to as Dev A. The procedure will be treated with fewer details, as specific aspects of the modeling approach were reported in the pre-

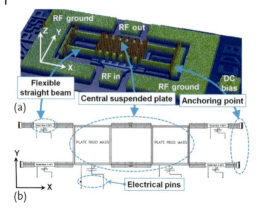

Figure 7.22 (a) Three-dimensional measured profile of the series ohmic switch discussed in this section. The intrinsic device is based on a central rigid gold plate connected to four straight flexible suspensions. The DC biasing electrodes are electrically separated from the input/output RF lines. For this reason, an additional DC biasing pad is visible on the right. (b) Spectre schematic of the series ohmic switch. Since the biasing electrode has an interdigitated configuration under the suspended plate (see Figure 7.23), the whole transducer is split in two halves in order not to overestimate the attractive force.

vious two sections. However, since the topology of Dev A is rather different from that of the varactors previously reported, the example that will be discussed here will provide the reader with additional tips and tricks. The 3D measured profile of the sample under study is shown in Figure 7.22a. Also, this sample was manufactured with the FBK RF-MEMS technology platform, but it does not belong to the same batch of variable capacitors already discussed.

The suspended plate is realized in thick gold (5.5 µm thick) and its in-plane dimensions are 260 µm (along the X axis) and 140 µm (along the Y axis). It presents three by six (20×20 µm^2) holes, and four straight beams (165 µm long, 10 µm wide, and 2.5 µm thick) are connected to its corners. In this case the DC bias line and electrode (in polysilicon) are electrically separated by the RF lines (in gold and multimetal). The biasing electrodes and RF lines underneath the suspended bridge are interdigitated (see Section 2.3) as Figure 7.23 reports.

The RF lines are interrupted, that is, they do not connect the input and output pads, so the switch is open when the suspended plate is not actuated, while it is closed when the membrane pulls in, as it shorts the RF input to the output (gold/multimetal ohmic contact). In order to prevent the RF signal being shorted to RF ground through the suspending beams when the plate is actuated, high-resistivity polysilicon serpentines connect the suspended gold structure to the RF ground lines, decoupling the DC and the RF ground.

Because of the interdigitated topology just mentioned, the effective area of the polysilicon biasing electrodes covers only half of the suspended bridge surface. Since the rigid plate transducer implemented in the MEMS library always assumes an underlying biasing electrode area equal to the bridge area, this would result in an overestimation of the electrostatic force in the simulations, and in a predicted pull-

7.4 Electromechanical Modeling of an RF-MEMS Series Ohmic Switch (Dev A) with Compact Models | 215

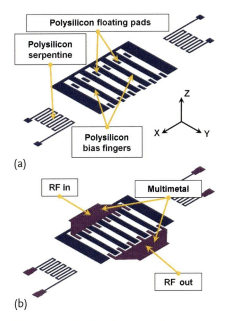

Figure 7.23 (a) Three-dimensional schematic of the polysilicon layer (Dev A) realizing the fingers of the DC bias electrode, the floating pads for elevating the RF contacts, and the serpentine-like resistors decoupling the RF signal and the DC signal. (b) Three-dimensional schematic where the multimetal layer realizing the RF input/output contacts is also visible.

in voltage significantly lower than the actual one. To circumvent this problem, in the simulation schematic diagram the 260 µm long (X axis) rigid plate is split into two 130 µm plates connected together, as shown in Figure 7.22b. Only one plate is biased (the left one), and both include three by three holes on their surface. This allows one, on one hand, to properly calculate the electrostatic attraction force and, on the other hand, also to keep the right description of the dynamic effects (inertia due to the mass and viscous damping influenced by the area and by the number of holes). Since the air gap model with four DOFs is adopted (see Sections 2.3 and 5.2.1), there are no torque contributions around the Y axis due to asymmetric vertical forces (acting only on half of the whole bridge area), and the two smaller plates always remain parallel to the substrate. In this fabrication batch, a thicker sacrificial layer is deployed aiming at a larger air gap, expected to be around 4 µm. Moreover, the RF input/output multimetal lines under the bridge are placed on electrically floating polysilicon pads, leading to the presence of a residual air gap between the suspended gold and the polysilicon biasing electrodes of about 840 nm in the actuated state (see Figure 7.23).

The pull-in/pull-out quasi-static characteristic is measured as in the variable capacitor case, by applying a 20 Hz triangular voltage ranging this time from −70 to +70 V. The whole experimental characteristic is compared with the results of the

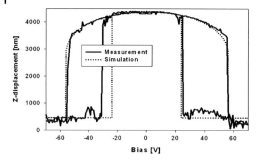

Figure 7.24 Experimental versus simulated pull-in/pull-out characteristic of the MEMS series ohmic switch. The controlling signal is a 20 Hz, zero mean value, 140 V peak-to-peak triangular voltage. The shift between the first and the second pull-out is visible, similar to the variable capacitor case (see Figure 7.5).

DC simulation in Figure 7.24, where a correct prediction of the pull-in, taking place at ±56 V, is obtained by means of compact models.

Concerning the pull-out, the remarks made for the variable capacitor in the previous section apply here as well. The first measured pull-out takes place at −33.5 V, while the second one takes place at +25.2 V, the latter being in good agreement with the simulated one of ±23.8 V. Recalling the considerations in Section 7.2, we find there is a significant shift between the two measured pull-outs, the second one being closer to ideality. Moreover, since the MEMS suspended plate discussed here has a very good planarity with respect to the variable capacitors previously analyzed, the accurate simulated results in Figure 7.24 were obtained by accounting for the nominal residual air gap mentioned above (840 nm) due to the particular topology of the MEMS switch (see Figures 7.22 and 7.23) without the need to fit its value.

We now focus on the dynamic behavior of the ohmic switch. In this case, the set of experimental data refer to the RF power envelope versus time. An RF signal at a single frequency (4 dBm at 12 GHz) is applied to the input port of the switch by using the VNA as a power source in single-port and single-frequency mode (continuous wave [206]). The power flowing to the switch output port is converted into an equivalent voltage by means of a power detector (Agilent 432B) [207] and is displayed on an oscilloscope triggered by the bias signal controlling the switch and supplied by an arbitrary waveform generator. When the switch is not actuated (i.e., open), the input power does not reach the output and the oscilloscope displays a nearly zero voltage. On the other hand, when the switch pulls in, the input power, attenuated by the cables, probes, and switch ohmic losses, reaches the output and is displayed on the oscilloscope as a nonzero (negative) voltage. The applied biasing signal waveform, displayed in Figure 7.25a, is rectangular with 0 V low voltage, 60 V high voltage, 300 Hz frequency, 700 μs rise and fall times, and 1 ms pulse width. The plot in Figure 7.25b reports the vertical displacement (transient analysis) of the schematic in Figure 7.22b, while Figure 7.25c shows the measured behavior of the detector output voltage. A very good correspondence between the down/up transitions of the measured output voltage and the simulated mechanical actuation/release instants of the ohmic switch is proven.

7.4 Electromechanical Modeling of an RF-MEMS Series Ohmic Switch (Dev A) with Compact Models

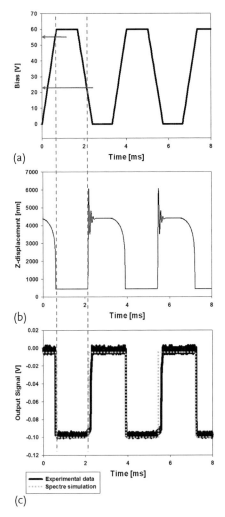

Figure 7.25 (a) Biasing signal applied to the series ohmic switch. The pulse frequency is 300 Hz with 0 V low value, 60 V high value, 700 μs rise and fall time, and 1 ms pulse width. (b) Simulated vertical displacement of the ohmic switch (transient analysis). (c) Output power envelope of the ohmic switch, measured through the power detector output voltage. A very good match between the output voltage transitions and the mechanical actuation/release of the switch is observed. Vertical dashed lines help to identify the connection between the plate actuation/release and the pull-in/pull-out voltages.

A final validation is now reported concerning a MEMS ohmic switch with the same geometry as that reported in Figure 7.22, although included in another fabrication batch with different characteristics (e.g., air gap, suspended gold thickness). As the modeling approach for the variable capacitors considered in Section 7.3 and for the ohmic switch considered in this section has been already discussed in detail, only the measured versus simulated results are reported. Figure 7.26 reports the

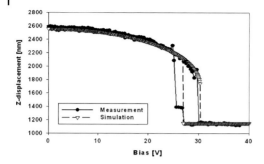

Figure 7.26 Experimental versus simulated pull-in/pull-out characteristic of a MEMS series ohmic switch. Good agreement between the simulated and measured pull-in (about 30 V) and pull-out (about 27 V) voltages is visible.

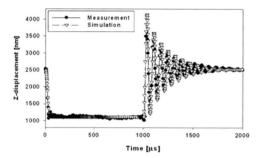

Figure 7.27 Experimental versus simulated transient behavior of the MEMS series ohmic switch. The applied bias is a square pulse (50% duty cycle) with a period of 2 ms, 0 V low value, and 33 V high value.

comparison of the simulated and measured pull-in/pull-out characteristic, showing a good prediction of the pull-in voltage (about 30 V) and pull-out voltage (about 27 V). Figure 7.27 reports a comparison of the simulated and experimental transient response when a square voltage (50% duty cycle with a period of 2 ms, 0 V low value, and 33 V high value) is applied. An accurate prediction of the actuation and release transitions (also of the plate oscillations concerning the latter) is achieved in Spectre, proving once again the appropriateness of the compact models used in this chapter and discussed in Chapter 5.

7.5
Electromagnetic Modeling and Simulation of an RF-MEMS Impedance-Matching Network (Dev E) for a GSM CMOS Power Amplifier

In this section the RF-MEMS-based reconfigurable impedance-matching network referred to as Dev E in Chapter 2 will be analyzed. The impedance tuner is conceived to be interfaced with a CMOS class E power amplifier (PA) for the GSM standard. The design, performance, and measured characteristics of the two parts as well as those of the hybrid RF-MEMS/CMOS block are reported and discussed.

7.5.1
Introduction

Telecommunication standards impose the need for cell phone transceivers to work in different frequency bands (the GSM protocol, as an example, employs four uplink bands); hence, PAs have to deal with a wide frequency range. Designing PAs with bandwidth exceeding 100 MHz is not an issue, and commercial solutions for GSM usually employ two PAs, whose operation frequencies are centered at approximately 900 MHz and approximately 1.8 GHz to allow signal amplification in the lower or the upper band, respectively. Despite this solution being effective, it imposes duplicating circuitry while having one PA in idle mode. In contrast, a reconfigurable PA would allow both the lower and the upper band to be covered, enabling the reduction of chip area. In this section, a hybrid RF-MEMS/CMOS PA demonstrator for the GSM standard is reported [208]. The PA is based on standard CMOS technology, while the impedance-matching network is entirely implemented with the RF-MEMS technology exploited for the examples discussed in this book (i.e., the RF-MEMS technology available at FBK in Italy). The latter synthesizes a purely real 12 Ω impedance on the PA side, and transforms it into a 50 Ω real impedance at the other end in order to match the antenna impedance. MEMS ohmic switches reconfigure the matching network in order to ensure such input/output impedance match at both GSM operation frequencies (900 MHz and 1.8 GHz), reducing the need for hardware redundancy as the same PA can work in both required bands. The CMOS PA and the RF-MEMS impedance matching network are mounted on a printed circuit board specifically designed to provide the DC bias, and are wire-bonded to each other. RF probe measurements of the RF-MEMS/CMOS PA are rather promising as it delivers 20 dBm power with drain efficiencies of 38 and 26% at 900 MHz and 1.8 GHz, respectively. RF simulations of the MEMS network are reported, comparing the expected results with experimental measurements of stand-alone MEMS matching network samples. On the basis of such comparisons, nonidealities of the MEMS network are highlighted and consideration is given as to how the design could be improved with the aim of better performance.

7.5.2
Electromagnetic Design and Optimization of the RF-MEMS Impedance-Matching Network

In this subsection we focus on the RF design of the MEMS-based impedance-matching network discussed here. The layout of the MEMS matching network is depicted in Figure 7.28 while its lumped element network is reported in Figure 7.29.

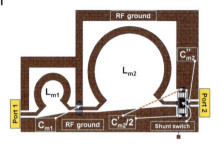

Figure 7.28 Layout of the MEMS-based reconfigurable impedance-matching network. It is based on two LC sections in order to obtain the desired input/output impedance transformation with 3/4-a-circle planar inductors and metal–insulator–metal shunt capacitors. The capacitance of the second LC section is reconfigured by means of a MEMS ohmic shunt switch.

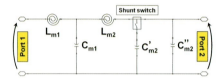

Figure 7.29 Lumped element scheme of the MEMS network reported in Figure 7.28. It is simulated in Agilent Advanced Design System in order to extract the optimum value of each reactive component, aiming at the impedance transformation required by the specifications of the power amplifier (PA).

It is based on two LC sections where the inductors are 3/4-a-circle gold planar ones, while all the shunt-to-ground capacitors are implemented with superposed multimetal and gold layers with low-temperature oxide in between (metal–insulator–metal capacitors). The realization of planar inductors is preferred to that based on suspended (in air) coils because, despite the latter solution enabling smaller area occupation and a larger quality factor when compared with the planar implementation, the presence of residual stress within the gold layer might significantly deform (out of plane) the suspended coil, thus jeopardizing the inductor performance. The MEMS shunt switch, whose topology is that referred to in this book as Dev A, reconfigures the capacitance of the second LC section. When the switch is not actuated (1.8 GHz GSM operating frequency) such a capacitance is defined by C''_{m2} (see Figures 7.28 and 7.29). In a different way, when the PA operates at 900 MHz, the ohmic switch is actuated and it connects C'_{m2} in parallel with C''_{m2}. Note that C'_{m2} is determined by two metal–insulator–metal capacitors, whose single value is $C'_{m2}/2$, located at the top and bottom ends of the switch, and connected in parallel. The capacitance reconfiguration just mentioned has been assigned to an ohmic shunt switch because it suits such a task better than a series one. Indeed, the reconfiguration is performed by loading the RF line with an additional shunt capacitance (C'_{m2}), which is exactly what a shunt switch does when actuated. Otherwise, a dual implementation of the two-state capacitor discussed above featuring a MEMS series ohmic switch would require a T-junction on the RF lines connected, through a series switch, to the C'_{m2} capacitor, leading to a more complex network

7.5 Electromagnetic Modeling and Simulation of an RF-MEMS Impedance-Matching Network...

realization. By reconfiguration of the capacitance of the second LC section, the input/output impedance transformation, that is, 12 Ω at port 1 and 50 Ω at port 2 at both GSM operating frequencies, is achieved.

The analysis of the RF behavior of the network is performed in Agilent ADS [203] and is based on the schematic lumped element network in Figure 7.29. Given the network topology, values of all the inductors and capacitors have to be extracted from simulations. In order to get a better prediction of the parasitic effects, such as resistive losses and capacitive coupling, the inductors are modeled by using the MSIND (microstrip round spiral inductor) element, available in the ADS TLines-Microstrip library [203]. This element models the circular inductor by accounting for its series resistive loss and capacitive coupling to ground. The substrate element (MSub) has also been defined and is associated with the two inductors, accounting for the dielectric constant of the silicon substrate and the conductivity of the gold layer realizing the actual inductors. The advantage of using the MSIND elements is that it provides a straightforward reading of the geometrical parameters needed to design the layout of the inductors. In particular, the two constraints imposed in the schematic in Figure 7.29 are that the conductor width of the planar inductor is 200 µm and that, since the 3/4-a-circle topology is chosen, the number of turns is 0.75. Differently, concerning the capacitors, once their optimum values have been extracted from ADS simulations, the layout features are calculated by using the well-known formula for a parallel plate capacitor (see Chapter 5). The behavior of the ohmic shunt switch is accounted for by linking the measured S-parameter dataset (in Touchstone format [209]) in the ADS schematic. A stand-alone switch sample (such as the one discussed in Section 7.4), fabricated in the same batch, was measured in both the actuated and the nonactuated position, corresponding to the inclusion and exclusion, respectively, of $C'_{m2}/2$ in the network in Figure 7.29. The influence of the CPW-like frame around the movable device has been de-embedded, not being present in the matching network (see Figure 7.28). For this purpose, dummy CPW pads without any MEMS device in between were included in the same fabrication batch and S-parameter measurements of such structures were performed. By use of the ADS optimization tool, a lumped element network reproducing the measured RF characteristic of the dummy CPW structure is extracted. The S-parameter characteristic of dummy pads is accurately reproduced by a 26 fF shunt capacitance in the frequency range of interest. Hence, the switch symbol, visible in Figure 7.29, is defined in the matching network schematic in ADS. It features a two-port data block linked to the proper MEMS switch measured S-parameter dataset, and a negative shunt capacitance (−26 fF) in input and output to compensate for the RF behavior of the CPW pads. Finally, with the target of the input/output impedance transformation previously mentioned, the values of the inductors and capacitors included in the schematic in Figure 7.29 are extracted by means of an S-parameter optimization in ADS. The optimum values determined for the inner radius of L_{m1} and L_{m2} are 545 and 1254 µm, respectively. Their inductance can be calculated by applying the

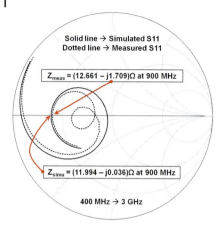

Figure 7.30 Smith chart of the simulated and measured S11 parameter of the MEMS network in Figure 7.28 configured for operation at 900 MHz. The experimental real part of the impedance is about 12.6 Ω, while the CMOS PA specification requires 12 Ω.

well-known formula for a spiral air-core coil planar inductor [210]:

$$L = \frac{R^2 N^2}{(2R + 2.8D)10^5} \quad (7.2)$$

where R is the inner radius, reported above, N is the number of turns, and D is the coil width, the latter two parameters being 0.75 and 200 μm, respectively. Consequently, the L_{m1} and L_{m2} values are 1.01 and 2.88 nH, respectively. Finally, the simulated capacitance values of C_{m1}, C'_{m2}, and C''_{m2} are 4.69, 3.8, and 1.44 pF, respectively. These extracted values ensure the requested input/output impedance transformation at 900 MHz and 1.8 GHz (see Section 7.5.3). Fabricated samples of the MEMS impedance-matching network were measured on a probe station before being wire-bonded to the CMOS PA, and the experimental S-parameter characteristic was compared against ADS simulations in both GSM bands. Figures 7.30 and 7.31 report the measured versus simulated S11 characteristic on a Smith chart for the MEMS network operating at 900 MHz and 1.8 GHz, respectively.

Concerning the low-frequency band, the experimental input impedance (real part) is rather close to the desired 12 Ω. Indeed, there is a difference of about 0.6 Ω between the simulation and the measurement. Differently, the measured imaginary part of the impedance shows a larger difference compared with the simulations. On the other hand, the experimental results referring to the high-frequency band show a more significant impedance mismatch, as the measured real part of the impedance is about 14.6 Ω instead of the expected 12 Ω. In the Smith chart in Figure 7.31, we see a circle in the phase rotation for the measured trace in the frequency range around 1.8 GHz which is not predicted by the ADS simulation. Such behavior relates to a resonance very likely due to the reactive parasitic effects of the MEMS network that are not properly taken into account by the schematic in Figure 7.29. Indeed, the additional inductive contribution due to the gold lines

7.5 Electromagnetic Modeling and Simulation of an RF-MEMS Impedance-Matching Network ...

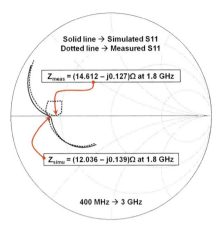

Figure 7.31 Smith chart of the simulated and measured S11 parameter of the MEMS network in Figure 7.28 configured for operation at 1.8 GHz. The experimental real part of the impedance is about 14.6 Ω, while the CMOS PA specification requires 12 Ω.

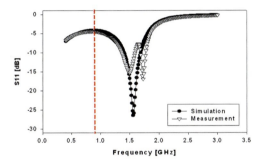

Figure 7.32 S11 rectangular plot for the simulated and measured MEMS impedance-matching network operating at 900 MHz.

connecting the two actual planar inductors, together with the input and output CPW lines, that are coupled with the shunt parasitic distributed capacitance they introduce (not accounted for in the simulated network) could be responsible for the observed behavior. Such a resonance around 1.8 GHz is clearly visible also when the network is configured to work at 900 MHz (see Figure 7.30), even if it is slightly shifted in frequency as C'_{m2} is disconnected (see Figure 7.29). Discrepancies between the measured and the simulated S11 characteristic around 1.8 GHz are also visible in the rectangular plots reported in Figure 7.32 (network working at 900 MHz) and Figure 7.33 (network working at 1.8 GHz). The vertical dashed lines help to locate the central frequency of the bands. These plots show that the measured impedance matching at 900 MHz is much closer to the desired 12 Ω as it is still far from the unwanted resonant frequency. A rather significant impedance mismatch at 1.8 GHz will mainly affect the efficiency of the PA, as will be discussed in the next subsections.

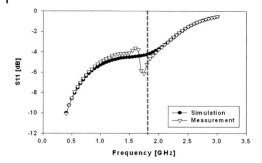

Figure 7.33 S11 rectangular plot for the simulated and measured MEMS impedance-matching network operating at 1.8 GHz.

7.5.3
Deign of a Reconfigurable Class E PA

In this subsection the design of a CMOS class E PA for the GSM standard is discussed. Class E PAs are switched-mode amplifiers and were presented for the first time in 1975 [211]. Their components, shown in Figure 7.34, are sized according to simple design equations [212]:

$$R_L = \frac{\alpha V_{DD}^2}{P_{OUT}} \; ; \quad C_{sh} = \frac{\beta}{\omega_0 R_L} \; ; \quad C_2 = \frac{\gamma}{\omega_0 R_L} \; ; \quad L_2 = \frac{Q_{out} R_L}{\omega_0} \quad (7.3)$$

where α, β, and γ are parameters that depend on the output network quality factor Q_{out}; P_{OUT} is the output power with an ideal efficiency of 100%; V_{DD} is the voltage supply; ω_0 is the working frequency; R_L is the load resistance. V_{DD} cannot exceed the maximum limit of approximately 3 V due to reliability constraints [213], and therefore increasing P_{OUT} lowers R_L. For a maximum output power of approximately 21 dBm in both the lower and the upper band, we calculated $R_L = 12\,\Omega$, assuming a real efficiency of approximately 40%, $Q_{out} = 2.5$, and $V_{DD} = 3$ V.

In order to obtain this load impedance from a 50 Ω antenna resistance in both the lower and the upper band, a reconfigurable matching network is needed. In this design, such a network is the MEMS-based one discussed, modeled, and characterized in Section 7.5.2. The efficiency and quality factor were intentionally kept rather

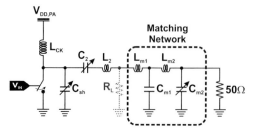

Figure 7.34 Schematic of a reconfigurable class E PA. The matching network down-transforms the 50 Ω antenna to the R_L optimum load.

7.5 Electromagnetic Modeling and Simulation of an RF-MEMS Impedance-Matching Network ...

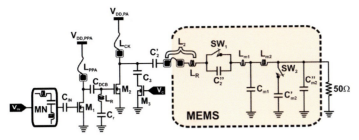

Figure 7.35 Schematic of the proposed reconfigurable class E PA. The MEMS-based matching network (circled by the dashed line) implements the LC series filter and the matching network, composed of two LC sections.

low in the PA design phase in order to make it more robust against impedance detuning arising from the MEMS network. The matching network comprises two LC ladders which down-transform the antenna resistance at approximately 900 MHz and approximately 1.8 GHz using a minimum number of reconfigurable components, namely, C_2 and C_{m2}, and avoiding switching inductors to reduce the power losses. In order to switch between the upper and the lower band while keeping a high operation efficiency, output network components (C_{sh}, C_2, and L_2) have to be reconfigured according to (7.3). Since the switch losses are one of the most significant power loss contributions, we reduced the number of switches used to reconfigure the network, keeping L_2 constant regardless of the operation frequency. Thus, the output network quality factor Q_{out} doubles at 1.8 GHz as shown in (7.3). Selecting a low Q_{out} is crucial in order to reduce the output network power losses [214]; hence, we chose $Q_{out} = 2.5$ at 900 MHz. In this way, the capacitance values at the low (L) and high (H) bands are $C_{sh,L} = 3.2\,\text{pF}$, $C_{2,L} = 14.9\,\text{pF}$, $C_{sh,H} = 1.6\,\text{pF}$, and $C_{2,H} = 2\,\text{pF}$, respectively. In Figure 7.35, a schematic of circuit of the reconfigurable PA, including the external MEMS network, is shown.

The PA comprises two common-source stages. In order to avoid gate oxide breakdown, we used MOSFETs with high breakdown voltage (7.5 V) and limited the supply voltage of the PA driving stage to $V_{DD,PPA} = 1.25\,\text{V}$, thus limiting the maximum voltage across the gate oxide stress below the DC breakdown limit [213]. In the implementation of the reconfigurable shunt capacitance C_{sh}, switch M_3 is employed in order to ground C_3 while using the M_2 (C_{dd,M_2}) and M_3 (C_{dd,M_3}) drain parasitic capacitances (see Figure 7.35). For the lower band, approximately 900 MHz, M_3 is switched on in order to increase the shunt capacitance, implementing $C_{sh,L}$. In contrast, when the PA is working at approximately 1.8 GHz, M_3 is switched off to obtain $C_{sh,H}$. In order to find values for the drain MOSFET capacitances and C_3 we solved the following system of equations:

$$C_{sh,L} = C_{dd,M_2} + C_3$$
$$C_{sh,H} = C_{dd,M_2} + \frac{C_{dd,M_3} C_3}{C_{dd,M_3} + C_3} \tag{7.4}$$

Assuming that C_3 is k times the M_3 drain parasitic capacitance ($C_3 = k C_{dd,M_3}$) and applying $C_{sh,L} = 2 C_{sh,H}$ allows one to calculate the capacitance values after some mathematical manipulations:

$$C_{dd,M_2} = \frac{k-1}{2k} C_{sh,L}$$

$$C_{dd,M_3} = \frac{k+1}{2k^2} C_{sh,L}$$

$$C_3 = \frac{k+1}{2k} C_{sh,L} \tag{7.5}$$

where $k > 1$ has to be selected to maximize the efficiency. The MOSFET power loss ($P_{L,MOS}$) depends inversely on the transistor width (W); hence, a large W is desirable to minimize the on-state resistance [214]. Unfortunately, this increases the drain capacitance, which is proportional to $1/P_{L,MOS}$. Thus, minimization of the MOSFET power losses requires increasing both C_{dd,M_2} and C_{dd,M_3}, but this cannot be done by means of k, as they have opposite dependence on k. The optimum $k = 3.56$ is found by imposing $C_{dd,M_2} = 2 C_{dd,M_3}$, which corresponds to equalizing the power loss due to M_2 to half of the M_3 one, as M_3 is switched on only for the lower-frequency band, that is half of the time. Substituting $k = 3.56$ into (7.5) allows one to calculate the optimum M_2 ($W = 6000$ µm) and M_3 ($W = 4500$ µm) widths and $C_3 = 9.5$ pF. The finite RF choke inductor $L_{CK} = 2.1$ nH is realized by a bond-wire inductor. To drive efficiently the M_2 gate, we used a double resonant matching network synthesizing high impedance at both 900 MHz and 1.8 GHz; it comprises the bond-wire inductor $L_{PPA} = 1.8$ nH, the planar inductor $L_R = 3.1$ nH, $C_r = 4.2$ pF, and the M_2 gate-source capacitance. The MEMS impedance-matching network to be interfaced with the CMOS PA also implements the $L_R C_2''$ series section (see Figure 7.35) as shown in the layout reported in Figure 7.36.

In this implementation, the series section just mentioned is interposed between the PA (which will be wire-bonded at port 1) and the actual MEMS matching network previously described. The L_R inductor radius is modeled according to (7.2), while the C_2'' capacitance (2.29 pF) is implemented with two $C_2''/2$ metal–insulator–metal capacitors in parallel. However, in this case C_2'' is connected in series on the RF line and a series ohmic switch selects/deselects such a capacitor.

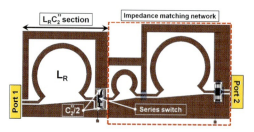

Figure 7.36 Layout of the complete MEMS network interfaced with the CMOS PA. An additional LC series section is interposed between the PA and the actual impedance-matching network shown in Figure 7.28.

The behavior of this switch is opposite to that of the one employed in the matching network, it being open when the switch is not actuated. In this case, C_2'' loads the RF lines. On the other hand, when the switch pulls in, it forms a low-resistance input/output electric path that shorts out the C_2'' capacitor. In the following subsection, the hybrid MEMS/CMOS class E PA demonstrator will be discussed together with its measured RF performance.

7.5.4
Experimental Results for the Hybrid RF-MEMS/CMOS PA

A microphotograph of the circuit of the PA prototype realized is shown in Figure 7.37. Two dies manufactured with CMOS and RF-MEMS technologies are glued on top of a DC biasing board on an FR4 substrate [215].

The dies are connected by means of bond wires in a system-on-chip fashion. The critical bond-wire inductors are L_{CK}, L_{PPA}, and the output one that synthesizes part of L_2 (see Figure 7.35). The chip manufactured with AMS (formerly Austriamicrosystems; www.austriamicrosystems.com (accessed 1 March 2008)) 0.35 μm technology implements the active part of the PA circuit. Grounding is provided by means of 10 bond wires, while two dedicated ground bond wires are used to connect the MEMS output network to ground. The CMOS die area is 1.44 mm² including pads, while the MEMS network area is 29.88 mm². The MEMS matching network input/output and the PA RF input/output are configured for GSG probing, also enabling the characterization of the matching network alone, as reported above. The complete PA was measured using microprobes in order to connect the input and output GSG pads. In order to make the PA work in the low-frequency band, $V_C = 3$ V (see Figure 7.35) is imposed on the M_3 gate, while the MEMS switches are actuated supplying 60 V DC bias (slightly larger than the pull-in voltage). The input-matching network is realized using external components. Figure 7.38 shows the output power and the drain efficiency of the PA prototype measured at 900 MHz and 1.8 GHz.

With the maximum $V_{DD,PA} = 3$ V supply, the PA delivers 20 dBm with drain efficiencies of 38 and 26% at 900 MHz and 1.8 GHz, respectively. The difference with respect to the simulations (44 and 40% at 900 and 1800 MHz, respectively) is partially caused by the behavior of the MEMS matching network that forces the PA to

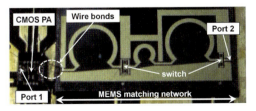

Figure 7.37 Microphotograph of the whole PA prototype. The CMOS and MEMS dies, implementing the active part of the circuit and the output network, respectively, are wire-bonded onto a DC testing board and configured for GSG probing.

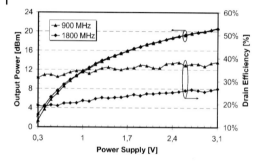

Figure 7.38 Output power and drain efficiency of the hybrid MEMS/CMOS PA (see Figure 7.37) measured at 900 MHz and 1.8 GHz.

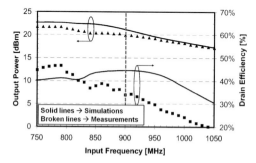

Figure 7.39 Output power (triangles) and drain efficiency (squares) measured in the 900 MHz frequency band. The solid lines represent the simulations performed considering the real behavior of the output network obtained from the MEMS network characterization (see Figures 7.30 and 7.32).

operate with a mistuned output network, thus degrading the circuit efficiency and performance. Such a mistuning is particularly significant in the 1.8 GHz band, as reported in the previous section. Indeed, in Figure 7.38, a drain efficiency decrease of about 10% is visible when the PA is working at 1.8 GHz over all the power supply voltage range. On the other hand, the impedance mistuning does not show a significantly different effect on the output power in the two frequency ranges. Nevertheless, the unavoidable uncertainties on the L_{PPA} and L_{CK} bond-wire inductances plus the ones connecting the CMOS to the MEMS parts also contribute to reducing the performance of the PA. This is particularly evident in Figures 7.39 and 7.40, which report the PA drain efficiency and output power versus frequency at 900 MHz and 1.8 GHz, respectively.

In both plots the broken lines represent the measured PA performance, while the solid lines refer to the simulation of the ideal PA in which the lumped network of the MEMS part highlighted in Figure 7.35 is replaced with the experimental S-parameter datasets (Touchstone format) of the network in Figure 7.36 measured at 900 MHz (both switches actuated) and 1.8 GHz (none of the switches actuated). Since in these simulations nonidealities of the MEMS matching network are accounted for while all the rest is ideal, discrepancies with the PA measured perfor-

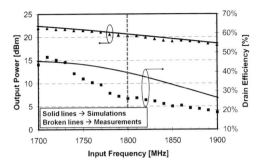

Figure 7.40 Output power (triangles) and drain efficiency (squares) measured in the 1.8 GHz frequency band. The solid lines represent the simulations performed considering the real behavior of the output network obtained from the MEMS network characterization (see Figures 7.31 and 7.33).

mance, visible in Figures 7.39 and 7.40, can only be explained by a rather significant difference between the inductances realized with the bond wires and their expected value plus, even if in smaller amount, the nonidealities of the CMOS PA. In particular, the drain efficiency decrease between simulated and measured curves is about 10% at both 900 MHz and 1.8 GHz. Differently, the inductance mistuning does not significantly affect the power delivered by the PA (as illustrated in Figures 7.39 and 7.40). Despite such technology and assembly issues, the low output network quality factor we selected reduces the sensitivity of the circuit to component variations, allowing the PA to deliver more than 18 dBm in both the lower and the upper band.

7.6 Electromagnetic Simulation of an RF-MEMS Capacitive Switch (Dev B1) in ANSYS HFSS™

In this section the use of the ANSYS HFSS electromagnetic FEM simulator [153], introduced in Chapter 4, is shown concerning the simulation of the S-parameter characteristic of an RF-MEMS variable capacitor (Dev B1 in Chapter 2) in both the on configuration and the off configuration, and the comparison of the simulated data against experimental curves. The aim of this example is concentrated on the identification of nonidealities of the physical device due to technology aspects and their inclusion in the simulations in order to obtain a better prediction of the real characteristics of a fabricated device. From the measured 3D profile of the fabricated Dev B1 sample reported in Figure 7.41, it is clear that some consideration has to be given to the vertical profile of the real device.

From the observation by means of an optical profilometer, the movable surface of the RF-MEMS variable capacitor is not planar, but presents an arching of the gold membrane (concerning both the central plate and the four suspensions) induced by the residual stress gradient built within the metallization during the fabrication process. Apart from the effect that such a bowing has on the electromechanical

Figure 7.41 Three-dimensional measured profile of the RF-MEMS variable capacitor (Dev B1 in Chapter 2) obtained by means of an interferometry-based optical profiling system.

properties of the device (i.e., on the overall elastic constant and, in turn, on the pull-in and pull-out levels), the scarce planarity of the central plate clearly affects the high capacitance value of the RF-MEMS device (i.e., when the movable membrane is pulled in). A variable residual air gap between the actuated gold plate and the underlying insulating layer (and fixed electrode) will be distributed between the surfaces of capacitor electrodes, resulting in a significant decrease of the on-state capacitance. This issue was discussed in Sections 7.2 and 7.3. Having a high capacitance value lower than expected in a shunt-to-ground capacitive switch results in poorer isolation (S21 parameter) and lower reflection (S11 parameter) as a larger portion of the RF signal will pass to the output terminal [216]. The presence of a residual air gap in the on state (due to the residual stress) can be easily accounted for in the electromagnetic simulation, by calculating an effective averaged thickness of air, to be uniformly inserted underneath the movable plate in the on state (see Section 7.2). In this case, by means of profile measurements performed with the optical profilometer, a sensible averaged value for the residual air gap is determined to be around 150 nm. The vertical profile of the movable plate can be easily modified in HFSS by parameterizing the vertical position of the block corresponding to the gold of the suspended plate (see Section 4.5). In the simulation of the on state, instead of considering a zero distance between the suspended plate and the insulating layer, a residual distance of 150 nm (of air) is implemented. On the other hand, the effect of the arching of the plate can be disregarded concerning the off state (i.e., rest position) of the RF-MEMS movable membrane. In this case, a more important aspect has to be accounted for: the actual extent of the air gap, which in some cases can be considerably different from the nominal expected value (i.e., 3 μm) [139]. Such a difference can be related to multiple and diverse reasons. First of all, the spinning of the photoresist used as the sacrificial layer can lead to an initial thickness of the patterned layer different from the nominal value. Moreover, the photoresist can be more or less effective in reproducing the vertical profiles of the underlying layers on its upper surface. Such a characteristic is called planarization, and the better this feature is, the more insensitive the profile of the spun photoresist is to the vertical steps of the underlying layers it covers. Figure 7.42

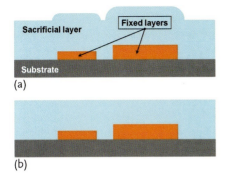

Figure 7.42 (a) Schematic cross section of a nonplanarizing photoresist material adopted as the sacrificial layer. The photoresist reproduces the vertical profile of the underlying deposited layers, presenting a uniform thickness above all the devices. (b) Schematic cross section of a planarizing photoresist layer. Its profile is flat regardless of the diverse profiles of the underlying layers, resulting in a thickness less than the nominal value with regard to the layers deposited above the substrate.

shows schematically the difference between a nonplanarizing (Figure 7.42a) and a planarizing (Figure 7.42b) sacrificial layer, highlighting the effect on the nominal and effective thickness of the spun material and, in turn, on the air gap of the final RF-MEMS devices.

In this case, the photoresist used tends to planarize the underlying profile, and, since the height of the area underneath the suspended plate is greater than that of the rest of the device (due to the presence of the fixed electrode), it is reasonable to expect a final air gap smaller than 3 μm. In order to extract an accurate indication of the actual air gap of the DUT, the electromechanical pull-in/pull-out characteristic is measured and the vertical position of the RF-MEMS device is observed against the applied bias. Figure 7.43 shows the experimental pull-in/pull-out characteristic of a Dev B1 sample obtained by means of a dynamic measurement system based (once again) on an interferometer equipped with a stroboscopic illuminator [30].

From the plot an effective air gap of 2.2 μm is extracted as the difference between the vertical thickness of the RF-MEMS device in the rest position (no applied bias)

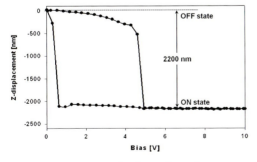

Figure 7.43 Experimental pull-in/pull-out characteristic of a physical Dev B1 sample obtained with a dynamic profiling system. The actual air gap is less than the nominal expected value.

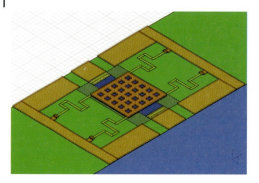

Figure 7.44 Three-dimensional schematic in ANSYS HFSS of the RF-MEMS variable capacitor (Dev B1) discussed in this section.

and after the pull-in has been reached (at around 5 V). Consequently, in the HFSS an air gap of 2.2 μm rather than 3 μm is implemented for the simulation of the RF-MEMS device in the off position (i.e., closed switch). Having a rest position air gap smaller than that expected means realizing a larger low capacitance value, resulting in slightly poorer transmission (S21 parameter) and reflection (S11 parameter), especially in the high-frequency range. After extraction of the effective values for the on-state and off-state air gap (residual air gap of 150 nm and 2.2 μm, respectively) the HFSS simulations are launched, referring to the 3D schematic in Figure 7.44, as reported and discussed in Chapter 4.

The comparison of the measured and simulated S-parameter characteristics (from 400 MHz to 13.5 GHz) of the RF-MEMS variable capacitor are reported in the Figures 7.45–7.48. Figures 7.45 and 7.46 report the S11 (reflection) and S21 (transmission) characteristics, respectively, of the DUT in the off state (closed switch). Figures 7.47 and 7.48 report the S11 (reflection) and S21 (isolation) characteristics, respectively, of the DUT in the on state (open switch). The good match of the measured and simulated curves in both RF-MEMS device configurations proves that the extracted values concerning the air gap in the device rest position, as well as the effective residual air gap thickness in the actuated position (due to

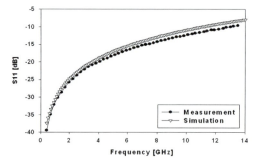

Figure 7.45 Measured versus simulated (in HFSS) S11 characteristic (reflection) of Dev B1 in the off state from 400 MHz to 13.5 GHz.

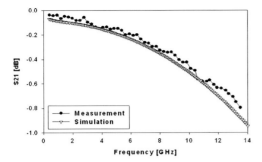

Figure 7.46 Measured versus simulated (in HFSS) S21 characteristic (transmission) of Dev B1 in the off state from 400 MHz to 13.5 GHz.

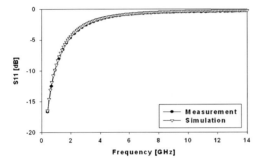

Figure 7.47 Measured versus simulated (in HFSS) S11 characteristic (reflection) of Dev B1 in the on state from 400 MHz to 13.5 GHz.

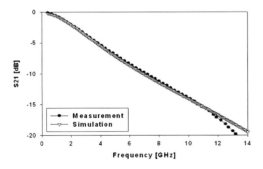

Figure 7.48 Measured versus simulated (in HFSS) S21 characteristic (isolation) of Dev B1 in the on state from 400 MHz to 13.5 GHz.

the presence of residual stress), are sensible and based on consistent deductions and considerations based on the physical observation of fabricated samples.

7.7
Electromagnetic Simulation of a MEMS-Based Reconfigurable RF Power Attenuator (Dev D) in ANSYS HFSS

In this section the use of HFSS to simulate the S-parameter characteristic of the RF-MEMS-based reconfigurable power attenuator introduced in Chapter 2 and referred to as Dev D is described. The RF-MEMS network is described in detail in [71], and it features a few polysilicon resistors that can load (or not) the RF line, depending on the configuration (on or off) of the MEMS cantilever-type series ohmic switches. Depending on the overall resistive load on the RF line, the network changes the level of attenuation applied to the input RF signal. A microphotograph of the network (already presented in Chapter 1) is reported in Figure 7.49.

The network features two parallel branches (each comprising three resistors), and one or both of the split lines can be selected by the two leftmost RF-MEMS ohmic switches, indeed doubling the number of implemented resistive loads (i.e., a single resistance or two equal resistances in parallel). An HFSS 3D schematic of the whole network is reported in Figure 7.50, and was obtained by importing the 3D model in SAT format, previously generated with the layout editing software tool (see Chapters 3 and 4).

A close-up of the microphotograph reported in Figure 7.51 shows in detail the split RF lines, each featuring three polysilicon resistors cascaded in series (R_1, R_2, R_3) and the RF-MEMS cantilever-type ohmic switches for their selection (or shorting). The two leftmost RF-MEMS switches are needed to select one or both of the RF lines and, differently from the other microrelays, can be controlled independently.

As visible, the 3D structure is already set for the simulation (e.g., the rectangular ports for imposing the excitations are already defined). For all the details on the simulation settings, refer to the example reported in Chapter 4. Figures 7.52

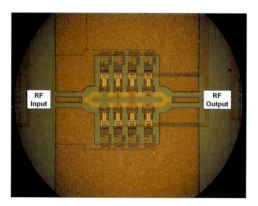

Figure 7.49 Microphotograph of a MEMS-based reconfigurable RF power attenuator fabricated sample (Dev D).

7.7 Electromagnetic Simulation of a MEMS-Based Reconfigurable RF Power Attenuator ...

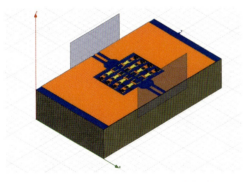

Figure 7.50 HFSS 3D schematic of the MEMS-based reconfigurable RF power attenuator discussed in this section (Dev D) imported in the SAT file format.

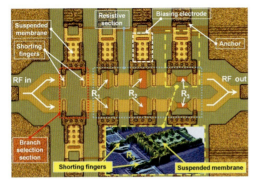

Figure 7.51 Microphotograph (close-up) of the central part of the network reported in Figure 7.49. The RF line shows two parallel branches, each loaded by three resistors of different value (R_1, R_2, R_3). The leftmost switching stage (composed of two microswitches) selects one or both of the branches, while all the other switches short the resistors, reconfiguring the resistance of the whole network.

and 7.53 report two close-ups of the HFSS 3D model, where the most relevant parts of the networks are labeled and indicated.

It can be seen that the suspended gold membranes of all the switches are plain and do not contain the holes needed for technology reasons (the holes are visible in Figures 7.49 and 7.51). Typically, exporting complicated geometries, like in this case where the holes are circular, makes the SAT file generation phase rather critical, which might lead to errors, jeopardizing the model generation itself. For this reason, some layout simplifications have to be performed, however without changing significantly the behavior of the structure of interest. For example, neglecting the presence of holes on the MEMS surface can influence the electromechanical characteristics of the device, e.g., the pull-in and pull-out voltages as well as the dynamic behavior (air viscous damping). Nonetheless, in this case we are solely observing the RF characteristics of the network. If an RF electrode was deployed underneath the suspended gold membrane, like it was for Dev B1 discussed in

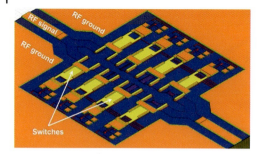

Figure 7.52 Close-up of the HFSS 3D model of the RF-MEMS network, highlighting the central part of the device comprising the eight microswitches, and the two banks of polysilicon resistors.

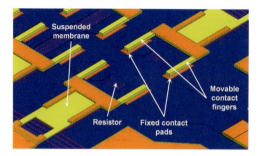

Figure 7.53 Close-up of the HFSS 3D model of the RF-MEMS network, highlighting the details of two MEMS ohmic switches and the resistors they control.

the previous sections, including or not including holes would influence both the on-state and the off-state capacitance, and consequently the S-parameter behavior. In the case of Dev D, no RF electrodes are placed underneath the suspended membrane, so we can neglect the presence of holes without affecting the electromagnetic characteristics of the whole RF-MEMS network. Again, concerning the reconfiguration of the switches (on/off), the only parts involved in opening and closing the short circuit bypassing the polysilicon resistors are the MEMS ohmic switch fingers on the free end of the cantilevers (see Figure 7.53). For this reason, in order to simplify the parameterization of the HFSS structure, just the contact fingers are allowed to move upward and downward, opening or closing the ohmic contact with the gold contact pads underneath.

With regard to the HFSS simulation settings, a nominal sheet resistance of 100 $\Omega/\square$ is expected for the polysilicon layer realizing the resistive loads. According to the form factor of the three polysilicon resistors (see Figure 7.51), their resistance is expected to be 10, 40, and 100 Ω, respectively. However, by measurement of the on-purpose test structure included in the same wafer layout of the RF-MEMS attenuators, a definitely larger effective sheet resistance was extracted for the polysilicon layer, it being 250 $\Omega/\square$ (i.e., 2.5 times greater than the nominal one). Consequently, the actual resistance of the three loads becomes 25, 100,

7.7 Electromagnetic Simulation of a MEMS-Based Reconfigurable RF Power Attenuator...

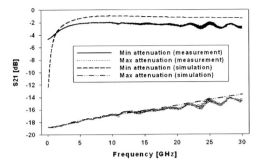

Figure 7.54 Comparison of the measured and simulated (in HFSS) minimum and maximum attenuation levels (i.e., S21 parameter) of the RF-MEMS-based reconfigurable attenuator (Dev D) from 100 MHz to 30 GHz.

and 250 Ω, respectively, and these values are accounted for in the HFSS simulation. Among the $2^4 = 16$ possible attenuation configurations realized by the RF-MEMS network, the minimum and maximum ones are simulated in HFSS. The minimum attenuation is reached when all the switches shorting the polysilicon resistor are actuated, and also when both branches are selected (i.e,. all eight switches comprising the network are on). On the other hand, the maximum attenuation is reached when none of the polysilicon resistors are shorted (i.e., all the corresponding switches are in the rest position), and just one of the two branches is selected (i.e., just one switch is actuated, it being one of the two leftmost ones in Figures 7.49 and 7.51). The comparison of the measured and simulated attenuation levels in the frequency range from 100 MHz to 30 GHz is reported in the plot in Figure 7.54. The plot shows good qualitative agreement between the simulated and measured S21 parameter characteristics, proving that the effective value of the polysilicon resistivity is correct.

However, looking in more detail at the plot in Figure 7.54, we find that additional considerations are necessary. The HFSS prediction of the maximum attenuation configuration matches better the measured characteristic than does the minimum one, the simulated curve for the low attenuation underestimating the losses of about 1 dB over most of the observed frequency range. The reason for this (however acceptable) disagreement is very likely due to the contact ohmic resistance of the actuated RF-MEMS switches with the underlying fixed gold pads. The HFSS simulation, indeed, considers an ideal metal-to-metal contact, while in the real device the contact resistance of the actuated switches can be a bit larger (e.g., 1–5 Ω), due, for instance, to the gold surface roughness. Since in the minimum attenuation configuration all the switches are actuated, an unwanted small resistance is inserted in series on the RF line per switch stage, building in the end a parasitic resistance of 10–20 Ω that is compatible with about 1 dB of additional losses affecting the S21 characteristic. This issue is not present in the maximum attenuation configuration, as the RF-MEMS switches are not actuated, and consequently the small parasitic resistors just mentioned are not inserted on the RF line (loaded by the polysilicon resistors).

Another qualitative disagreement is visible in the measured versus simulated characteristics concerning the very low frequency range (up to 1 GHz), once again looking at the minimum attenuation level. The simulated characteristic starts from a low loss level (about 12 dB) and rapidly increases to a few decibels, while the measured characteristic begins from about 5 dB. This difference is very likely due to parasitic large series capacitances associated with the gold to multimetal (and multimetal to polysilicon) transitions comprised in each calibrated resistor. This issue is the same as that discussed in Section 7.3 referring to Dev B2, and it is responsible for the additional losses seen in the measured characteristic, and not accounted for in the HFSS simulation. The same disagreement is not seen when looking at the measured versus simulated maximum attenuation curves, as the switches are all actuated, and consequently the layer-to-layer vertical transitions (and the linked parasitic effects) are shorted.

All these nonideal effects can be accounted for in the HFSS simulations, for example, by changing the resistivity of the layers used for the ohmic contacts of the RF-MEMS switches (see Figure 4.29), as well as by adding a very thin insulating layer with regard to the vertical layer-to-layer transitions. Of course, the level of accuracy that the MEMS scientist wants to achieve strongly depends (once again) on the targets of his/her analysis. Typically, the inclusion of second-order effects implies a complication of the model, which is not always sensible if, for example, like in this case, the target is to have a reasonably accurate prediction of the attenuation levels implemented by the RF-MEMS-based reconfigurable network.

7.8
Conclusions

This chapter reported several simulations of different RF-MEMS devices performed with various approaches and aimed at the prediction of both electromechanical and electromagnetic aspects of such case studies. The simulations presented were treated by accounting for the nonidealities of the devices arising from technology aspects. Modeling techniques for the inclusion of such deviations from the expected device characteristics were presented and used so that the accuracy of the simulation approach the RF-MEMS scientist chooses can be significantly increased once the specific technology platform for the manufacturing of such devices has been determined.

Acknowledgments The author would like to acknowledge Dr. Paola Farinelli (RF Microtech Srl, Perugia, Italy) and all the staff from the DIEI Department at the University of Perugia (Italy) for making available access to the ANSYS HFSS 3D FEM electromagnetic simulation tool used for the examples reported in this chapter. The author would like to acknowledge the Advanced Research Center on Electronic Systems for Information and Communication Technologies E. De Castro (ARCES) of the University of Bologna (Italy) for making available access to the optical profiling system (Wyko NT1100 DMEMS from Veeco) used for the experimental measurement of the RF-MEMS samples discussed in this chapter, and for making available access to the Agilent ADS software environment. Finally, the author would like to acknowledge Prof. Luca Larcher (DiSMI Department, University of Modena and Reggio Emilia, Italy) for collaboration in the development of the hybrid CMOS/RF-MEMS class E PA for the GSM standard reported in this chapter.

Appendix A
Rigid Plate Electromechanical Transducer (Complete Model)

Abstract: The focus of this appendix is the detailed description of the multiphysical model of the rigid plate electromechanical transducer discussed in Chapter 5 and implemented in the Verilog-A [101] programming language. The information that is discussed does not represent an integration with respect to the features reported in Chapter 5, but is rather a description from scratch of the rigid plate model at a significantly deeper level of detail. Consequently, this appendix is intended for those readers who are interested in a close examination of the implementation details related to the microelectromechanical (MEMS) system rigid plate model.

The appendix is arranged in sections, each of which discusses different features of the MEMS rigid plate transducer. First of all, the mechanical model is reported, it being developed both for a plate with four degrees of freedom (DOFs; that is, a plate always parallel to the substrate) and for a plate with six DOFs. Subsequently, the electrostatic model, the transduction between the electrical and mechanical domains, and the viscous damping model are discussed according to the simplified plate model with four DOFs and according to the complete one with six DOFs.

A.1
Introduction

First of all, it is necessary to introduce the description of a rigid plate in a suitable reference system. A schematic of the rigid plate is reported in Figure A.1. The reference frame origin is defined in the center, and the local reference system (xyz) is assumed to coincide with the global one (XYZ). The critical dimensions of the plate are the width (W), length (L), and thickness (T) along the x, y, and z axes, respectively. Since in the cases of interest the thickness of the plate is much smaller than the width and length $(T \ll W, L)$, the specific point along T chosen as the reference system origin is irrelevant. We will assume that such a point corresponds to the center of mass of the plate, under the hypothesis that the material is homogeneous. In other words, the origin is chosen in the center of W, L, and T, as shown in Figure A.2.

The mechanical model of the rigid plate will now be described in detail, firstly, referring to a simplified model with four degrees of freedom (DOFs), and subsequently introducing the complete mechanical model with six DOFs.

Practical Guide to RF-MEMS, First Edition. Jacopo Iannacci.
© 2013 WILEY-VCH Verlag GmbH & Co. KGaA. Published 2013 by WILEY-VCH Verlag GmbH & Co. KGaA.

Appendix A Rigid Plate Electromechanical Transducer (Complete Model)

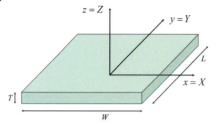

Figure A.1 Rigid plate with the origin of the reference system defined in the center of the three axes.

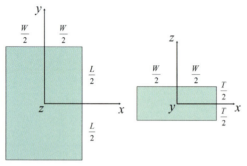

Figure A.2 The origin of the local reference system is placed at the center of mass of the plate.

A.2
Mechanical Model of the Rigid Plate with Four DOFs

The mechanical model of the rigid plate with four DOFs allows three linear displacements and one rotation. In more detail, the plate can move along the three coordinate axes (xyz) and can rotate around the vertical axis (z). The plate with four DOFs does not represent just a simple example to explain the approach chosen for its implementation. Indeed, a very common realization of a microelectromechanical system (MEMS) switch or varactor is based on a rigid plate suspended above a fixed electrode by means of flexible structures (e.g., deformable beams). In this case, it can be reasonably assumed that when a biasing voltage is applied between the plate and the underlying electrode, the former moves downward by keeping a parallel position with respect to the substrate. Therefore, no rotation will occur around the x and y axes of the plate, and consequently the model with four DOFs is appropriate at least for a first-stage analysis of the entire topology.

The approach chosen to define the mechanical behavior of the plate is rather simple. The main physical effect involved is the inertia of the body, expressed by the well known formula

$$F = ma \tag{A1}$$

where F is the force applied to the body, m is the mass, and a is its acceleration. Thus, the main issue to be solved is how to compose the set of external forces and

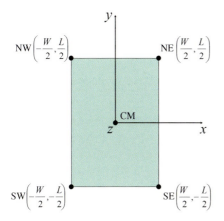

Figure A.3 Rigid plate vertexes with their coordinates in the local reference system. "CM" refers to the center of mass.

torques applied to the plate in order to define their resultant contributions. Let us assume that the only points in which external stimuli can be applied to the plate are its four vertexes (nodes). Looking at Figure A.3, and starting from the top-right vertex (clockwise direction), one can address the nodes as NE (northeast), SE (southeast), SW (southwest), and NW (northwest). All the external forces and torques are then referred to the center of mass, addressed as the point CM in Figure A.3, and defined as the center point of W, L, and T in Figure A.2. Consequently, the forces applied to each node have to be decomposed into their contributions along the X and Y directions. Finally, the arms of each node with respect to the center on the XY plane must be derived in order to transform the imposed force into the corresponding torques applied at the center of mass. It is important to highlight that the force components and arms are referred to the global system (XYZ). This is necessary because, when describing different components joined together, the consistency between them has to be maintained by choosing a common and unique reference system.

Let us now assume the plate rotates by a generic angle θ_z around the vertical Z axis, as shown in Figure A.4. The contributions determining the angle θ_z will be defined later. Let us now consider that a force F_{SE} on the XY plane is applied to the SE node. Its two components along the X and Y directions are $F_{X_{SE}}$ and $F_{Y_{SE}}$, respectively. First of all, the contribution of the force components applied along the X and Y axes, involved in the calculation of the torque around the Z axis must be determined. In other words, a relationship between the forces applied to the SE node and the forces applied to the center of mass has to be defined.

Taking into account the inertia effects expressed by (A1) and applied to the center of mass, we define the previously mentioned relationships as follows:

$$F_{\Delta X_{SE}} = F_{X_{SE}} - F_{X_{CM}} \tag{A2}$$

$$F_{\Delta Y_{SE}} = F_{Y_{SE}} - F_{Y_{CM}} \tag{A3}$$

where $F_{\Delta X_{SE}}$ and $F_{\Delta Y_{SE}}$ are the effective forces applied to the SE node along the X and Y directions, respectively. The latter take into account the inertia effect applied to the center of mass, and defined as $F_{X_{CM}}$ and $F_{Y_{CM}}$ in (A2) and (A3):

$$F_{X_{CM}} = -m\frac{d^2}{dt^2}(P_{X_{CM}}) = -ma_{X_{CM}} \tag{A4}$$

$$F_{Y_{CM}} = -m\frac{d^2}{dt^2}(P_{Y_{CM}}) = -ma_{Y_{CM}} \tag{A5}$$

where m is the mass of the plate (assuming a body made of homogeneous material), while $P_{X_{CM}}$ and $P_{Y_{CM}}$ describe the position of the center of mass along the X and Y directions, respectively. Their double time derivative leads to the acceleration, expressed by $a_{X_{CM}}$ and $a_{Y_{CM}}$ in (A4) and (A5). It must be noted that the second terms of such equations are negative (minus sign). Indeed, according to the principles of dynamics, the inertia of a body stimulated with a certain external applied force tends to counteract the imposed stimulus [217]. Finally, as the contribution along the Y direction of the external force applied to the SE node is oriented in the negative direction of the axis (see Figure A.4), the actual expression of (A3) is

$$F_{A_{Y_{SE}}} = -F_{Y_{SE}} - F_{Y_{CM}} \tag{A6}$$

Let us focus now on Figure A.4 once again. The two segments addressed as X_{SE} and Y_{SE} are the projections of the SE corner in the global reference system (XYZ), and depend on the coordinates of the SE point in the local frame (xyz) and on the angle θ_Z. Such segments can be easily calculated by using the rotation matrix [169]. Indeed, the coordinates of the SE corner being equal to $W/2$ and $-L/2$ along the x and y axes, respectively, we have

$$X_{SE} = \left(\frac{W}{2}\right)\cos\theta_Z + \left(\frac{L}{2}\right)\sin\theta_Z \tag{A7}$$

$$Y_{SE} = \left(\frac{W}{2}\right)\sin\theta_Z - \left(\frac{L}{2}\right)\cos\theta_Z \tag{A8}$$

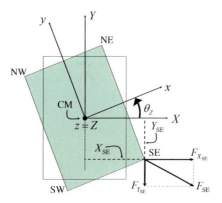

Figure A.4 Rigid plate after a rotation θ_z around the Z axis. A force is applied to the bottom-right (SE) corner.

Once the two arms X_{SE} and Y_{SE} have been determined, the contributions to the Z-axis torque of the center of mass due to the external forces applied to the SE corner are calculated as follows:

$$T_{\theta_{Z_{CM}}} = |Y_{SE}|F_{\Delta X_{SE}} - |X_{SE}|F_{\Delta Y_{SE}} \tag{A9}$$

In (A9) $T_{\theta_{Z_{CM}}}$ represents the Z-axis torque. Moreover, the absolute value of the X_{SE} and Y_{SE} arms is taken into account as, being projections of a generic point, they can assume a negative value. In the previous equation such magnitudes are used as arms (i.e., segments) in the torque calculation, and consequently they must always have a positive sign. On the other hand, the force components define the sign of each term on the right in (A9). According to the right-hand rule, when the product of the force multiplied by the arm induces a counterclockwise rotation around the Z axis, its contribution to the torque is positive. This is the case for the first term on the right in (A9). Differently, as the force applied to the SE node along the Y axis is oriented along its negative direction, this brings about a clockwise rotation of the Z axis. Such a consideration explains the negative sign of the second term on the right in (A9). Equation (A9) must now be completed by including the moment of inertia of the plate itself. In this case it has the form

$$T_{\theta_{Z_{CM}}} = -I_Z \frac{d^2\theta_Z}{dt^2} + |Y_{SE}|F_{\Delta X_{SE}} - |X_{SE}|F_{\Delta Y_{SE}} \tag{A10}$$

where I_Z is the moment of inertia of the plate with respect to the Z axis, expressed as follows [217]:

$$I_Z = \frac{1}{12}m(L^2 + W^2) \tag{A11}$$

In (A10) the moment of inertia is multiplied by the double derivative of θ_Z with respect to time. This represents the angular acceleration of the plate around its vertical axis. Furthermore, the sign of the first term in the second part of (A10) is negative because the principles of dynamics, already mentioned when referring to (A4) and (A5), are valid when dealing with the moment of inertia as well. Moving the focus now to the rotation angle θ_Z, we must define its effective value, which is expressed by two terms:

$$\theta_Z = \theta_{Z_{INIT}} + \theta_{Z_{INST}} \tag{A12}$$

where $\theta_{Z_{INIT}}$ is the initial rotation of the plate around the Z axis, as it could be not aligned to the global frame (XYZ) before applying any external force or torque, and $\theta_{Z_{INST}}$ is the instantaneous rotation of the center of mass around the Z axis, it being dependent on all the forces and torques applied to the four vertexes (nodes) of the plate.

Before proceeding with further explanations concerning the mechanical nodes of the plate, we must stress an important aspect. As the rigid plate allows translations along the three coordinate axes, the origin of its local frame (xyz) might not be

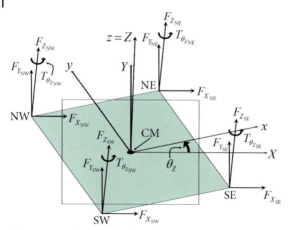

Figure A.5 Rigid plate with all the forces and torques applied to the four vertexes.

superimposed on the origin of the global system ($X\,Y\,Z$). However, this aspect can be disregarded, as we are interested in the relative displacements of each element after a stimulus has been applied, and not in the absolute position of each device. On the other hand, it is critical to maintain a consistent description of all the magnitudes along the three axes, which must be the same for all devices. This is the reason why the global frame $X\,Y\,Z$ is adopted and all the forces and displacements are referred to it.

Let us now extend what we have discussed thus far to the most general case, in which forces with three components along the X, Y, and Z axes, as well as torques around the Z axis, are applied to all four nodes (NE, SE, SW, and NW), like it is depicted in Figure A.5.

We now have to rewrite the equations previously reported in order to include all the stimuli mentioned above. Starting from the relationships among the forces applied to the four nodes and the internal forces related to the center of mass, we must integrate (A2) and (A3) with other expressions, as follows:

$$F_{\Delta X_{NE}} = F_{X_{NE}} - F_{X_{CM}} \tag{A13}$$

$$F_{\Delta Y_{NE}} = F_{Y_{NE}} - F_{Y_{CM}} \tag{A14}$$

$$F_{\Delta Z_{NE}} = F_{Z_{NE}} - F_{Z_{CM}} \tag{A15}$$

$$F_{\Delta X_{SE}} = F_{X_{SE}} - F_{X_{CM}} \tag{A16}$$

$$F_{\Delta Y_{SE}} = F_{Y_{SE}} - F_{Y_{CM}} \tag{A17}$$

$$F_{\Delta Z_{SE}} = F_{Z_{SE}} - F_{Z_{CM}} \tag{A18}$$

$$F_{\Delta X_{SW}} = F_{X_{SW}} - F_{X_{CM}} \tag{A19}$$

$$F_{\Delta Y_{SW}} = F_{Y_{SW}} - F_{Y_{CM}} \tag{A20}$$

$$F_{\Delta Z_{SW}} = F_{Z_{SW}} - F_{Z_{CM}} \tag{A21}$$

A.2 Mechanical Model of the Rigid Plate with Four DOFs

$$F_{\Delta X_{NW}} = F_{X_{NW}} - F_{X_{CM}} \tag{A22}$$

$$F_{\Delta Y_{NW}} = F_{Y_{NW}} - F_{Y_{CM}} \tag{A23}$$

$$F_{\Delta Z_{NW}} = F_{Z_{NW}} - F_{Z_{CM}} \tag{A24}$$

where $F_{X_{CM}}$ and $F_{Y_{CM}}$ are expressed by (A4) and (A5), while $F_{Z_{CM}}$ is given by

$$F_{Z_{CM}} = -m \frac{d^2}{dt^2}(P_{Z_{CM}}) = -m a_{Z_{CM}} \tag{A25}$$

In the previous equation, $P_{Z_{CM}}$ is the position of the center of mass along the Z direction. Now, a set of similar equations has to be introduced concerning the moment of inertia around the Z axis. Such relationships are expressed as follows:

$$T_{\Delta \theta_{Z_{NE}}} = T_{\theta_{Z_{NE}}} - T_{\theta_{Z_{CM}}} \tag{A26}$$

$$T_{\Delta \theta_{Z_{SE}}} = T_{\theta_{Z_{SE}}} - T_{\theta_{Z_{CM}}} \tag{A27}$$

$$T_{\Delta \theta_{Z_{SW}}} = T_{\theta_{Z_{SW}}} - T_{\theta_{Z_{CM}}} \tag{A28}$$

$$T_{\Delta \theta_{Z_{NW}}} = T_{\theta_{Z_{NW}}} - T_{\theta_{Z_{CM}}} \tag{A29}$$

where $T_{\theta_{Z_{NE}}}$, $T_{\theta_{Z_{SE}}}$, $T_{\theta_{Z_{SW}}}$, and $T_{\theta_{Z_{NW}}}$ are the external torques applied to the NE, SE, SW, and NW nodes, respectively. $T_{\theta_{Z_{CM}}}$ corresponds to (A10). In this case it is expressed in the following form:

$$\begin{aligned} T_{\theta_{Z_{CM}}} = -I_Z \frac{d^2 \theta_Z}{dt^2} &- \text{sgn}(Y_{NE})|Y_{NE}|F_{\Delta X_{NE}} \\ &+ \text{sgn}(X_{NE})|X_{NE}|F_{\Delta Y_{NE}} \\ &- \text{sgn}(Y_{SE})|Y_{SE}|F_{\Delta X_{SE}} \\ &+ \text{sgn}(X_{SE})|X_{SE}|F_{\Delta Y_{SE}} \\ &- \text{sgn}(Y_{SW})|Y_{SW}|F_{\Delta X_{SW}} \\ &+ \text{sgn}(X_{SW})|X_{SW}|F_{\Delta Y_{SW}} \\ &- \text{sgn}(Y_{NW})|Y_{NW}|F_{\Delta X_{NW}} \\ &+ \text{sgn}(X_{NW})|X_{NW}|F_{\Delta Y_{NW}} \end{aligned} \tag{A30}$$

where X_{NE}, Y_{NE}, X_{SW}, Y_{SW}, X_{NW}, and Y_{NW}, are the arms along the X and Y axes for the NE, SW, and NW nodes, respectively, calculated with the same approach used to obtain (A7) and (A8). The signs of the eight terms related to the forces applied on the XY plane in (A30) are determined with the right-hand rule, already mentioned for (A9).

To complete the mechanical model of the rigid plate with four DOFs, we now define the relative displacement of each node with respect to the center of mass in the global frame. The relative displacement of a generic point P in an arbitrary instant is defined as the difference between its instantaneous position P_{INST} and the initial position P_{INIT}:

$$\Delta P = P_{INST} - P_{INIT} \tag{A31}$$

As mentioned before, the angle θ_Z is composed of two contributions, namely, the initial rotation of the plate $\theta_{Z_{INIT}}$ and the instantaneous rotation $\theta_{Z_{INST}}$ (see (A12)). Consequently, when the angle $\theta_{Z_{INIT}}$ has a nonzero value, we have to calculate the position of each node in the global frame and we have to consider this as the initial position. Referring to the NE node and using once again the rotation matrix [169], we find the initial abscissas along the X and Y directions are

$$X_{NE_{INIT}} = \left(\frac{W}{2}\right)\cos\theta_{Z_{INIT}} - \left(\frac{L}{2}\right)\sin\theta_{Z_{INIT}} \tag{A32}$$

$$Y_{NE_{INIT}} = \left(\frac{W}{2}\right)\sin\theta_{Z_{INIT}} + \left(\frac{L}{2}\right)\cos\theta_{Z_{INIT}} \tag{A33}$$

The instantaneous position of the NE node is given by

$$X_{NE_{INST}} = \left(\frac{W}{2}\right)\cos\theta_Z - \left(\frac{L}{2}\right)\sin\theta_Z \tag{A34}$$

$$Y_{NE_{INST}} = \left(\frac{W}{2}\right)\sin\theta_Z + \left(\frac{L}{2}\right)\cos\theta_Z \tag{A35}$$

where θ_Z is expressed by (A12). Finally, the relative displacement of the NE node with respect to the center of mass, generally expressed by (A31), now becomes

$$\Delta X_{NE} = X_{NE_{INIT}} - X_{NE_{INST}} \tag{A36}$$

$$\Delta Y_{NE} = Y_{NE_{INIT}} - Y_{NE_{INST}} \tag{A37}$$

Following the same approach, one can effortlessly derive the proper set of equations defining the relative displacements along the XY plane for the SE, SW, and NW nodes. Since the simplified model discussed here does not allow any rotation of the plate around the X and Y axes (four DOFs), a relative displacement of each node with respect to the center of mass must never occur. To ensure this condition is met, the subsequent constraints must be imposed:

$$\Delta Z_{NE} = 0 \tag{A38}$$

$$\Delta Z_{SE} = 0 \tag{A39}$$

$$\Delta Z_{SW} = 0 \tag{A40}$$

$$\Delta Z_{NW} = 0 \tag{A41}$$

Eventually, another condition must be introduced to complete the definition of the plate as a rigid body, it being that any relative rotation must never occur between each node and the center of mass. Such a constraint results in the following conditions:

$$\theta_{Z_{NE}} - \theta_{Z_{CM}} = 0 \tag{A42}$$

$$\theta_{Z_{SE}} - \theta_{Z_{CM}} = 0 \tag{A43}$$

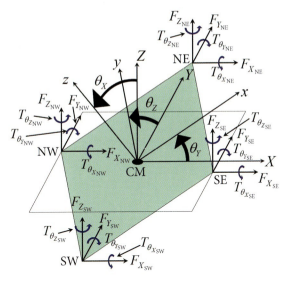

Figure A.6 Rigid plate (six degrees of freedom, DOFs) with all the forces and torques applied to the four corners.

$$\theta_{Z_{SW}} - \theta_{Z_{CM}} = 0 \tag{A44}$$

$$\theta_{Z_{NW}} - \theta_{Z_{CM}} = 0 \tag{A45}$$

The latter set of equations completes the mechanical model of the rigid plate with four DOFs. All the formulas presented so far can be easily implemented in the Verilog-A programming language, as discussed in Chapter 5.

A.3
Extension of the Mechanical Model of the Rigid Plate to Six DOFs

The extension of the mechanical model of the rigid plate with four DOFs to the complete one, featuring six DOFs, is rather straightforward. The model described in the previous pages has to be modified in order to account for the rotations around the X and Y axes at each node. Figure A.6 shows a rigid plate with three translational and three rotational stimuli (forces and torques, respectively) applied to the four nodes.

Starting from the relationships among the external translational forces at the four nodes and the internal forces on the center of mass, we find (A13)–(A15) and (A16)–(A24) are still valid. However, (A26)–(A29) must be extended as follows:

$$T_{\Delta\theta_{X_{NE}}} = T_{\theta_{X_{NE}}} - T_{\theta_{X_{CM}}} \tag{A46}$$

$$T_{\Delta\theta_{Y_{NE}}} = T_{\theta_{Y_{NE}}} - T_{\theta_{Y_{CM}}} \tag{A47}$$

$$T_{\Delta\theta_{Z_{NE}}} = T_{\theta_{Z_{NE}}} - T_{\theta_{Z_{CM}}} \tag{A48}$$

Appendix A Rigid Plate Electromechanical Transducer (Complete Model)

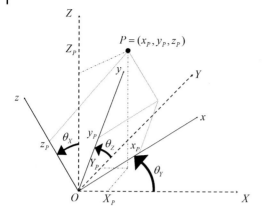

Figure A.7 Rotation of the local reference system around the axes of the global (fixed) one, identified by the angles θ_X, θ_Y, and θ_Z.

$$T_{\Delta\theta_{X_{SE}}} = T_{\theta_{X_{SE}}} - T_{\theta_{X_{CM}}} \tag{A49}$$

$$T_{\Delta\theta_{Y_{SE}}} = T_{\theta_{Y_{SE}}} - T_{\theta_{Y_{CM}}} \tag{A50}$$

$$T_{\Delta\theta_{Z_{SE}}} = T_{\theta_{Z_{SE}}} - T_{\theta_{Z_{CM}}} \tag{A51}$$

$$T_{\Delta\theta_{X_{SW}}} = T_{\theta_{X_{SW}}} - T_{\theta_{X_{CM}}} \tag{A52}$$

$$T_{\Delta\theta_{Y_{SW}}} = T_{\theta_{Y_{SW}}} - T_{\theta_{Y_{CM}}} \tag{A53}$$

$$T_{\Delta\theta_{Z_{SW}}} = T_{\theta_{Z_{SW}}} - T_{\theta_{Z_{CM}}} \tag{A54}$$

$$T_{\Delta\theta_{X_{NW}}} = T_{\theta_{X_{NW}}} - T_{\theta_{X_{CM}}} \tag{A55}$$

$$T_{\Delta\theta_{Y_{NW}}} = T_{\theta_{Y_{NW}}} - T_{\theta_{Y_{CM}}} \tag{A56}$$

$$T_{\Delta\theta_{Z_{NW}}} = T_{\theta_{Z_{NW}}} - T_{\theta_{Z_{CM}}} \tag{A57}$$

where $T_{\theta_{X_{CM}}}$ and $T_{\theta_{Y_{CM}}}$ are the torques at the center of mass around the X and Y axes, respectively. Let us now extend the definition of the arms by taking into account the XY out-of-plane displacements of the nodes. We basically have three segments which identify each node in the global frame (XYZ), as shown in Figure A.7.

Taking for instance the NE corner, and assuming its local abscissa along the z axis equals zero, as was already discussed for Figure A.2, we find its local coordinates are

$$NE(x, y, z) = \left(\frac{W}{2}, \frac{L}{2}, 0\right) \tag{A58}$$

In order to derive the three arms X_{NE}, Y_{NE}, and Z_{NE}, we have to use the rotation matrix [169]:

$$X_{NE} = (C_Y C_Z)\frac{W}{2} + (-C_X S_Z + S_X S_Y C_Z)\frac{L}{2} \tag{A59}$$

$$Y_{NE} = (C_Y S_Z)\frac{W}{2} + (C_X C_Z + S_X S_Y S_Z)\frac{L}{2} \tag{A60}$$

$$Z_{NE} = -S_Y\frac{W}{2} + (S_X C_Y)\frac{L}{2} \tag{A61}$$

The rotation matrix elements appearing in (A59)–(A61) are

$$C_X = \cos\theta_X \tag{A62}$$

$$S_X = \sin\theta_X \tag{A63}$$

$$C_Y = \cos\theta_Y \tag{A64}$$

$$S_Y = \sin\theta_Y \tag{A65}$$

$$C_Z = \cos\theta_Z \tag{A66}$$

$$S_Z = \sin\theta_Z \tag{A67}$$

Following the same approach, and substituting in (A59)–(A61) the proper coordinates of the other three nodes, we easily obtain all the arms, they being (X_{SE}, Y_{SE}, Z_{SE}), (X_{SW}, Y_{SW}, Z_{SW}), and (X_{NW}, Y_{NW}, Z_{NW}). Once all the arms are available, the equations for the torque at the center of mass around the X and Y axes ($T_{\theta_{X_{CM}}}$, $T_{\theta_{Y_{CM}}}$) can be written. Such equations complete (A30), which is still valid for the model with six DOFs, and they are expressed as follows:

$$\begin{aligned} T_{\theta_{X_{CM}}} = &-I_X\frac{d^2\theta_X}{dt^2} - \text{sgn}(Z_{NE})|Z_{NE}|F_{\Delta Y_{NE}} \\ &+ \text{sgn}(Y_{NE})|Y_{NE}|F_{\Delta Z_{NE}} \\ &- \text{sgn}(Z_{SE})|Z_{SE}|F_{\Delta Y_{SE}} \\ &+ \text{sgn}(Y_{SE})|Y_{SE}|F_{\Delta Z_{SE}} \\ &- \text{sgn}(Z_{SW})|Z_{SW}|F_{\Delta Y_{SW}} \\ &+ \text{sgn}(Y_{SW})|Y_{SW}|F_{\Delta Z_{SW}} \\ &- \text{sgn}(Z_{NW})|Z_{NW}|F_{\Delta Y_{NW}} \\ &+ \text{sgn}(Y_{NW})|Y_{NW}|F_{\Delta Z_{NW}} \end{aligned} \tag{A68}$$

$$\begin{aligned} T_{\theta_{Y_{CM}}} = &-I_Y\frac{d^2\theta_Y}{dt^2} + \text{sgn}(Z_{NE})|Z_{NE}|F_{\Delta X_{NE}} \\ &- \text{sgn}(X_{NE})|X_{NE}|F_{\Delta Z_{NE}} \\ &+ \text{sgn}(Z_{SE})|Z_{SE}|F_{\Delta X_{SE}} \\ &- \text{sgn}(X_{SE})|X_{SE}|F_{\Delta Z_{SE}} \\ &+ \text{sgn}(Z_{SW})|Z_{SW}|F_{\Delta X_{SW}} \\ &- \text{sgn}(X_{SW})|X_{SW}|F_{\Delta Z_{SW}} \\ &+ \text{sgn}(Z_{NW})|Z_{NW}|F_{\Delta X_{NW}} \\ &- \text{sgn}(X_{NW})|X_{NW}|F_{\Delta Z_{NW}} \end{aligned} \tag{A69}$$

where I_X and I_Y represent the moment of inertia of the plate around the X and Y axes, respectively, and are expressed as

$$I_X = \frac{1}{12} m(L^2 + T^2) \tag{A70}$$

$$I_Y = \frac{1}{12} m(W^2 + T^2) \tag{A71}$$

In (A68) and (A69), I_X and I_Y are multiplied by the angular acceleration of the center of mass around the X and Y axes, respectively. A further step is the calculation of the relative displacements of each node with respect to the center of mass. As reported in (A12) for θ_Z, the angles θ_X and θ_Y include an initial and instantaneous contribution, their complete expression being as follows:

$$\theta_X = \theta_{X_{INIT}} + \theta_{X_{INST}} \tag{A72}$$

$$\theta_Y = \theta_{Y_{INIT}} + \theta_{Y_{INST}} \tag{A73}$$

Equation (A31) is applied also in this case, and the instantaneous and initial positions of each node depend on the three rotation angles. The relationships are similar to (A59)–(A61), and taking as an example the NE node, we find the initial position along the three coordinate axes $(X\,Y\,Z)$ is given by

$$X_{NE_{INIT}} = (C_{Y_{INIT}} C_{Z_{INIT}}) \frac{W}{2}$$
$$+ (-C_{X_{INIT}} S_{Z_{INIT}} + S_{X_{INIT}} S_{Y_{INIT}} C_{Z_{INIT}}) \frac{L}{2} \tag{A74}$$

$$Y_{NE_{INIT}} = (C_{Y_{INIT}} S_{Z_{INIT}}) \frac{W}{2}$$
$$+ (C_{X_{INIT}} C_{Z_{INIT}} + S_{X_{INIT}} S_{Y_{INIT}} S_{Z_{INIT}}) \frac{L}{2} \tag{A75}$$

$$Z_{NE_{INIT}} = -S_{Y_{INIT}} \frac{W}{2} + (S_{X_{INIT}} C_{Y_{INIT}}) \frac{L}{2} \tag{A76}$$

The instantaneous position of the NE node is given by

$$X_{NE_{INST}} = (C_{Y_{INST}} C_{Z_{INST}}) \frac{W}{2}$$
$$+ (-C_{X_{INST}} S_{Z_{INST}} + S_{X_{INST}} S_{Y_{INST}} C_{Z_{INST}}) \frac{L}{2} \tag{A77}$$

$$Y_{NE_{INST}} = (C_{Y_{INST}} S_{Z_{INST}}) \frac{W}{2}$$
$$+ (C_{X_{INST}} C_{Z_{INST}} + S_{X_{INST}} S_{Y_{INST}} S_{Z_{INST}}) \frac{L}{2} \tag{A78}$$

$$Z_{NE_{INST}} = -S_{Y_{INST}} \frac{W}{2} + (S_{X_{INST}} C_{Y_{INST}}) \frac{L}{2} \tag{A79}$$

Finally, the relative displacements of the NE node with respect to the center of mass are

$$\Delta X_{NE} = X_{NE_{INIT}} - X_{NE_{INST}} \tag{A80}$$

$$\Delta Y_{NE} = Y_{NE_{INIT}} - Y_{NE_{INST}} \tag{A81}$$

$$\Delta Z_{NE} = Z_{NE_{INIT}} - Z_{NE_{INST}} \tag{A82}$$

By applying the same procedure, we can straightforwardly derive the relative displacements of the other three nodes. In the case of six DOFs, the conditions expressed by (A38)–(A41) must not be imposed as each node of the plate is free to have a different abscissa along the Z axis due to the rotations θ_X and θ_Y. Furthermore, in order to ensure that the plate with six DOFs behaves as a rigid body, (A42)–(A45) have to be extended as follows:

$$\theta_{X_{NE}} - \theta_{X_{CM}} = 0 \tag{A83}$$

$$\theta_{Y_{NE}} - \theta_{Y_{CM}} = 0 \tag{A84}$$

$$\theta_{Z_{NE}} - \theta_{Z_{CM}} = 0 \tag{A85}$$

$$\theta_{X_{SE}} - \theta_{X_{CM}} = 0 \tag{A86}$$

$$\theta_{Y_{SE}} - \theta_{Y_{CM}} = 0 \tag{A87}$$

$$\theta_{Z_{SE}} - \theta_{Z_{CM}} = 0 \tag{A88}$$

$$\theta_{X_{SW}} - \theta_{X_{CM}} = 0 \tag{A89}$$

$$\theta_{Y_{SW}} - \theta_{Y_{CM}} = 0 \tag{A90}$$

$$\theta_{Z_{SW}} - \theta_{Z_{CM}} = 0 \tag{A91}$$

$$\theta_{X_{NW}} - \theta_{X_{CM}} = 0 \tag{A92}$$

$$\theta_{Y_{NW}} - \theta_{Y_{CM}} = 0 \tag{A93}$$

$$\theta_{Z_{NW}} - \theta_{Z_{CM}} = 0 \tag{A94}$$

A.3.1
Placement of Nodes along the Edges of a Rigid Plate

In order to complete the mechanical model for rigid plates with four and six DOFs, let us now briefly describe an additional feature included in both of them. Up to now we took into account four nodes at the corners, through which external stimuli are imposed on the plate. However, four additional nodes are included in the rigid plate model implemented in Verilog-A in order to extend its usability. In several radio frequency MEMS switch realizations, the suspending springs are not connected to the four corners, but are placed in arbitrary positions along the edges of a rigid plate. The additional nodes mentioned above correspond to the four middle

254 | Appendix A Rigid Plate Electromechanical Transducer (Complete Model)

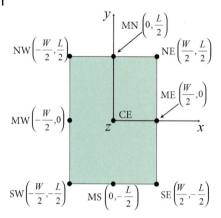

Figure A.8 Rigid plate with eight nodes. The middle point along the four edges is also introduced.

points along the edges of the plate, as shown in Figure A.8 starting from the right edge and proceeding clockwise, and they are named middle-east (ME), middle-south (MS), middle-west (MW), and middle-north (MN), respectively. Moreover, the possibility of having each of the eight nodes displaced with respect to the four corners and four middle points, along the local x or y axis, is also implemented. This allows more flexibility in designing a complete MEMS structure to be simulated as, for instance, it is possible to connect flexible beams everywhere along the edges of the plate. As an example, in Figure A.9, an offset D_y is applied to the NE node. When the four additional middle nodes are included, the mechanical model of the rigid plate discussed so far is still valid. It is then necessary to extend the previous equations (e.g., torques around the axes and displacements) with suitable terms taking into account their presence. Moreover, when an offset is applied to

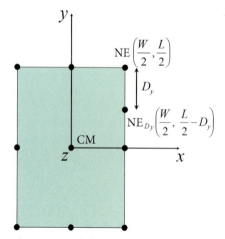

Figure A.9 Rigid plate with an offset D_y applied to the NE node along the local y axis.

one or several nodes, we have to calculate its (their) position, considering the actual local coordinates also including such a displacement. For example, referring to Figure A.9, we have to calculate (A59)–(A61) of the NE node by substituting its original set of coordinates (A58) with the set including the D_y offset:

$$NE_{D_y}(x, y, z) = \left(\frac{W}{2}, \frac{L}{2} - D_y, 0\right) \tag{A95}$$

A.4
Contact Model for Rigid Plates with Four and Six DOFs

An important extension of the mechanical model for rigid plates with four and six DOFs is the inclusion of the effect of collision with the substrate lying underneath the suspended rigid plate, it being a typical functional condition occurring in MEMS switches (i.e., pull-in). For this purpose a suitable feature in the mechanical model of the plate considering the presence of a surface underneath it has to be defined. As already mentioned in the previous sections, the collision model is firstly introduced for the rigid plate with four DOFs, and is subsequently extended to the complete model with six DOFs.

The approach to the contact modeling is based on the introduction of a suitable mechanical constraint when a certain condition on the vertical displacement of the plate is verified. For this, an initial value of the distance between the plate and the plane underneath (i.e., gap) must be defined. In this way, the relative vertical displacement of the plate, previously introduced in the mechanical model, is turned into an absolute displacement with respect to the floor plane. When the gap is equal to zero, the collision occurs. In order not to allow any further downward movement of the plate, which would imply a physically nonsensical negative gap, an appropriate force must counteract its motion. This is obtained by defining a material with a fictitious elastic constant, which is set to a very large value. When the collision occurs, the reaction force is applied to the center of mass. As a consequence, any further small downward displacement of the plate generates a very large force, counteracting its lowering, and eventually stopping the movement. Let us now write in mathematical notation what has just been discussed. The contact force $F_{Z_{TOUCH}}$ must be added to (A25) as follows:

$$F_{Z_{CM}} = -ma_{Z_{CM}} + F_{Z_{TOUCH}} \tag{A96}$$

In order to distinguish the case in which the collision either does or does not occur, a binary flag TC_{FLAG} (touch flag) is defined as follows:

$$TC_{FLAG} = \begin{cases} 1 & \text{if } (Z_{CM_{INST}} < 0) \text{ and } (|Z_{CM_{INST}}| \geq |Z_{GAP}|) \\ 0 & \text{otherwise} \end{cases} \tag{A97}$$

where $Z_{CM_{INST}}$ is the instantaneous displacement of the center of mass along the Z direction, while Z_{GAP} is the initial gap between the plate and the substrate. A

Appendix A Rigid Plate Electromechanical Transducer (Complete Model)

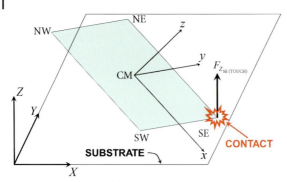

Figure A.10 Collision of the SE corner of the plate with the bottom substrate plane.

further step consists in defining the value of $F_{Z_{TOUCH}}$ depending on the state of TC_{FLAG}:

$$F_{Z_{TOUCH}} = \begin{cases} K_{TOUCH}(Z_{GAP} + Z_{CM_{INST}}) & \text{if } TC_{FLAG} = 1 \\ 0 & \text{if } TC_{FLAG} = 0 \end{cases} \quad (A98)$$

where K_{TOUCH} is a fictitious elastic constant with a very large value, and models the lower contact surface (i.e., the underneath electrode). In the Verilog-A implementation of the contact model, its magnitude is

$$K_{TOUCH} = 10^9 \left[\frac{N}{m}\right] \quad (A99)$$

Extending the contact model to the rigid plate with six DOFs is rather straightforward. The main difference is that the collision condition must be defined for each node, rather than for solely the center of mass. Indeed, the collision might occur at one or two nodes, depending on the orientation of the plate around the three axes. Let us suppose that the collision occurs just to the SE corner, as depicted in Figure A.10.

The first step is to define a flag variable similar to (A97) in order to detect when the collision occurs at the SE corner. It is expressed in the form

$$TC_{SE_{FLAG}} = \begin{cases} 1 & \text{if } (Z_{SE_{INST}} < 0) \text{ and } (|Z_{SE_{INST}}| \geq |Z_{GAP} + Z_{SE_{INIT}}|) \\ 0 & \text{otherwise} \end{cases}$$

$$(A100)$$

where $Z_{SE_{INIT}}$ and $Z_{SE_{INST}}$ are the initial and instantaneous positions of the SE corner along the Z axis, and are obtained through (A76) and (A79), respectively, calculated for the proper coordinates of the SE corner in the local frame (xyz). The definition of $Z_{SE_{INIT}}$ is shown in Figure A.11, where the lateral view of the initial position for a plate rotated around the Y axis (angle $\theta_{Y_{INIT}}$) is reported.

A.4 Contact Model for Rigid Plates with Four and Six DOFs

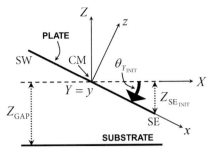

Figure A.11 Lateral view of a plate with an initial nonzero orientation angle $\theta_{Y_{INIT}}$.

Subsequently, the contact force $F_{Z_{SE(TOUCH)}}$ in Figure A.10 is determined, and it is expressed, similarly to (A98), in the form

$$F_{Z_{SE(TOUCH)}} = \begin{cases} K_{TOUCH}(Z_{GAP} + Z_{SE_{INST}} + Z_{SE_{INIT}}) & \text{if } TC_{SE_{FLAG}} = 1 \\ 0 & \text{if } TC_{SE_{FLAG}} = 0 \end{cases} \quad (A101)$$

Finally, the contribution of $F_{Z_{SE(TOUCH)}}$ due to the collision of the SE corner acts on the balance of torques at the center of mass. In particular, $F_{Z_{SE(TOUCH)}}$ contributes to the torques around the X and Y axes, and consequently (A68) and (A69) have to be rewritten as follows:

$$\begin{aligned}
T_{\theta_{X_{CM}}} = &-I_X \frac{d^2 \theta_X}{dt^2} - \text{sgn}(Z_{NE})|Z_{NE}|(F_{\Delta Y_{NE}}) \\
&+ \text{sgn}(Y_{NE})|Y_{NE}|(F_{\Delta Z_{NE}}) \\
&- \text{sgn}(Z_{SE})|Z_{SE}|(F_{\Delta Y_{SE}}) \\
&+ \text{sgn}(Y_{SE})|Y_{SE}|(F_{\Delta Z_{SE}} + F_{Z_{SE(TOUCH)}}) \\
&- \text{sgn}(Z_{SW})|Z_{SW}|(F_{\Delta Y_{SW}}) \\
&+ \text{sgn}(Y_{SW})|Y_{SW}|(F_{\Delta Z_{SW}}) \\
&- \text{sgn}(Z_{NW})|Z_{NW}|(F_{\Delta Y_{NW}}) \\
&+ \text{sgn}(Y_{NW})|T_{NW}|(F_{\Delta Z_{NW}}) \quad (A102)
\end{aligned}$$

$$\begin{aligned}
T_{\theta_{Y_{CM}}} = &-I_Y \frac{d^2 \theta_Y}{dt^2} + \text{sgn}(Z_{NE})|Z_{NE}|(F_{\Delta X_{NE}}) \\
&- \text{sgn}(X_{NE})|X_{NE}|(F_{\Delta Z_{NE}}) \\
&+ \text{sgn}(Z_{NE})|Z_{SE}|(F_{\Delta X_{SE}}) \\
&- \text{sgn}(X_{NE})|X_{SE}|(F_{\Delta Z_{SE}} + F_{Z_{SE(TOUCH)}}) \\
&+ \text{sgn}(Z_{NE})|Z_{SW}|(F_{\Delta X_{SW}}) \\
&- \text{sgn}(X_{NE})|X_{SW}|(F_{\Delta Z_{SW}}) \\
&+ \text{sgn}(Z_{NE})|Z_{NW}|(F_{\Delta X_{NW}}) \\
&- \text{sgn}(X_{NE})|X_{NW}|(F_{\Delta Z_{NW}}) \quad (A103)
\end{aligned}$$

Following the same approach, we calculate the contact forces for the other three nodes ($F_{Z_{NE(TOUCH)}}$, $F_{Z_{SW(TOUCH)}}$, and $F_{Z_{NW(TOUCH)}}$) and the complete expressions of (A102) and (A103) become

$$\begin{aligned} T_{\theta_{X_{CM}}} = -I_X \frac{d^2 \theta_X}{dt^2} &- \operatorname{sgn}(Z_{NE})|Z_{NE}|(F_{\Delta Y_{NE}}) \\ &+ \operatorname{sgn}(Y_{NE})|Y_{NE}|(F_{\Delta Z_{NE}} + F_{Z_{NE(TOUCH)}}) \\ &- \operatorname{sgn}(Z_{SE})|Z_{SE}|(F_{\Delta Y_{SE}}) \\ &+ \operatorname{sgn}(Y_{SE})|Y_{SE}|(F_{\Delta Z_{SE}} + F_{Z_{SE(TOUCH)}}) \\ &- \operatorname{sgn}(Z_{SW})|Z_{SW}|(F_{\Delta Y_{SW}}) \\ &+ \operatorname{sgn}(Y_{SW})|Y_{SW}|(F_{\Delta Z_{SW}} + F_{Z_{SW(TOUCH)}}) \\ &- \operatorname{sgn}(Z_{NW})|Z_{NW}|(F_{\Delta Y_{NW}}) \\ &+ \operatorname{sgn}(Y_{NW})|T_{NW}|(F_{\Delta Z_{NW}} + F_{Z_{NW(TOUCH)}}) \quad (A104) \end{aligned}$$

$$\begin{aligned} T_{\theta_{Y_{CM}}} = -I_Y \frac{d^2 \theta_Y}{dt^2} &+ \operatorname{sgn}(Y_{NE})|Z_{NE}|(F_{\Delta X_{NE}}) \\ &- \operatorname{sgn}(Y_{NE})|X_{NE}|(F_{\Delta Z_{NE}} + F_{Z_{NE(TOUCH)}}) \\ &+ \operatorname{sgn}(Y_{SE})|Z_{SE}|(F_{\Delta X_{SE}}) \\ &- \operatorname{sgn}(Y_{SE})|X_{SE}|(F_{\Delta Z_{SE}} + F_{Z_{SE(TOUCH)}}) \\ &+ \operatorname{sgn}(Y_{SW})|Z_{SW}|(F_{\Delta X_{SW}}) \\ &- \operatorname{sgn}(Y_{SW})|X_{SW}|(F_{\Delta Z_{SW}} + F_{Z_{SW(TOUCH)}}) \\ &+ \operatorname{sgn}(Y_{NW})|Z_{NW}|(F_{\Delta X_{NW}}) \\ &- \operatorname{sgn}(Y_{NW})|X_{NW}|(F_{\Delta Z_{NW}} + F_{Z_{NW(TOUCH)}}) \quad (A105) \end{aligned}$$

To complete the description of the collision effects, a simple model accounting for the dynamic friction of the plate when it is touching the lower substrate and moving (i.e., sliding) onto the XY plane is implemented. It is not shown here for the sake of brevity.

A.5
Electrostatic Model of the Rigid Plate

We now introduce the electromechanical transduction performed via the suspended rigid plate when a biasing voltage is imposed between the movable and the fixed (underlying) electrode. As in previous sections, the simplest case will be discussed first, and then more and more features will be added.

A.5.1
The Four DOFs Condition

When the rigid plate allows only four DOFs, the two electrodes, that is, the fixed one and the movable plate itself, are always in a parallel position. In this configura-

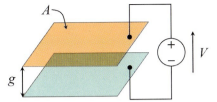

Figure A.12 Electrostatic model for two parallel plates. A voltage V is applied between them.

tion the electrostatic problem is very simple, and the capacitance and electrostatic attraction force are calculated with the following well-known formulas [92]:

$$C = \varepsilon \frac{A}{g} \tag{A106}$$

$$F_{EL} = \frac{1}{2}\varepsilon V^2 \frac{A}{g^2} \tag{A107}$$

In (A106) and (A107), C is the capacitance, F_{EL} is the electrostatic attraction force, A is the area of the electrodes, g is the vertical distance between them, V is the applied voltage, and ε is the dielectric constant of the insulator (e.g., air) (see Figure A.12).

Let us focus on the capacitance. Referring to the rigid plate model with four DOFs already discussed, one can rewrite (A106) as follows:

$$C = \varepsilon \frac{WL}{(Z_{GAP} + Z_{CM_{INST}})} \tag{A108}$$

where the denominator represents the instantaneous gap between the plate and the substrate. In other words, (A108) is the expression for the variable capacitance depending on the vertical displacement of the plate. The current flowing through the suspended plate is then defined as the time derivative of the total accumulated charge on the two electrodes:

$$I = \frac{dQ}{dt} \tag{A109}$$

where

$$Q = CV \tag{A110}$$

For the electrostatic force, (A107) can be rewritten by taking into account the instantaneous gap between the two plates:

$$F_{Z_{EL}} = \frac{1}{2}\varepsilon V^2 \frac{A}{(Z_{GAP} + Z_{CM_{INST}})^2} \tag{A111}$$

With (A111) the dependence of the electrostatic force on the instantaneous gap is introduced. In (A111) we introduced $F_{Z_{EL}}$ instead of F_{EL} because it is oriented

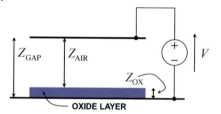

Figure A.13 Parallel plates with an oxide layer deposited on the lower electrode. Its thickness Z_{OX} is enlarged to make it easily visible.

along the Z axis. Finally, the electromechanical transduction is accounted for by introducing $F_{Z_{EL}}$ in the balance of forces at the center of mass. This means rewriting (A96) as

$$F_{Z_{CM}} = -MA_{Z_{CM}} + F_{Z_{TOUCH}} - F_{Z_{EL}} \tag{A112}$$

Very often, during the fabrication of MEMS suspended switches, a thin insulating layer is deposited onto the lower electrode. This is necessary in order to prevent the short circuit that would occur after the collapse of the suspended plate onto the substrate (i.e., physical contact of two conductive materials). The influence of the dielectric layer on the capacitance and electrostatic force is negligible when the gap is large. However, as the air gap between the two plates decreases, the presence of the insulating layer affects the C and $F_{Z_{EL}}$ values, especially in the limit case of the pull-in. The model discussed thus far is extended by also including the deposited insulating layer. A schematic view of the suspended plate with an insulating layer on the lower electrode is shown in Figure A.13.

In such a schematic, an oxide (i.e., silicon dioxide) layer is included, it being a material very often used in the fabrication of MEMS switches. The first aspect considered is that the gap between the plates is reduced because of the presence of the oxide layer. This means that in all the previous formulas related to the mechanical model, we have to replace the initial gap Z_{GAP} with Z_{AIR}, which is expressed in the form (see Figure A.13)

$$Z_{AIR} = Z_{GAP} - Z_{OX} \tag{A113}$$

Subsequently, the capacitance and electrostatic force must be rewritten keeping in mind the oxide layer. Starting from the first one, we determine the total capacitance C of the entire device by connecting in series the two capacitors, namely, one with air as dielectric material (C_{AIR}), and the second one with oxide (C_{OX}), as shown in Figure A.14.

C is then defined with the well-known formula

$$\frac{1}{C} = \frac{1}{C_{AIR}} + \frac{1}{C_{OX}} \tag{A114}$$

which can be rewritten as

$$C = \frac{C_{AIR} + C_{OX}}{C_{AIR} C_{OX}} \tag{A115}$$

A.5 Electrostatic Model of the Rigid Plate

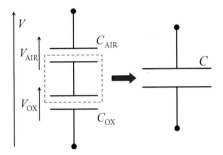

Figure A.14 Because of the oxide layer, an additional capacitance is inserted in series with the one realized by the air gap.

where

$$C_{AIR} = \varepsilon_{AIR} \frac{WL}{(Z_{AIR} + Z_{CM_{INST}})} \quad (A116)$$

and

$$C_{OX} = \varepsilon_{OX} \frac{WL}{Z_{OX}} \quad (A117)$$

In (A116) and (A117), ε_{AIR} and ε_{OX} represent the dielectric constants of air and oxide, respectively.

Now moving the focus onto the electrostatic force, we have to rewrite (A111) as a function of the voltage drop across the air gap (V_{AIR} in Figure A.14) instead of the total voltage V. The proper expression for V_{AIR} can be derived by means of the series capacitors mentioned above. According to the principle of charge conservation and looking at Figure A.14, we find the total charge accumulated on one plate must be equal to the charge on the other plate within the dashed box. This means that

$$Q_{AIR} = Q_{OX} \quad (A118)$$

According to (A116), (A117) can be written as

$$C_{AIR} V_{AIR} = C_{OX} V_{OX} \quad (A119)$$

However,

$$V = V_{AIR} + V_{OX} \quad (A120)$$

and transforming this equation, we obtain

$$V_{OX} = V - V_{AIR} \quad (A121)$$

Substituting (A121) in (A119), we obtain the following expression for V_{AIR}:

$$V_{AIR} = \frac{C_{OX}}{C_{AIR} + C_{OX}} V \quad (A122)$$

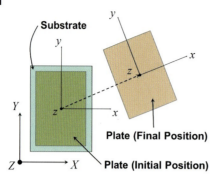

Figure A.15 When large displacements and rotations occur, the plate projection on the underlying substrate could fall (partially or totally) outside its surface.

Finally, the actual electrostatic force accounting for the oxide layer is obtained by replacing V in (A111) with (A122), resulting in

$$F_{Z_{EL}} = \frac{1}{2} \varepsilon_{AIR} V_{AIR}^2 \frac{WL}{(Z_{AIR} + Z_{CM_{INST}})^2} \tag{A123}$$

Additionally, the total capacitance and electrostatic force accounting for the oxide layer, expressed by (A115) and (A123), respectively, can be expressed as a function of an effective dielectric constant ε_{EFF}. The latter depends on both ε_{AIR} and ε_{OX}, and its expression including the scheme of Figure A.14 is

$$\varepsilon_{EFF} = \frac{\varepsilon_{OX} \varepsilon_{AIR} Z_{GAP}}{Z_{OX} \left(\varepsilon_{AIR} + \varepsilon_{OX} \frac{Z_{AIR}}{Z_{OX}} \right)} \tag{A124}$$

Equation (A124) substituted into (A108) and (A111) gives the complete electrical effects also accounting for the oxide layer. A further consideration must now be highlighted. The rigid plate models with four and six DOFs can move along the three axes (X, Y, Z), and also rotate around them (the plate with four DOFs can rotate only around Z, while the plate with six DOFs can rotate around all of them). It might happen then that due to large displacements and rotations, the suspended plate ends up in a configuration in which it partially overlaps, or does not overlap at all, the electrode underneath, as shown in Figure A.15.

In the latter case, the previous formulas for the capacitance and electrostatic force would not be valid anymore. However, we assume that the typical displacements and rotations the plate undergoes are small enough and the underlying electrode is large enough that the suspended plate projection (or shadow) lies always and entirely within its edges.

A.5.2
Extension to the Case of Six DOFs

Following the usual approach, we now extend the electrostatic model to the rigid plate with six DOFs. For the sake of brevity, we directly mention the presence of

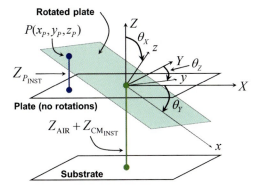

Figure A.16 Definition of the vertical distance between a generic point P on the rotated plate area and the substrate.

the oxide layer on the lower electrode. The main feature that must be considered is that when all the rotation angles (θ_X, θ_Y, θ_Z) have nonzero values, every point on the surface of the plate has a different distance from the substrate. In Figure A.16 a generic point P with coordinates x_P, y_P, z_P in the local frame of the plate is taken as an example.

Its distance from the substrate is the composition of two contributions. One is the instantaneous distance of the center of mass of the plate from the substrate, defined as $Z_{AIR} + Z_{CM_{INST}}$. The second segment is the projection along the global Z axis of the point P, and is expressed by the same formula as (A79):

$$Z_{P_{INST}} = -\sin\theta_Y x_P + (\sin\theta_X \cos\theta_Y) y_P + (\cos\theta_X \cos\theta_Y) z_P \quad (A125)$$

where θ_X and θ_Y are defined as in (A72) and (A73), respectively. Moreover, in (A125)

$$z_P = -\frac{T}{2} \quad (A126)$$

Indeed, in Figure A.2, we see the local reference system origin is superimposed on the center of mass of the plate. However, the electrical effects must be calculated at the lower face of the plate, which has a local coordinate as indicated by (A126). From (A125), we see that a rotation of the plate around the global Z axis does not yield a contribution to $Z_{P_{INST}}$.

It is now possible to rewrite the equation for the capacitance and electrostatic force, in an integral form, over the whole area of the plate (see Figure A.3). In this way, the actual gap for each point on the surface of the plate is considered. Equation (A116) then becomes

$$C_{AIR} = \varepsilon_{AIR} \int_{-\frac{W}{2}}^{\frac{W}{2}} \int_{-\frac{L}{2}}^{\frac{L}{2}} \frac{dx\,dy}{(Z_{AIR} + Z_{CM_{INST}} + Z_{P_{INST}})} \quad (A127)$$

Similarly, the electrostatic attraction force of (A123) becomes

$$F_{Z_{EL}} = \frac{1}{2}\varepsilon_{AIR} V_{AIR}^2 \int_{-\frac{W}{2}}^{\frac{W}{2}} \int_{-\frac{L}{2}}^{\frac{L}{2}} \frac{dx\,dy}{(Z_{AIR} + Z_{CM_{INST}} + Z_{P_{INST}})^2} \qquad (A128)$$

Both (A127) and (A128) are integrated in a closed form, which is implemented in Verilog-A without particular issues. Equation (A128) represents the electrostatic attraction force expressed in the local frame of the plate (xyz). Before applying its contribution to the center of mass, we must decompose it into three corresponding forces in the global frame, $F_{X_{EL}}$, $F_{Y_{EL}}$, and $F_{Z_{EL}}$, according to the rotation matrix [169], as follows:

$$F_{X_{EL}} = (\sin\theta_X \sin\theta_Z + \cos\theta_X \sin\theta_Y \cos\theta_Z) F_{z_{EL}} \qquad (A129)$$

$$F_{Y_{EL}} = (-\sin\theta_X \cos\theta_Z + \cos\theta_X \sin\theta_Y \sin\theta_Z) F_{z_{EL}} \qquad (A130)$$

$$F_{Z_{EL}} = (\cos\theta_X \cos\theta_Y) F_{z_{EL}} \qquad (A131)$$

Now it is possible by applying the contribution of the electrostatic force to the force balance at the center of mass to rewrite (A4), (A5), and (A96) as

$$F_{X_{CM}} = -MA_{X_{CM}} - F_{X_{EL}} \qquad (A132)$$

$$F_{Y_{CM}} = -MA_{Y_{CM}} + F_{Y_{EL}} \qquad (A133)$$

$$F_{Z_{CM}} = -MA_{Z_{CM}} + F_{Z_{TOUCH}} - F_{Z_{EL}} \qquad (A134)$$

The same approach is used in order to derive the torque contribution due to the electrostatic force. It is sufficient to include in the integration of (A128) the arm along the local x and y directions in order to get the proper expressions for the torque:

$$T_{\theta x_{EL}} = \frac{1}{2}\varepsilon_{AIR} V_{AIR}^2 \int_{-\frac{W}{2}}^{\frac{W}{2}} \int_{-\frac{L}{2}}^{\frac{L}{2}} \frac{y_P}{(Z_{AIR} + Z_{CM_{INST}} + Z_{P_{INST}})^2} dx\,dy \qquad (A135)$$

$$T_{\theta y_{EL}} = \frac{1}{2}\varepsilon_{AIR} V_{AIR}^2 \int_{-\frac{W}{2}}^{\frac{W}{2}} \int_{-\frac{L}{2}}^{\frac{L}{2}} \frac{x_P}{(Z_{AIR} + Z_{CM_{INST}} + Z_{P_{INST}})^2} dx\,dy \qquad (A136)$$

where $T_{\theta x_{EL}}$ and $T_{\theta y_{EL}}$ are the torques due to the electrostatic force around the local x and y axes, respectively, and y_P and x_P are the arms of the generic point P on the surface of the plate, along the y and x directions, respectively. Also in this case, the torques must be decomposed into their contributions in the global frame (XYZ). Using the rotation matrix notation mentioned before, we rewrite them as follows:

$$T_{\theta X_{EL}} = C_Y C_Z T_{\theta x_{EL}} + (-C_X S_Z + S_X S_Y C_Z) T_{\theta y_{EL}} \qquad (A137)$$

A.5 Electrostatic Model of the Rigid Plate

$$T_{\theta Y_{EL}} = C_Y S_Z T_{\theta x_{EL}} + (C_X C_Z + S_X S_Y S_Z) T_{\theta y_{EL}} \tag{A138}$$

$$T_{\theta Z_{EL}} = -S_Y T_{\theta x_{EL}} + S_X C_Y T_{\theta y_{EL}} \tag{A139}$$

Finally, (A137)–(A139) have to be applied to the torque balance equations referred to the center of mass. Thus, (A30), (A104), and (A105) become

$$\begin{aligned}
T_{\theta Z_{CM}} = -I_Z \frac{d^2 \theta_Z}{dt^2} &- \text{sgn}(Y_{NE})|Y_{NE}|F_{\Delta X_{NE}} \\
&+ \text{sgn}(X_{NE})|X_{NE}|F_{\Delta Y_{NE}} \\
&- \text{sgn}(Y_{SE})|Y_{SE}|F_{\Delta X_{SE}} \\
&+ \text{sgn}(X_{SE})|X_{SE}|F_{\Delta Y_{SE}} \\
&- \text{sgn}(Y_{SW})|Y_{SW}|F_{\Delta X_{SW}} \\
&+ \text{sgn}(X_{SW})|X_{SW}|F_{\Delta Y_{SW}} \\
&- \text{sgn}(Y_{NW})|Y_{NW}|F_{\Delta X_{NW}} \\
&+ \text{sgn}(X_{NW})|X_{NW}|F_{\Delta Y_{NW}} \\
&- T_{\theta Z_{EL}}
\end{aligned} \tag{A140}$$

$$\begin{aligned}
T_{\theta X_{CM}} = -I_X \frac{d^2 \theta_X}{dt^2} &- \text{sgn}(Z_{NE})|Z_{NE}|(F_{\Delta Y_{NE}}) \\
&+ \text{sgn}(Y_{NE})|Y_{NE}|(F_{\Delta Z_{NE}} + F_{Z_{NE(TOUCH)}}) \\
&- \text{sgn}(Z_{SE})|Z_{SE}|(F_{\Delta Y_{SE}}) \\
&+ \text{sgn}(Y_{SE})|Y_{SE}|(F_{\Delta Z_{SE}} + F_{Z_{SE(TOUCH)}}) \\
&- \text{sgn}(Z_{SW})|Z_{SW}|(F_{\Delta Y_{SW}}) \\
&+ \text{sgn}(Y_{SW})|Y_{SW}|(F_{\Delta Z_{SW}} + F_{Z_{SW(TOUCH)}}) \\
&- \text{sgn}(Z_{NW})|Z_{NW}|(F_{\Delta Y_{NW}}) \\
&+ \text{sgn}(Y_{NW})|Y_{NW}|(F_{\Delta Z_{NW}} + F_{Z_{NW(TOUCH)}}) \\
&- T_{\theta X_{EL}}
\end{aligned} \tag{A141}$$

$$\begin{aligned}
T_{\theta Y_{CM}} = -I_Y \frac{d^2 \theta_Y}{dt^2} &+ \text{sgn}(Y_{NE})|Z_{NE}|(F_{\Delta X_{NE}}) \\
&- \text{sgn}(Y_{NE})|X_{NE}|(F_{\Delta Z_{NE}} + F_{Z_{NE(TOUCH)}}) \\
&+ \text{sgn}(Y_{SE})|Z_{SE}|(F_{\Delta X_{SE}}) \\
&- \text{sgn}(Y_{SE})|X_{SE}|(F_{\Delta Z_{SE}} + F_{Z_{SE(TOUCH)}}) \\
&+ \text{sgn}(Y_{SW})|Z_{SW}|(F_{\Delta X_{SW}}) \\
&- \text{sgn}(Y_{SW})|X_{SW}|(F_{\Delta Z_{SW}} + F_{Z_{SW(TOUCH)}}) \\
&+ \text{sgn}(Y_{NW})|Z_{NW}|(F_{\Delta X_{NW}}) \\
&- \text{sgn}(Y_{NW})|X_{NW}|(F_{\Delta Z_{NW}} + F_{Z_{NW(TOUCH)}}) \\
&- T_{\theta Y_{EL}}
\end{aligned} \tag{A142}$$

It is important to stress that the contributions of the electrical torques in (A140)–(A142) are negative because the electrostatic force is attractive, and consequently is oriented along the z axis in the negative direction.

A.5.3
Curved Electric Field Lines Model

An important detail neglected up to now is that when the plate assumes uneven positions in space (i.e., not parallel to the substrate), the electric field lines are not straight lines anymore. In the model discussed thus far, such lines are supposed to be straight segments. This is a reasonable assumption when the plate orientation angles ($\theta_X, \theta_Y, \theta_Z$) are very small [170]. However, it is possible to enhance the accuracy of the electrostatic model of the plate for larger angles by taking into account the curvature of the field lines. The hypothesis we assume is that the field lines are orthogonal to both the coming-out and coming-in surfaces (i.e., the plate area and the substrate) [218]. The solution satisfying this condition is to assume the field lines are arcs of circumference. A simplified case in which only θ_Y has a nonzero value is shown in Figure A.17.

Given a generic point $P(x_P, y_P, z_P = -T/2)$ on the surface of the plate, two new reference systems related to it are defined. One is the local system, with the axes oriented as xyz and with origin at the point P, with axes x_P^1, y_P^1, z_P^1 (see Figure A.17). The second reference system is oriented as the global frame XYZ, and its origin corresponds to the vertical projection, along the Z axis, of the point P on the XY plane. This system is X_P^1, Y_P^1, Z_P^1, as reported in Figure A.17. Such additional systems are introduced in order to unlink the calculation of the field arc length with respect to the particular set of coordinates for the point P on the xy plane. Indeed, all we need to know is the angle θ_Y and the segment PI_P. The coherence among the global (XYZ) and local (xyz) systems is maintained in any case. Indeed, the information about the local coordinates of the point P is already included in Z_P (see Figure A.17), defined in Figure A.16 as

$$Z_P = Z_{AIR} + Z_{CM_{INST}} + Z_{P_{INST}} \tag{A143}$$

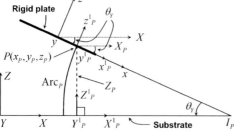

Figure A.17 Example of a curved electric field line, when only θ_Y has a nonzero value.

According to the properties of the triangles, the segment PI_P in Figure A.17 is defined as

$$PI_P = \frac{Z_P}{\sin \theta_Y} \tag{A144}$$

and finally, the field line Arc_P is

$$\text{Arc}_P = PI_P \theta_Y \tag{A145}$$

If we now introduce the coefficient σ_{Arc} defined as

$$\sigma_{\text{Arc}} = \frac{\text{Arc}_P}{Z_P} \tag{A146}$$

and if we replace Arc_P with (A145), σ_{Arc} becomes independent of Z_P, and consequently on the set of coordinates (x_P, y_P) on the surface of the plate

$$\sigma_{\text{Arc}} = \frac{\theta_Y}{\sin \theta_Y} \tag{A147}$$

This allows us to define the arc length of each point on the surface of the plate as

$$\text{Arc}_P = \sigma_{\text{Arc}} Z_P \tag{A148}$$

Finally, in order to include the curvature of the electric field lines in the capacitance and electrostatic force calculation, the denominator of (A127), (A128), (A135), and (A136) has to be multiplied by σ_{Arc}, which is constant with respect to the integration variables:

$$C_{\text{AIR}} = \frac{\varepsilon_{\text{AIR}}}{\sigma_{\text{Arc}}} \int_{-\frac{W}{2}}^{\frac{W}{2}} \int_{-\frac{L}{2}}^{\frac{L}{2}} \frac{dx\,dy}{(Z_{\text{AIR}} + Z_{\text{CM}_{\text{INST}}} + Z_{P_{\text{INST}}})} \tag{A149}$$

$$F_{z_{\text{EL}}} = \frac{1}{2} \frac{\varepsilon_{\text{AIR}} V_{\text{AIR}}^2}{\sigma_{\text{Arc}}^2} \int_{-\frac{W}{2}}^{\frac{W}{2}} \int_{-\frac{L}{2}}^{\frac{L}{2}} \frac{dx\,dy}{(Z_{\text{AIR}} + Z_{\text{CM}_{\text{INST}}} + Z_{P_{\text{INST}}})^2} \tag{A150}$$

$$T_{\theta x_{\text{EL}}} = \frac{1}{2} \frac{\varepsilon_{\text{AIR}} V_{\text{AIR}}^2}{\sigma_{\text{Arc}}^2} \int_{-\frac{W}{2}}^{\frac{W}{2}} \int_{-\frac{L}{2}}^{\frac{L}{2}} \frac{y_P}{(Z_{\text{AIR}} + Z_{\text{CM}_{\text{INST}}} + Z_{P_{\text{INST}}})^2} dx\,dy \tag{A151}$$

$$T_{\theta y_{\text{EL}}} = \frac{1}{2} \frac{\varepsilon_{\text{AIR}} V_{\text{AIR}}^2}{\sigma_{\text{Arc}}^2} \int_{-\frac{W}{2}}^{\frac{W}{2}} \int_{-\frac{L}{2}}^{\frac{L}{2}} \frac{x_P}{(Z_{\text{AIR}} + Z_{\text{CM}_{\text{INST}}} + Z_{P_{\text{INST}}})^2} dx\,dy \tag{A152}$$

Now we have to extend the curved field lines model to the most general case, in which all of the orientation angles have nonzero values. In Figure A.18, both θ_X

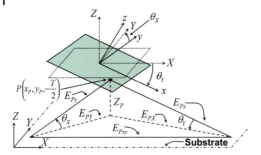

Figure A.18 When θ_X and θ_Y have nonzero values, each point P admits two segments (E_{P_x} and E_{P_y}) to determine the arc length.

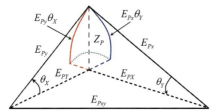

Figure A.19 The two segments E_{P_x} and E_{P_y} generate two different arc lengths for the same point P.

and θ_Y have nonzero values, while θ_Z is supposed to be zero as the definition of the arc length does not depend on it.

Given a generic point $P(x_P, y_P)$ on the surface of the plate, it now admits two extensions along the local x and y directions until the intersections with the substrate are reached, which are E_{P_x} and E_{P_y}. This leads to two possible arcs for the same point P, namely, $E_{P_x}\theta_Y$ and $E_{P_y}\theta_X$, reported in Figure A.19.

It is necessary to find a solution in order to identify the proper arc for the point P. Following an empirical approach, we define the field line arc for the point P as the one described by the bisector $E_{P_{\text{Arc}}}$ of the 90° angle formed by E_{P_x} and E_{P_y}, multiplied by the angle θ_{Arc} shown in Figure A.20. Both $E_{P_{\text{Arc}}}$ and θ_{Arc} are determined, with simple trigonometric calculations, on the triangles formed by the edges E_{P_x}, E_{P_y}, $E_{P_{xy}}$ and E_{PX}, E_{PY}, $E_{P_{xy}}$, which are not reported here for the sake of brevity. Also in this case, it is possible to define a suitable coefficient σ_{Arc} independent of Z_P, for which (A148) is still valid, as are (A149)–(A152).

A.6
Electrostatic Model of the Plate with Holes

The processing of MEMS based on surface micromachining employs the use of sacrificial layers for the release of suspended structures, such as rigid plates. In order to achieve a homogeneous removal, small openings (holes) are usually de-

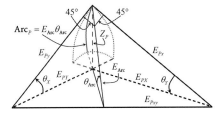

Figure A.20 The field arc related to the point P is empirically determined with simple trigonometric considerations.

Figure A.21 Scanning electron microscopy microphotograph of a suspended plate with holes distributed on its surface. Such openings allow the etching solution to homogeneously reach the sacrificial layer underneath the plate.

fined on the surface of the plate in order to allow the etching solution to reach the underlying sacrificial layer over all the surface of the suspended membrane. A scanning electron microscopy microphotograph of a suspended plate with holes on its surface is shown in Figure A.21. The presence of the holes influences the electric field, and must be taken into account. The electrostatic model of the rigid plate presented thus far describes this effect.

Let us start as usual from the case of a rigid plate with four DOFs. The approach chosen to determine the influence of the holes on the capacitance and electrostatic force is based on the superposition principle, which is allowed by the electric field [219]. Firstly, the capacitance and electrostatic force are defined by (A115) and (A123), respectively. Subsequently, the contribution to the capacitance and electrostatic force due to the *negative* hole surface is calculated and subtracted from (A115) and (A123). This is possible because both the capacitance and the electrostatic force are linearly dependent on the area, indeed making the principle of superposition effects valid. Finally, the effective value for the two electric magnitudes, due to the holes, is determined.

The width and length of each hole along the x and y axes, labeled as W_H and L_H, respectively, must now be defined, as shown in Figure A.22. If the numbers of holes along the local x and y axes are m and n, the total area of the holes A_{Holes} is

$$A_{\text{Holes}} = W_H L_H m n \tag{A153}$$

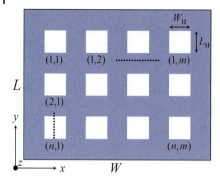

Figure A.22 Schematic of a plate with holes. There are supposed to be *m* holes along the local *x* axis and *n* holes along the local *y* axis.

First of all, the presence of holes is accounted for in the mechanical model of the rigid plate by replacing the mass *m* in (A1) with an effective value M_{EFF}. Within it, the total volume of the holes is subtracted from the value calculated for a rigid plate without openings on its surface. The expression is

$$M_{EFF} = \rho(W\,LT - W_H L_H T m n) \tag{A154}$$

where ρ is the density of the material the rigid plate is made of. The next step is the calculation of the capacitance and electrostatic force due to the area expressed by (A153). Thus, (A116), (A117), and (A123) are rewritten as follows:

$$C_{AIR} = \varepsilon_{AIR} \left[\frac{W L}{(Z_{AIR} + Z_{CM_{INST}})} - \frac{A_{Holes}}{(Z_{AIR} + Z_{CM_{INST}})} \right] \tag{A155}$$

$$C_{OX} = \varepsilon_{OX} \left[\frac{W L}{Z_{OX}} - \frac{A_{Holes}}{Z_{OX}} \right] \tag{A156}$$

$$F_{Z_{EL}} = \frac{1}{2} \varepsilon_{AIR} V_{AIR}^2 \left[\frac{W L}{(Z_{AIR} + Z_{CM_{INST}})^2} - \frac{A_{Holes}}{(Z_{AIR} + Z_{CM_{INST}})^2} \right] \tag{A157}$$

Let us now apply the same approach to the rigid plate model with six DOFs. In this case, there is an important aspect to be taken into account. Since the plate can assume any position (as shown in Figure A.16), each hole gives a different contribution to the capacitance and electrostatic force. Indeed, since the distance from the substrate is unique for each point on the surface of the plate, it is not possible to consider the total area of the holes, but each of them must be treated separately.

Figure A.23 represents the most general case, in which the holes are placed unevenly on the surface of the plate. In this case, the lateral distance of the holes from the four edges of the plate (left, right, top, bottom) are defined as Δ_{xL}, Δ_{xR}, Δ_{yT}, and Δ_{yB}, respectively.

A.6 Electrostatic Model of the Plate with Holes

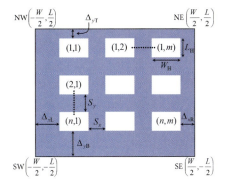

Figure A.23 Schematic of a plate with six DOFs with holes. Differently from the case with four DOFs, each hole gives a different contribution to the electric magnitudes, depending on its position.

Given these four parameters, the spacing between adjacent holes along the local x and y axes, identified by S_x and S_y, respectively, are derived as follows:

$$S_x = \frac{W - \Delta_{xL} - \Delta_{xR} - W_H m}{m - 1} \tag{A158}$$

$$S_y = \frac{L - \Delta_{yT} - \Delta_{yB} - L_H n}{n - 1} \tag{A159}$$

Now, the contribution of the electrical magnitudes related to the holes is calculated by performing an integration on the area of each hole in the local coordinates xy. By using the two indexes i and j, one can *sweep* the whole area of the plate and integrate only on the area of the holes. To do this, a double sum must be introduced in the integral formulas for the capacitance, electrostatic force, and torques. Let us now rewrite (A149)–(A152) by replacing the denominators with (A143). Taking into consideration the contribution of the holes, we obtain

$$C_{\text{AIR}} = \frac{\varepsilon_{\text{AIR}}}{\sigma_{\text{Arc}}} \left(\int_{-\frac{W}{2}}^{\frac{W}{2}} \int_{-\frac{L}{2}}^{\frac{L}{2}} \frac{dx\,dy}{Z_P} - \sum_{i=0}^{m-1} \sum_{j=0}^{n-1} \int_{a(i)}^{b(i)} \int_{c(j)}^{d(j)} \frac{dx\,dy}{Z_P} \right) \tag{A160}$$

$$F_{z\text{EL}} = \frac{1}{2} \frac{\varepsilon_{\text{AIR}} V_{\text{AIR}}^2}{\sigma_{\text{Arc}}^2} \left(\int_{-\frac{W}{2}}^{\frac{W}{2}} \int_{-\frac{L}{2}}^{\frac{L}{2}} \frac{dx\,dy}{Z_P^2} - \sum_{i=0}^{m-1} \sum_{j=0}^{n-1} \int_{a(i)}^{b(i)} \int_{c(j)}^{d(j)} \frac{dx\,dy}{Z_P^2} \right) \tag{A161}$$

$$T_{\theta x\text{EL}} = \frac{1}{2} \frac{\varepsilon_{\text{AIR}} V_{\text{AIR}}^2}{\sigma_{\text{Arc}}^2} \left(\int_{-\frac{W}{2}}^{\frac{W}{2}} \int_{-\frac{L}{2}}^{\frac{L}{2}} \frac{y_P}{Z_P^2} dx\,dy - \sum_{i=0}^{m-1} \sum_{j=0}^{n-1} \int_{a(i)}^{b(i)} \int_{c(j)}^{d(j)} \frac{y_P}{Z_P^2} dx\,dy \right)$$

$$\tag{A162}$$

Appendix A Rigid Plate Electromechanical Transducer (Complete Model)

$$T_{\theta\,\text{YEL}} = \frac{1}{2}\frac{\varepsilon_{\text{AIR}}V_{\text{AIR}}^2}{\sigma_{\text{Arc}}^2}\left(\int_{-\frac{W}{2}}^{\frac{W}{2}}\int_{-\frac{L}{2}}^{\frac{L}{2}}\frac{x_P}{Z_P^2}dx\,dy - \sum_{i=0}^{m-1}\sum_{j=0}^{n-1}\int_{a(i)}^{b(i)}\int_{c(j)}^{d(j)}\frac{x_P}{Z_P^2}dx\,dy\right)$$

(A163)

In the previous equations the integral indexes $a(i)$, $b(i)$, $c(j)$, and $d(j)$ are expressed as follows:

$$a(i) = -\frac{W}{2} + \Delta_{xL} + i(W_H + S_x) \qquad \text{(A164)}$$

$$b(i) = -\frac{W}{2} + \Delta_{xL} + i(W_H + S_x) + W_H \qquad \text{(A165)}$$

$$c(j) = \frac{L}{2} - \Delta_{yT} - L_H - j(L_H + S_y) \qquad \text{(A166)}$$

$$d(j) = \frac{L}{2} - \Delta_{yT} - j(L_H + S_y) \qquad \text{(A167)}$$

Finally, the C_{OX} expression to be associated with (A160) in order to get the total capacitance C is (A156). The edges of the plate, as well as the edges of the holes, introduce boundary effects due to the distortion of the electric field lines (i.e., fringing) that must be kept in view in order to get an accurate prediction of the electrical effects. This is what will be discussed in the following sections.

A.7
Electrostatic Model of the Fringing Effect

In order to complete the electrostatic model for the rigid plate, the fringing effect must be kept in mind. Such an effect is due to the distortion of the electric field lines near the boundaries of the plate [219]. Figure A.24 shows the bent field lines due to the fringing effect close to the edges of two parallel plates.

Such distorted lines give a contribution to the capacitance and electrostatic force, which is rather limited when the plate dimensions are large and the gap in between the two electrodes is very small. Because of this, some authors decided not

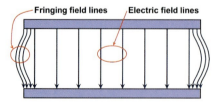

Figure A.24 Behavior of fringing field lines near the edges of the two plates.

to include the fringing effect in their models [170]. However, in order to produce an electrostatic model suitable in a wider range of cases, encompassing larger gaps and holes on surface of the plate as well, it was decided to include the fringing effect in the present work. Several approaches to face this issue are available in the literature, ranging from fully empirical to fully analytical solutions [220, 221].

The solution we propose is *semianalytical*. This means that it is based on a rather simple analytical model, and includes a few empirically determined parameters, in order to get an accurate prediction of the fringing effect for a wide range of topologies, on one hand, and a simple software implementation, on the other. Moreover, the model proposed considers two different fringing contributions, which are kept separated. The first one is due to the xy in-plane boundary effects near the edges of the plate, and will be discussed in this section. The second effect comes from the vertical faces of the plate, when its orientation angles assume nonzero values, which is the topic of the next section. First of all, we determine the proper curvature of the fringing field lines by means of a coefficient γ, as is shown in Figure A.25.

The coefficient γ ranges within the interval $]0, 1]$. When $\gamma = 1$, the fringing line $\text{Arc} F_P$ is half a circumference with diameter Z_P, while when $\gamma \to 0$, $\text{Arc} F_P \to Z_P$ (see Figure A.25). The proper value of γ is empirically determined by means of FEM simulations [163]. Given γ, by means of simple trigonometric considerations (not shown here) it is possible to determine the coefficient $\sigma_{\text{Arc}F}$ as follows:

$$\sigma_{\text{Arc}F} = \frac{1+\gamma^2}{2\gamma} \arcsin\left(\frac{2\gamma}{1+\gamma^2}\right) \tag{A168}$$

by means of which the fringing field arc length is

$$\text{Arc} F_P = \sigma_{\text{Arc}F} Z_P \tag{A169}$$

Once $\text{Arc} F_P$ has been determined, another coefficient must be introduced. This is ζ and it represents the in-plane depth in which the fringing field is supposed to act, as shown in Figure A.26.

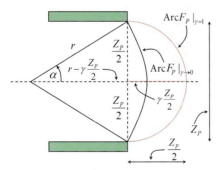

Figure A.25 Definition of the curvature of the fringing lines as a function of the gap Z_P and an empirical coefficient γ.

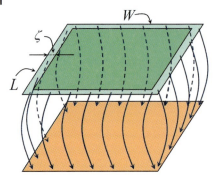

Figure A.26 The fringing field lines are supposed to be confined within a *belt* of width ζ, starting from the four edges of the plate and moving inward.

Furthermore, ζ is empirically determined by means of FEM simulations as well. Finally, it is possible to define the fringing contribution to the capacitance and electrostatic force by determining their values per unit length on each edge, and subsequently multiplying them by ζ. Let us start with the case of the plate with four DOFs. The total capacitance and electrostatic force accounting for the fringing effect in Figure A.26, are expressed by rewriting (A116) and (A123) as

$$C_{AIR} = \varepsilon_{AIR} \left[\frac{WL}{(Z_{AIR} + Z_{CM_{INST}})} + \frac{\zeta(2W + 2L)}{\sigma_{ArcF}(Z_{AIR} + Z_{CM_{INST}})} \right] \quad (A170)$$

$$F_{Z_{EL}} = \frac{1}{2} \varepsilon_{AIR} V_{AIR}^2 \left[\frac{WL}{(Z_{AIR} + Z_{CM_{INST}})^2} + \frac{\zeta(2W + 2L)}{\sigma_{ArcF}^2(Z_{AIR} + Z_{CM_{INST}})^2} \right] \quad (A171)$$

where the second term in the brackets represents the fringing capacitance and the fringing electrostatic force, respectively.

If there are holes on the surface of the plate, the same approach is applied to each of them. Thus, there is a region characterized by width ζ around the four edges of each hole within which the fringing field is supposed to be confined [222]. In this case, (A155) and (A157), also including the fringing term of (A170) and (A171), can be expressed as follows:

$$\begin{aligned}
C_{AIR} &= \varepsilon_{AIR} \frac{WL}{(Z_{AIR} + Z_{CM_{INST}})} \\
&\quad - \varepsilon_{AIR} \frac{A_{Holes}}{(Z_{AIR} + Z_{CM_{INST}})} \\
&\quad + \varepsilon_{AIR} \frac{\zeta(2W + 2L)}{\sigma_{ArcF}(Z_{AIR} + Z_{CM_{INST}})} \\
&\quad + \varepsilon_{AIR} \frac{\zeta(2W_H + 2L_H)mn}{\sigma_{ArcF}(Z_{AIR} + Z_{CM_{INST}})}
\end{aligned} \quad (A172)$$

$$F_{Z_{EL}} = \frac{1}{2}\varepsilon_{AIR} V_{AIR}^2 \frac{WL}{(Z_{AIR} + Z_{CM_{INST}})^2}$$
$$-\frac{1}{2}\varepsilon_{AIR} V_{AIR}^2 \frac{A_{Holes}}{(Z_{AIR} + Z_{CM_{INST}})^2}$$
$$+\frac{1}{2}\varepsilon_{AIR} V_{AIR}^2 \frac{\zeta(2W + 2L)}{\sigma_{ArcF}^2(Z_{AIR} + Z_{CM_{INST}})^2}$$
$$+\frac{1}{2}\varepsilon_{AIR} V_{AIR}^2 \frac{\zeta(2W_H + 2L_H)mn}{\sigma_{ArcF}^2(Z_{AIR} + Z_{CM_{INST}})^2} \qquad (A173)$$

When dealing with the model with six DOFs, we have to determine the fringing contributions one by one for each edge because, as pointed out earlier, each point on the surface of the plate has a unique distance from the substrate. The fringing capacitance and electrostatic force per unit length are calculated at each of the four edges of the plate, and are then multiplied by ζ. The latter one is supposed to be small enough to make the vertical distance variations within it negligible. In other words, Z_P of each point along an edge is supposed to be constant when moving inward on the surface of the plate for a distance equal to ζ. Given this assumption, we now proceed to calculate the fringing capacitance and electrostatic force, referring to the four edges of the plate, starting from the lower one and going clockwise, that is, SW–SE, SE–NE, NE–NW, and finally NW–SW. From Figure A.3, we have

$$y_{(SW-SE)} = -\frac{L}{2} \qquad (A174)$$

$$x_{(SE-NE)} = +\frac{W}{2} \qquad (A175)$$

$$y_{(NE-NW)} = +\frac{L}{2} \qquad (A176)$$

$$x_{(NW-SW)} = -\frac{W}{2} \qquad (A177)$$

By using (A143), in which we expressed the dependence on the local coordinates (xyz) of each point (which appears in (A125)), we can rewrite (A149) and (A150) as follows:

$$C_{AIR} = \frac{\varepsilon_{AIR}}{\sigma_{Arc}} \int_{-\frac{W}{2}}^{\frac{W}{2}} \int_{-\frac{L}{2}}^{\frac{L}{2}} \frac{dx\,dy}{Z_P\left(x, y, -\frac{T}{2}\right)}$$

$$+\zeta\left[\frac{\varepsilon_{AIR}}{\sigma_{ArcF}} \int_{-\frac{W}{2}}^{\frac{W}{2}} \frac{dx}{Z_P\left(x, -\frac{L}{2}, -\frac{T}{2}\right)}\right] + \zeta\left[\frac{\varepsilon_{AIR}}{\sigma_{ArcF}} \int_{-\frac{L}{2}}^{\frac{L}{2}} \frac{dy}{Z_P\left(\frac{W}{2}, y, -\frac{T}{2}\right)}\right]$$

$$+\zeta\left[\frac{\varepsilon_{AIR}}{\sigma_{ArcF}} \int_{-\frac{W}{2}}^{\frac{W}{2}} \frac{dx}{Z_P\left(x, \frac{L}{2}, -\frac{T}{2}\right)}\right] + \zeta\left[\frac{\varepsilon_{AIR}}{\sigma_{ArcF}} \int_{-\frac{L}{2}}^{\frac{L}{2}} \frac{dy}{Z_P\left(-\frac{W}{2}, y, -\frac{T}{2}\right)}\right] \qquad (A178)$$

Appendix A Rigid Plate Electromechanical Transducer (Complete Model)

$$F_{ZEL} = \frac{1}{2}\frac{\varepsilon_{AIR} V_{AIR}^2}{\sigma_{Arc}^2} \int_{-\frac{W}{2}}^{\frac{W}{2}}\int_{-\frac{L}{2}}^{\frac{L}{2}} \frac{dx\,dy}{Z_P\left(x,y,-\frac{T}{2}\right)^2}$$

$$+ \zeta\left[\frac{1}{2}\frac{\varepsilon_{AIR} V_{AIR}^2}{\sigma_{ArcF}^2}\int_{-\frac{W}{2}}^{\frac{W}{2}}\frac{dx}{Z_P\left(x,-\frac{L}{2},-\frac{T}{2}\right)^2}\right] + \zeta\left[\frac{1}{2}\frac{\varepsilon_{AIR} V_{AIR}^2}{\sigma_{ArcF}^2}\int_{-\frac{L}{2}}^{\frac{L}{2}}\frac{dy}{Z_P\left(\frac{W}{2},y,-\frac{T}{2}\right)^2}\right]$$

$$+ \zeta\left[\frac{1}{2}\frac{\varepsilon_{AIR} V_{AIR}^2}{\sigma_{ArcF}^2}\int_{-\frac{W}{2}}^{\frac{W}{2}}\frac{dx}{Z_P\left(x,\frac{L}{2},-\frac{T}{2}\right)^2}\right] + \zeta\left[\frac{1}{2}\frac{\varepsilon_{AIR} V_{AIR}^2}{\sigma_{ArcF}^2}\int_{-\frac{L}{2}}^{\frac{L}{2}}\frac{dy}{Z_P\left(-\frac{W}{2},y,-\frac{T}{2}\right)^2}\right]$$

(A179)

To complete the fringing effect we now introduce the contribution due to the holes for a rigid plate with six DOFs. With an approach similar to the one just presented, we consider the contribution of all the bottom, right, top, and left edges of the holes. In more detail, in the next formulas the holes are treated row by row for the bottom and top edges, and column by column for the right and left edges. Thus, (A160) and (A161), also including (A178) and (A179), can be rewritten as follows:

$$C_{AIR} = \frac{\varepsilon_{AIR}}{\sigma_{Arc}}\int_{-\frac{W}{2}}^{\frac{W}{2}}\int_{-\frac{L}{2}}^{\frac{L}{2}}\frac{dx\,dy}{Z_P\left(x,y,-\frac{T}{2}\right)} - \frac{\varepsilon_{AIR}}{\sigma_{Arc}}\sum_{i=0}^{m-1}\sum_{j=0}^{n-1}\int_{a(i)}^{b(i)}\int_{c(j)}^{d(j)}\frac{dx\,dy}{Z_P\left(x,y,-\frac{T}{2}\right)}$$

$$+ \zeta\left[\frac{\varepsilon_{AIR}}{\sigma_{ArcF}}\int_{-\frac{W}{2}}^{\frac{W}{2}}\frac{dx}{Z_P\left(x,-\frac{L}{2},-\frac{T}{2}\right)}\right] + \zeta\left[\frac{\varepsilon_{AIR}}{\sigma_{ArcF}}\int_{-\frac{L}{2}}^{\frac{L}{2}}\frac{dy}{Z_P\left(\frac{W}{2},y,-\frac{T}{2}\right)}\right]$$

$$+ \zeta\left[\frac{\varepsilon_{AIR}}{\sigma_{ArcF}}\int_{-\frac{W}{2}}^{\frac{W}{2}}\frac{dx}{Z_P\left(x,\frac{L}{2},-\frac{T}{2}\right)}\right] + \zeta\left[\frac{\varepsilon_{AIR}}{\sigma_{ArcF}}\int_{-\frac{L}{2}}^{\frac{L}{2}}\frac{dy}{Z_P\left(-\frac{W}{2},y,-\frac{T}{2}\right)}\right]$$

$$+ \zeta\left[\frac{\varepsilon_{AIR}}{\sigma_{ArcF}}\sum_{i=0}^{m-1}\sum_{j=0}^{n-1}\int_{a(i)}^{b(i)}\frac{dx}{Z_P\left(x,c(j),-\frac{T}{2}\right)}\right]$$

$$+ \zeta\left[\frac{\varepsilon_{AIR}}{\sigma_{ArcF}}\sum_{i=0}^{m-1}\sum_{j=0}^{n-1}\int_{c(j)}^{d(j)}\frac{dy}{Z_P\left(b(i),y,-\frac{T}{2}\right)}\right]$$

$$+ \zeta\left[\frac{\varepsilon_{AIR}}{\sigma_{ArcF}}\sum_{i=0}^{m-1}\sum_{j=0}^{n-1}\int_{a(i)}^{b(i)}\frac{dx}{Z_P\left(x,d(j),-\frac{T}{2}\right)}\right]$$

$$+ \zeta\left[\frac{\varepsilon_{AIR}}{\sigma_{ArcF}}\sum_{i=0}^{m-1}\sum_{j=0}^{n-1}\int_{c(j)}^{d(j)}\frac{dy}{Z_P\left(a(i),y,-\frac{T}{2}\right)}\right]$$

(A180)

$$F_{z_{EL}} = \frac{1}{2} \frac{\varepsilon_{AIR} V_{AIR}^2}{\sigma_{Arc}^2} \int_{-\frac{W}{2}}^{\frac{W}{2}} \int_{-\frac{L}{2}}^{\frac{L}{2}} \frac{dx\,dy}{Z_P\left(x, y, -\frac{T}{2}\right)^2}$$

$$- \frac{1}{2} \frac{\varepsilon_{AIR} V_{AIR}^2}{\sigma_{Arc}^2} \sum_{i=0}^{m-1} \sum_{j=0}^{n-1} \int_{a(i)}^{b(i)} \int_{c(j)}^{d(j)} \frac{dx\,dy}{Z_P\left(x, y, -\frac{T}{2}\right)^2}$$

$$+ \zeta \left[\frac{1}{2} \frac{\varepsilon_{AIR} V_{AIR}^2}{\sigma_{ArcF}^2} \int_{-\frac{W}{2}}^{\frac{W}{2}} \frac{dx}{Z_P\left(x, -\frac{L}{2}, -\frac{T}{2}\right)^2} \right]$$

$$+ \zeta \left[\frac{1}{2} \frac{\varepsilon_{AIR} V_{AIR}^2}{\sigma_{ArcF}^2} \int_{-\frac{L}{2}}^{\frac{L}{2}} \frac{dy}{Z_P\left(\frac{W}{2}, y, -\frac{T}{2}\right)^2} \right]$$

$$+ \zeta \left[\frac{1}{2} \frac{\varepsilon_{AIR} V_{AIR}^2}{\sigma_{ArcF}^2} \int_{-\frac{W}{2}}^{\frac{W}{2}} \frac{dx}{Z_P\left(x, \frac{L}{2}, -\frac{T}{2}\right)^2} \right]$$

$$+ \zeta \left[\frac{1}{2} \frac{\varepsilon_{AIR} V_{AIR}^2}{\sigma_{ArcF}^2} \int_{-\frac{L}{2}}^{\frac{L}{2}} \frac{dy}{Z_P\left(-\frac{W}{2}, y, -\frac{T}{2}\right)^2} \right]$$

$$+ \zeta \left[\frac{1}{2} \frac{\varepsilon_{AIR} V_{AIR}^2}{\sigma_{ArcF}^2} \sum_{i=0}^{m-1} \sum_{j=0}^{n-1} \int_{a(i)}^{b(i)} \frac{dx}{Z_P\left(x, c(j), -\frac{T}{2}\right)^2} \right]$$

$$+ \zeta \left[\frac{1}{2} \frac{\varepsilon_{AIR} V_{AIR}^2}{\sigma_{ArcF}^2} \sum_{i=0}^{m-1} \sum_{j=0}^{n-1} \int_{c(j)}^{d(j)} \frac{dy}{Z_P\left(b(i), y, -\frac{T}{2}\right)^2} \right]$$

$$+ \zeta \left[\frac{1}{2} \frac{\varepsilon_{AIR} V_{AIR}^2}{\sigma_{ArcF}^2} \sum_{i=0}^{m-1} \sum_{j=0}^{n-1} \int_{a(i)}^{b(i)} \frac{dx}{Z_P\left(x, d(j), -\frac{T}{2}\right)^2} \right]$$

$$+ \zeta \left[\frac{1}{2} \frac{\varepsilon_{AIR} V_{AIR}^2}{\sigma_{ArcF}^2} \sum_{i=0}^{m-1} \sum_{j=0}^{n-1} \int_{c(j)}^{d(j)} \frac{dy}{Z_P\left(a(i), y, -\frac{T}{2}\right)^2} \right] \quad \text{(A181)}$$

The electrostatic force due to the fringing field has been assumed to be not large enough to give a significant contribution in terms of torques. This is the reason why expressions like (A181) corresponding to (A162) and (A163) are not reported here.

A.7.1
Fringing Effect on the Vertical Faces of the Plate

The very last effect to be accounted for is the definition of the capacitance and electrostatic force due to the vertical faces of the plate, when θ_X and (or) θ_Y have nonzero values. This is also considered as a fringing effect, since it is not dominant when determining the final capacitance and electrostatic force values. Moreover, for its nature, this contribution is applied only to the plate model with six DOFs. The case of $\theta_X = 0$ and $\theta_Y > 0$ is shown in Figure A.27.

The SE–NE vertical face, that is, between the SE and NE nodes, is exposed to the electrode underneath, and consequently it gives a contribution to the electrical magnitudes. On the other hand, the SW–NW vertical face is shadowed by the lower face of the plate, and consequently its area is not exposed to the substrate. Of course, the other two faces (SW–SE and NW–NE) do not contribute to the capacitance and electrostatic force, because when $\theta_X = 0$ they are perpendicular with respect to the lower electrode. The case of either $\theta_X > 0$ or $\theta_Y > 0$ is depicted in Figure A.28.

In this case, the SW–NW and NW–NE vertical faces are shadowed by the lower surface of the plate, while the other two contribute to the fringing field. Before introducing the mathematical formulation for the capacitance and electrostatic force related to the vertical faces, we will summarize which face (faces) is (are) involved in the fringing phenomenon, depending on the orientation angles. In Table A.1, depending on all the possible combinations of θ_X and θ_Y, the face (faces) contributing to the fringing and the others which are shadowed are listed.

Moreover, for each vertical face generating fringing effects, the direction of the local x or y axis (positive or negative) along with the electrostatic force is reported.

Let us now refer to the case of $\theta_X > 0$ and $\theta_Y > 0$. We proceed to calculate the capacitance and electrostatic force due to the fringing effect on the SW–SE and SE–NE vertical faces. Concerning the capacitance, these two contributions are

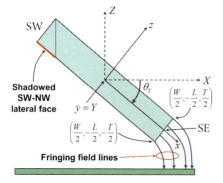

Figure A.27 Fringing effect due to the vertical faces when $\theta_Y > 0$. The SW–NW vertical face is shadowed by the lower surface of the plate.

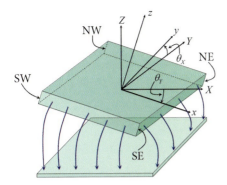

Figure A.28 Fringing effect due to the vertical faces when $\theta_X > 0$ and $\theta_Y > 0$. The SW–NW and NW–NE vertical faces are shadowed by the lower surface of the plate.

Table A.1 For each vertical face, it is stated if it gives or does not give a contribution to the fringing, depending on the value of θ_X and θ_Y, as well as whether the electrostatic force is oriented along the positive or the negative local x and y axes.

θ_X	θ_Y	SW–SE	SE–NE	NE–NW	NW–SW	x	y
$= 0$	$= 0$	Shadowed	Shadowed	Shadowed	Shadowed	No	No
> 0	$= 0$	Fringing	Shadowed	Shadowed	Shadowed	No	–
< 0	$= 0$	Shadowed	Shadowed	Fringing	Shadowed	No	+
$= 0$	> 0	Shadowed	Fringing	Shadowed	Shadowed	+	No
> 0	> 0	Fringing	Fringing	Shadowed	Shadowed	+	–
< 0	> 0	Shadowed	Fringing	Fringing	Shadowed	+	+
$= 0$	< 0	Shadowed	Shadowed	Shadowed	Fringing	–	No
> 0	< 0	Fringing	Shadowed	Shadowed	Fringing	–	–
< 0	< 0	Shadowed	Shadowed	Fringing	Fringing	–	+

called $C_{AIR(SW-SE)}$ and $C_{AIR(SE-NE)}$, respectively:

$$C_{AIR(SW-SE)} = \frac{\varepsilon_{AIR}}{\sigma_{ArcVF}} \int_{-\frac{W}{2}}^{\frac{W}{2}} \int_{-\frac{T}{2}}^{\frac{T}{2}} \frac{dx\,dz}{Z_P\left(x, -\frac{L}{2}, z\right)} \qquad (A182)$$

$$C_{AIR(SE-NE)} = \frac{\varepsilon_{AIR}}{\sigma_{ArcVF}} \int_{-\frac{L}{2}}^{\frac{L}{2}} \int_{-\frac{T}{2}}^{\frac{T}{2}} \frac{dy\,dz}{Z_P\left(\frac{W}{2}, y, z\right)} \qquad (A183)$$

where σ_{ArcVF} is a coefficient derived for the vertical faces with an approach similar to the one leading to (A147). Equations (A182) and (A183) must then be added to (A178). Let us now focus on the electrostatic force. The two contributions of the

SW–SE and SE–NE vertical faces are called $F_{Y(\text{SW-SE})}$ and $F_{X(\text{SE-NE})}$, respectively:

$$F_{Y(\text{SW-SE})} = -\frac{1}{2} \frac{\varepsilon_{\text{AIR}} V_{\text{AIR}}^2}{\sigma_{\text{ArcVF}}^2} \int_{-\frac{W}{2}}^{\frac{W}{2}} \int_{-\frac{T}{2}}^{\frac{T}{2}} \frac{dx\,dz}{Z_P\left(x, -\frac{L}{2}, z\right)^2} \quad (A184)$$

$$F_{X(\text{SE-NE})} = +\frac{1}{2} \frac{\varepsilon_{\text{AIR}} V_{\text{AIR}}^2}{\sigma_{\text{ArcVF}}^2} \int_{-\frac{L}{2}}^{\frac{L}{2}} \int_{-\frac{T}{2}}^{\frac{T}{2}} \frac{dy\,dz}{Z_P\left(\frac{W}{2}, y, z\right)^2} \quad (A185)$$

It must be noted that in (A184) and (A185) the proper sign of the force (according to Table A.1) has already been taken into account. Now, $F_{Y(\text{SW-SE})}$ and $F_{X(\text{SE-NE})}$ have to be decomposed into the contributions along the global X, Y, and Z axes, labeled as $F_{X(\text{Fring})}$, $F_{Y(\text{Fring})}$, and $F_{Z(\text{Fring})}$, respectively. Using the rotation matrix formulation, we obtain

$$F_{X(\text{Fring})} = C_Y C_Z F_{X(\text{SE-NE})} + (-C_X S_Z + S_X S_Y C_Z) F_{Y(\text{SW-SE})} \quad (A186)$$

$$F_{Y(\text{Fring})} = C_Y S_Z F_{X(\text{SE-NE})} + (C_X C_Z + S_X S_Y S_Z) F_{Y(\text{SW-SE})} \quad (A187)$$

$$F_{Z(\text{Fring})} = -S_Y F_{X(\text{SE-NE})} + S_X C_Y F_{Y(\text{SW-SE})} \quad (A188)$$

Finally, (A186)–(A188) must be added to (A132)–(A134) at the center of mass, becoming

$$F_{X\text{CM}} = -MA_{X\text{CM}} - F_{X\text{EL}} + F_{X(\text{Fring})} \quad (A189)$$

$$F_{Y\text{CM}} = -MA_{Y\text{CM}} + F_{Y\text{EL}} + F_{Y(\text{Fring})} \quad (A190)$$

$$F_{Z\text{CM}} = -MA_{Z\text{CM}} + F_{Z\text{TOUCH}} - F_{Z\text{EL}} + F_{Z(\text{Fring})} \quad (A191)$$

Still referring to the case of $\theta_X > 0$ and $\theta_Y > 0$, we now include the holes in our formulation. By applying the same considerations in Table A.1 for the SW–SE, SE–NE, NE–NW, and NW–SW edges of plate to the bottom, right, top, and left edges of the holes, respectively, and introducing $C_{\text{AIR(Bottom)}}$ and $C_{\text{AIR(Right)}}$, we get

$$C_{\text{AIR(Bottom)}} = \frac{\varepsilon_{\text{AIR}}}{\sigma_{\text{ArcVF}}} \sum_{i=0}^{m-1} \sum_{j=0}^{n-1} \int_{a(i)}^{b(i)} \int_{-\frac{T}{2}}^{\frac{T}{2}} \frac{dx\,dz}{Z_P(x, c(j), z)} \quad (A192)$$

$$C_{\text{AIR(Right)}} = \frac{\varepsilon_{\text{AIR}}}{\sigma_{\text{ArcVF}}} \sum_{i=0}^{m-1} \sum_{j=0}^{n-1} \int_{c(j)}^{d(j)} \int_{-\frac{T}{2}}^{\frac{T}{2}} \frac{dy\,dz}{Z_P(b(i), y, z)} \quad (A193)$$

Equations (A192) and (A193) together with (A182) and (A183) added to (A180) yield the total capacitance accounting for all the effects. Introducing a similar formulation for the electrostatic force, we have

$$F_{Y(\text{Bottom})} = -\frac{1}{2} \frac{\varepsilon_{\text{AIR}} V_{\text{AIR}}^2}{\sigma_{\text{ArcVF}}^2} \sum_{i=0}^{m-1} \sum_{j=0}^{n-1} \int_{a(i)}^{b(i)} \int_{-\frac{T}{2}}^{\frac{T}{2}} \frac{dx\,dz}{Z_P(x, c(j), z)^2} \quad (A194)$$

$$F_{x(\text{Right})} = \frac{1}{2} \frac{\varepsilon_{\text{AIR}} V_{\text{AIR}}^2}{\sigma_{\text{ArcVF}}^2} \sum_{i=0}^{m-1} \sum_{j=0}^{n-1} \int_{c(j)-\frac{I}{2}}^{d(j)} \int_{-\frac{I}{2}}^{\frac{I}{2}} \frac{dy\,dz}{Z_P} (b(i), y, z)^2 \qquad (A195)$$

Equations (A186)–(A188) become

$$\begin{aligned}F_{X(\text{Fring})} &= C_Y C_Z (F_{x(\text{SE-NE})} + F_{x(\text{Right})}) \\&\quad + (-C_X S_Z + S_X S_Y C_Z)(F_{y(\text{SW-SE})} + F_{y(\text{Bottom})})\end{aligned} \qquad (A196)$$

$$\begin{aligned}F_{Y(\text{Fring})} &= C_Y S_Z (F_{x(\text{SE-NE})} + F_{x(\text{Right})}) \\&\quad + (C_X C_Z + S_X S_Y S_Z)(F_{y(\text{SW-SE})} + F_{y(\text{Bottom})})\end{aligned} \qquad (A197)$$

$$\begin{aligned}F_{Z(\text{Fring})} &= -S_Y (F_{x(\text{SE-NE})} + F_{x(\text{Right})}) \\&\quad + S_X C_Y (F_{y(\text{SW-SE})} + F_{y(\text{Bottom})})\end{aligned} \qquad (A198)$$

Finally, (A196)–(A198) contribute to (A189)–(A191) at the center of mass, as already shown before.

A.8 Viscous Damping Model

Another physical effect related to the plate when it is moving within a fluid, for example, air, is viscous damping. This phenomenon is related to the viscosity of the medium within which the body is immersed as its particles hit the moving structure. Moreover, the viscous damping force depends also on the velocity of the body which is moving within the fluid [223]. In the currently discussed plate model, two contributions for the viscous damping have to be distinguished. One is due to the vertical movement of the plate, and relies on the so-called squeeze-film damping theory, based on the assumption that the fluid is incompressible [165]. The second contribution is given by the in-plane (xy directions) movement of the plate, and its description is based on the Couette-type damping theory, which employs a laminar flow regime [167].

A.8.1 Squeeze-Film Viscous Damping of the Rigid Plate with Holes

The squeeze-film damping model adopted for the rigid plate with holes on its surface is based on the approach proposed by Bao et al. [166]. Such a model uses a modified Reynolds equation, accounting for the presence of openings on the surface of the plate. The solution refers to a plate which is parallel to the substrate, and the total viscous damping force counteracting the z motion of the plate is expressed as follows:

$$F_{Z_{\text{VD}}} = -4Rl^2 \left(\frac{W}{2} - 1\right)\left(\frac{L}{2} - 1\right) \qquad (A199)$$

Here R represents the resistance of the viscous fluid surrounding the moving plate:

$$R = \frac{12\mu}{(Z_{AIR})^3} V_{Z_{CM}} \tag{A200}$$

In (A200), μ is the viscosity of the fluid, while $V_{Z_{CM}}$ is the velocity of the center of mass along the Z axis. In (A199), l is represented by the following form factor:

$$l = \sqrt{\frac{2T(Z_{AIR})^3 \eta(\beta)}{3\beta^2 \left(\frac{l_{hole}}{\sqrt{\pi}}\right)^2}} \tag{A201}$$

In (A201) there are two nondimensional parameters:

$$\eta(\beta) = \left[1 + \frac{3\left(\frac{l_{hole}}{\sqrt{\pi}}\right)^4 K(\beta)}{16 T(Z_{AIR})^3}\right] \tag{A202}$$

$$K(\beta) = 4\beta^2 - \beta^4 - 4\ln\beta - 3 \tag{A203}$$

where β is the ratio of the radius of the hole to the radius of the subelement containing it on the surface of the plate,

$$\beta = \frac{r_{hole}}{r_{cell}} \tag{A204}$$

Since we are dealing with squared or rectangular instead of circular holes, (A204) is adjusted in order to fit the β parameter to the model discussed here. Let us refer to the more general case of rectangular holes. It is possible to write a relationship in which the area of the circular hole is set equal to the area of the corresponding rectangular opening (see Figure A.23):

$$\pi (r_{hole})^2 = W_H L_H \tag{A205}$$

The same relationship is then written for the surface of the subelement:

$$\pi (r_{cell})^2 = \frac{W}{m} \frac{L}{n} \tag{A206}$$

Deriving the expressions for r_{hole} and r_{cell} from (A205) and (A206), and substituting them into (A204), we obtain for the latter

$$\beta = \sqrt{\frac{mn\, W_H L_H}{W L}} \tag{A207}$$

The latter equations complete the description of the model proposed by Bao et al. and are applied to the rigid plate with four DOFs with holes on its surface. Finally, the viscous damping force expressed in (A199) must be applied to the force balance at the center of mass along the Z axis, as done earlier in this chapter.

A.8 Viscous Damping Model

We also propose an extension of the model of Bao *et al.* to describe the viscous damping effect related to the rigid plate model with six DOFs. In this case, we calculate the local vertical velocity of each cell (or subelement) containing one hole on the surface of the plate. We assume a constant velocity over all area of the cell, and such a magnitude is determined at the middle point. The latter is composed of three contributions. The first is the velocity of the center of mass along the Z direction. The other two are tangential contributions, and are determined by the angular velocity with which the plate rotates around the X and Y axes ($\dot{\theta}_X$ and $\dot{\theta}_Y$), multiplied by the arms referred to the specific point of interest on the surface of the plate along the y and x axes, respectively. Given a generic point P on the surface of the plate (with coordinates x_P, y_P), its velocity along Z, accounting for both the linear movement of the center of mass of the plate and the tangential contributions, is expressed as

$$V_{Z_P}(x_P, y_P) = V_{Z_{CM}} - x_P \dot{\theta}_Y + y_P \dot{\theta}_X \tag{A208}$$

In the point P coordinates we did not introduce the one along z, usually accounted for in the electrostatic model as shown earlier in this chapter. This is because we assume a displacement of $-\frac{T}{2}$ (see Figure A.16) to give a negligible contribution to the tangential velocity. Moreover, the vertical distance of the point P from the substrate is still defined as

$$Z_P(x_P, y_P, z_P) = Z_{AIR} + Z_{CM_{INST}} + Z_{P_{INST}} \tag{A209}$$

where $Z_{P_{INST}}$ is defined by (A125). Assuming that each cell contains one hole quasi-parallel to the substrate, and also assuming a small velocity variation between adjacent cells, we can still apply the boundary conditions of the model of Bao *et al.* to the edges of the plate. Consequently, the model of Bao *et al.* can also be applied to the surface of the plate discretized in subelements. Given these considerations, the total viscous damping force, directed along the local z axis of the plate, can be expressed as follows:

$$F_{z_{VD}} = -\sum_{i=0}^{m-1} \sum_{j=0}^{n-1} \frac{V_{Z_P}\left\{\left[-\frac{W}{2} + \frac{W}{m}\left(i + \frac{1}{2}\right)\right], \left[\frac{L}{2} - \frac{L}{n}\left(j + 1 - \frac{1}{2}\right)\right]\right\}}{Z_P\left\{\left[-\frac{W}{2} + \frac{W}{m}\left(i + \frac{1}{2}\right)\right], \left[\frac{L}{2} - \frac{L}{n}\left(j + 1 - \frac{1}{2}\right)\right], -\frac{T}{2}\right\}^3} \Gamma \tag{A210}$$

where Γ is expressed as

$$\Gamma = 48\mu l^2 \left(\frac{W}{2m} - l\right) \left(\frac{L}{2n} - l\right) \tag{A211}$$

in which the parameters l and $\eta(\beta)$, expressed in (A201) and (A202), respectively, are rewritten with Z_{AIR} replaced by Z_P, multiplied by the same expression in the denominator of (A210). With the same approach already shown before, $F_{z_{VD}}$ is decomposed into its three contributions in the global frame (XYZ), and applied to the force balance at the center of mass. Following a similar approach, and using the arms on the xy plane for center point of each cell, we can derive the torques around the X and Y axes due to the viscous damping effect and apply them to the torque balance at the center of mass, as well. The latter contribution is not shown here for the sake of brevity.

A.8.1.1 Squeeze-Film Damping of the Rigid Plate without Holes

A model accounting for the viscous damping effect is implemented also for the case of the rigid plate without holes. Such a model is also based on the squeeze-film damping assumption, but it represents just a first-approach assessment of the viscous damping effect for the rigid plate. Indeed, it is based on the reduction of the Reynolds equation to the case of a structure characterized by the length much larger than the width ($L \gg W$). Although in the case of the rigid plate $L \sim W$, we assume that this model can be applied in order to have a reasonable assessment of the influence of viscous damping on the dynamics of this structures. However, since the simplified model mentioned above is more appropriate in the case of the flexible beam, for which the condition $L \gg W$ is fully satisfied, the squeeze-film damping model will be discussed in more detail in Appendix B. The total force counteracting the plate dynamics is expressed by (A188), and it is applied at the center of mass. By means of a similar approach, and by employing the same assumptions stated for the extension of the model of Bao et al. to the case of six DOFs, we implement a viscous damping model for the rigid plate with nonzero orientation angles. The squeeze-film damping model for rigid plates with four and six DOFs without holes is not described in detail here.

A.8.2
Viscous Damping Model for Lateral Movements

In the case of lateral plate movements on the xy plane, assuming a laminar behavior of the fluid underneath the plate, we can assume its velocity distribution, as the plate moves, to be as reported in Figure A.29. The fluid behaves as a linear slide-film damper, and the viscous damping increases proportionally to the ratio of the plate area and the distance from the substrate. This behavior of the fluid is called Couette-type flow [167].

In the case of the rigid plate with four DOFs, the viscous damping force due to its movement along the X and Y directions is then expressed as

$$F_{X_{VD}} = -\mu \frac{LW}{Z_{AIR}} V_{X_{CM}} \quad (A212)$$

$$F_{Y_{VD}} = -\mu \frac{LW}{Z_{AIR}} V_{Y_{CM}} \quad (A213)$$

These formulas are supposed to be valid for the rigid plate with six DOFs as the orientation angles are assumed to be not large enough to introduce a significant variation in the behavior of the lateral damping. Consequently, the Couette-type flow regime is assumed to be still valid. Such assumptions are compliant with the

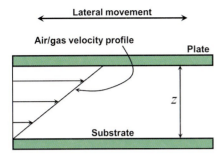

Figure A.29 When the plate moves laterally, the velocity distribution of the fluid underneath is described by Couette-type flow.

qualitative implementation of the viscous damping effect for the rigid plate proposed thus far. All the effects due to air layers on top of the rigid plate are disregarded in this viscous damping implementation.

A.8.3
Effect of the Mean Free Path of Gas Molecules

The description of the viscous damping phenomenon is completed by keeping in mind the effect of gas rarefaction. When the gap between the plate and the substrate is small enough to be comparable with the mean free path of the fluid molecules, the gas cannot be considered incompressible anymore [165]. This effect is included by replacing the constant value of the fluid viscosity μ with an expression dependent on the vertical distance between the plate and the substrate. The expression proposed by Veijola et al. [168] for the effective viscosity is

$$\mu_{\text{eff}}(z) = \frac{\mu}{1 + 9.638\, Kn^{1.159}} \tag{A214}$$

where Kn is the Knudsen number, defined as the ratio between the mean free path of gas molecules λ and the vertical distance of the plate from the substrate z:

$$Kn = \frac{\lambda}{z} \tag{A215}$$

When z is larger than λ, the Knudsen number is very small, and consequently $\mu_{\text{eff}}(z)$ can be considered constant and equal to μ. Differently, when z decreases significantly, indeed becoming comparable with the mean free path, the denominator of (A214) becomes larger than 1, and eventually $\mu_{\text{eff}}(z) < \mu$. Equation (A214) must replace μ in all the equations describing the vertical and lateral viscous damping previously shown if the gas rarefaction effect has to be considered.

A.9
Conclusions

The topic of this appendix was the detailed description of the MEMS rigid plate electromechanical transducers implemented in the Verilog-A programming language based library introduced in Chapter 5. The level of detail reported in this appendix was very deep, and for this reason, the information presented is intended for those readers who wish to know all the specific features of the model. The rigid plate model was discussed according to an incremental complexity approach. For this reason, the mechanical, electrostatic, and dynamic modeled features were presented first for a simplified model of a rigid plate with four DOFs, and were subsequently extended to the most general case of six DOFs.

Appendix B
Flexible Straight Beam (Complete Model)

Abstract: The focus of this appendix is the detailed description of the multiphysical model of the flexible straight beam introduced in Chapter 5 and implemented in the Verilog-A programming language [101]. The information here does not represent an integration with respect to the features reported in Chapter 5, but is rather a description from scratch of the flexible beam model at a significantly deeper level of detail. Consequently, this appendix is intended for those readers who are interested in a close examination of the implementation details related to the compact models of microelectromechanical systems.

Following an approach similar to the one used for the rigid plate in Appendix A, the mechanical model of the flexible beam is introduced first, starting from a very simple case. Subsequently, the electrostatic model and the viscous damping model are described.

B.1
Mechanical Model of the Flexible Beam with Two Degrees of Freedom

In this section, a very simple case is discussed in order to explain the approach adopted in defining the mechanical model of the flexible beam (also referred to as an Euler beam). A schematic of an Euler beam with dimensions W, L, and T, along the x, y, and z axes, respectively, is shown in Figure B.1. The two beam ends A and B are also indicated in the schematic. In this section we consider a beam with two degrees of freedom (DOFs), stimulated only along the x direction at the two ends (see Figure B.2).

The first aspect to be stressed is that the stimuli, that is, the forces F_{xA} and F_{xB}, are applied according to the superposition principle. This means that each stimulus is applied singularly, and all the others are considered to be equal to zero. Moreover, appropriate mechanical constraints are applied at the end of the beam which is not stimulated (see Figure B.2). The force F_{xA} applied at the beam end A induces a deformation equal to u_{xA}. On the other hand, the force F_{xB} deforms end B by u_{xB}. We now introduce the procedure to derive two matrices, called the *stiffness* matrix and the *mass* matrix, respectively. The first one defines the set of relationships between the forces and the corresponding deformations for each DOF of the flexible beam. In other words, the stiffness matrix describes the elastic be-

Practical Guide to RF-MEMS, First Edition. Jacopo Iannacci.
© 2013 WILEY-VCH Verlag GmbH & Co. KGaA. Published 2013 by WILEY-VCH Verlag GmbH & Co. KGaA.

Appendix B Flexible Straight Beam (Complete Model)

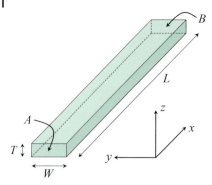

Figure B.1 Schematic of the Euler flexible beam. The beam length (along the x axis), width (along the y axis), and thickness (along the z axis) are L, W, and T, respectively.

havior of the deformable beam. On the other hand, the mass matrix describes the inertial behavior of the beam, defining the relationships between the forces and the accelerations associated with each DOF.

B.1.1
The Stiffness Matrix

Before focusing on the analytical details for the description of the elastic behavior of the beam, we must introduce an important hypothesis. The Euler beam model is based on the assumption that the length is much greater than the other two dimensions ($L \gg W, T$). Such a hypothesis means the beam falls in the range of validity of the plane stress [169]. Given this operating condition, let us express the set of forces and deformations, shown in Figure B.2, in the form of vectors as follows:

$$F = \begin{bmatrix} F_{xA} \\ F_{xB} \end{bmatrix} \tag{B1}$$

$$u = \begin{bmatrix} u_{xA} \\ u_{xB} \end{bmatrix} \tag{B2}$$

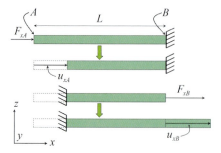

Figure B.2 Euler beam stimulated along the x direction at ends A and B. The end to which the stimulus is not applied has to be constrained (i.e., anchored).

B.1 Mechanical Model of the Flexible Beam with Two Degrees of Freedom

We now derive an expression taking the punctual deformation u_x due to F_{xA} and F_{xB} at a generic point x along the beam length. Assuming the local reference system origin is located at beam end A, and normalizing u_x with respect to the beam length L, we obtain

$$u_x = u_{xA} + (u_{xB} - u_{xA})\frac{x}{L} \tag{B3}$$

The first term on right in (B3) refers to end A, while the second term is taken at a generic point x along L. It should be noted that $x \in [0, L]$, and consequently $\frac{x}{L} \in [0, 1]$. Furthermore, by applying the stress–strain relationship [169] to (B3), we obtain

$$e_{xx} = \frac{\partial u_x}{\partial x} = \frac{1}{L}(u_{xB} - u_{xA}) = \frac{1}{L}\begin{bmatrix} -1 & 1 \end{bmatrix}\begin{bmatrix} u_{xA} \\ u_{xB} \end{bmatrix} \tag{B4}$$

Equation (B4) can be expressed in a compact form as

$$\boldsymbol{e} = \boldsymbol{b}\boldsymbol{u} \tag{B5}$$

where $\boldsymbol{e}$, $\boldsymbol{b}$, and $\boldsymbol{u}$ are, respectively,

$$\boldsymbol{e} = [e_{xx}] \tag{B6}$$

$$\boldsymbol{b} = \frac{1}{L}\begin{bmatrix} -1 & 1 \end{bmatrix} \tag{B7}$$

$$\boldsymbol{u} = \begin{bmatrix} u_{xA} \\ u_{xB} \end{bmatrix} \tag{B8}$$

The stiffness matrix $\boldsymbol{k}$ can be now introduced with the following expression:

$$\boldsymbol{k} = \int_v \boldsymbol{b}^T \boldsymbol{x} \boldsymbol{b} \, dV \tag{B9}$$

where the integral is calculated over the volume v. Moreover, $\boldsymbol{x}$ in this simple case is

$$\boldsymbol{x} = E \tag{B10}$$

where E is the Young's modulus. Finally, the stiffness matrix $\boldsymbol{k}$ can be expressed as

$$\boldsymbol{k} = \int_0^L \frac{1}{L}\begin{bmatrix} -1 \\ 1 \end{bmatrix}\frac{E}{L}\begin{bmatrix} -1 & 1 \end{bmatrix} A \, dx = \frac{AE}{L}\begin{bmatrix} 1 & -1 \\ -1 & 1 \end{bmatrix} \tag{B11}$$

where A is the transverse area of the beam, and

$$dV = A \, dx \tag{B12}$$

Once **k** has been determined, the relationship between the forces applied to the beam and the corresponding deformations is defined as

$$F = ku \tag{B13}$$

which written using the complete notation becomes

$$\begin{bmatrix} F_{xA} \\ F_{xB} \end{bmatrix} = \frac{AE}{L} \begin{bmatrix} 1 & -1 \\ -1 & 1 \end{bmatrix} \begin{bmatrix} u_{xA} \\ u_{xB} \end{bmatrix} \tag{B14}$$

B.1.2
The Mass Matrix

As mentioned earlier, the mass matrix defines the set of relationships between the forces applied to the beam and the accelerations related to all its DOFs. Translated into mathematical notation, this concept is expressed by the well-known expression

$$F = ma \tag{B15}$$

where the acceleration a is defined as the double time derivative of the displacement u:

$$a = \ddot{u} = \frac{\partial^2 u}{\partial t^2} \tag{B16}$$

We are now interested in extending (B15) to the case of multiple DOFs, which means expressing it in a vectorial form:

$$\boldsymbol{F} = m\boldsymbol{a} \tag{B17}$$

where **F** and **a**, in the simple case depicted in Figure B.2, are

$$\boldsymbol{F} = \begin{bmatrix} F_{xA} \\ F_{xB} \end{bmatrix} \tag{B18}$$

$$\boldsymbol{a} = \ddot{\boldsymbol{u}} = \begin{bmatrix} \ddot{u}_{xA} \\ \ddot{u}_{xB} \end{bmatrix} \tag{B19}$$

It is now clear that we have to look for a matrix form of the mass **m**, to be included in (B17). In order to derive **m**, we need to express all the nodal displacements as already done in Equation (B3). In this regard, let us introduce a nondimensional variable ξ, defined as follows

$$\xi = \frac{x}{L} \tag{B20}$$

that identifies the punctual abscissa x along the beam, normalized with respect to its length L [169]. By applying once again the superposition principle (see Figure B.2), and assuming a unitary deformation at both the nodes, that is, $u_{xA} = 1$

B.1 Mechanical Model of the Flexible Beam with Two Degrees of Freedom

and $u_{xB} = 1$, we can express the local displacement at a generic point x along the beam due to the two above-mentioned deformations by the vector a_m as

$$a_m = \begin{bmatrix} 1 - \xi & \xi \end{bmatrix} \tag{B21}$$

where the subscript m has been introduced in order not to confuse the vector a_m in (B21) with the acceleration in (B19). The interpretation of the geometrical meaning of (B21) is rather straightforward. When we consider the origin $x = 0$, $\xi = 0$, and consequently

$$a_m(\xi = 0) = \begin{bmatrix} 1 & 0 \end{bmatrix} \tag{B22}$$

This means that the deformation applied to node A exhibits its maximum value (1) at the origin, while it decreases as x increases, till it reaches 0 at end B. The opposite situation is

$$a_m(\xi = 1) = \begin{bmatrix} 0 & 1 \end{bmatrix} \tag{B23}$$

in which the deformation of end B does not give any contribution to node A, which of course is equal to 1 at end B. It is now possible to introduce the mass matrix m defined as

$$m = \int_v \rho a_m^T a_m \, dV \tag{B24}$$

where v is the volume of the object, ρ is the density of the material from which it is constituted, and a_m^T is the transposed vector a_m. Let us now expand (B24) as follows:

$$\begin{aligned} m &= \int_v \rho \begin{bmatrix} 1 - \xi \\ \xi \end{bmatrix} \begin{bmatrix} 1 - \xi & \xi \end{bmatrix} dV \\ &= \rho \int_v \begin{bmatrix} (1 - \xi)^2 & \xi(1 - \xi) \\ \xi(1 - \xi) & \xi^2 \end{bmatrix} dV \end{aligned} \tag{B25}$$

Integrating each element of the matrix and keeping in mind that

$$dV = A\,dx = AL\,d\xi \tag{B26}$$

we obtain

$$m = L \begin{bmatrix} \int_0^1 \rho A (1-\xi)^2 d\xi & \int_0^1 \rho A \xi (1-\xi) d\xi \\ \int_0^1 \rho A \xi (1-\xi) d\xi & \int_0^1 \rho A \xi^2 d\xi \end{bmatrix} \tag{B27}$$

Premultiplying m by $\rho A L$ and solving the integrals yields

$$m = \rho A L \begin{bmatrix} \left[\xi - \xi^2 + \frac{\xi^3}{3}\right]_0^1 & \left[\frac{\xi^2}{2} - \frac{\xi^3}{3}\right]_0^1 \\ \left[\frac{\xi^2}{2} - \frac{\xi^3}{3}\right]_0^1 & \left[\frac{\xi^3}{3}\right]_0^1 \end{bmatrix} \tag{B28}$$

and finally

$$m = \rho A L \begin{bmatrix} \frac{1}{3} & \frac{1}{6} \\ \frac{1}{6} & \frac{1}{3} \end{bmatrix} \tag{B29}$$

Finally, (B17) can be expressed as

$$\begin{bmatrix} F_{xA} \\ F_{xA} \end{bmatrix} = \rho A L \begin{bmatrix} \frac{1}{3} & \frac{1}{6} \\ \frac{1}{6} & \frac{1}{3} \end{bmatrix} \begin{bmatrix} \ddot{u}_{xA} \\ \ddot{u}_{xB} \end{bmatrix} \tag{B30}$$

Now that both the stiffness matrix k and mass matrix m have been introduced, we can combine the effects related to the elasticity and inertial behavior of the beam in a single constitutive equation, which is

$$F = ku + m\ddot{u} \tag{B31}$$

and finally expressed in an explicit matrix form this becomes

$$\begin{bmatrix} F_{xA} \\ F_{xB} \end{bmatrix} = \frac{AE}{L} \begin{bmatrix} 1 & -1 \\ -1 & 1 \end{bmatrix} \begin{bmatrix} u_{xA} \\ u_{xB} \end{bmatrix} + \rho A l \begin{bmatrix} \frac{1}{3} & \frac{1}{6} \\ \frac{1}{6} & \frac{1}{3} \end{bmatrix} \begin{bmatrix} \ddot{u}_{xA} \\ \ddot{u}_{xB} \end{bmatrix} \tag{B32}$$

B.2
Mechanical Model of the Flexible Beam with 12 DOFs

After the stiffness and mass matrices have been introduced, the next step is to extend their formulation, already derived for the Euler beam with two DOFs in the previous section, to the most general case of the beam with 12 DOFs. A schematic of such a model is shown in Figure B.3, where each beam end admits three linear deformations (along the x, y, and z axes), and three angular deformations (around the x, y, and z axes).

Before dealing with the analytical calculations, we must briefly discuss another matter. The model is addressed as a beam with 12 DOFs, while a Cartesian system

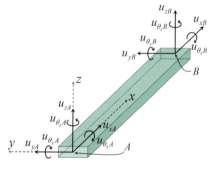

Figure B.3 Euler beam with 12 degrees of freedom (DOFs). Three linear and three angular deformations are allowed at end A and end B.

admits only six DOFs. Since the Euler beam is a flexible body, the deformations of each end are independent of the possible deformations at the other node. Because of this consideration, it is more straightforward to account for six DOFs per node instead of considering a unique local reference system. For the same reason, in the plate model presented in Appendix A, six overall DOFs instead of six DOFs for each node were considered (i.e., 24 DOFs), because in that case we were dealing with a rigid body.

B.2.1
The Stiffness Matrix for 12 DOFs

Let us now extend the stiffness matrix k in order to make it appropriate to describe the whole set of 12 DOFs. First of all, in this case (B1) and (B2) become

$$F^T = (F_{xA}, F_{yA}, F_{zA}, T_{\theta_x A}, T_{\theta_y A}, T_{\theta_z A}, F_{xB}, F_{yB}, F_{zB}, T_{\theta_x B}, T_{\theta_y B}, T_{\theta_z B}) \tag{B33}$$

$$u^T = (u_{xA}, u_{yA}, u_{zA}, u_{\theta_x A}, u_{\theta_y A}, u_{\theta_z A}, u_{xB}, u_{yB}, u_{zB}, u_{\theta_x B}, u_{\theta_y B}, u_{\theta_z B}) \tag{B34}$$

We now have to derive a set or relationships needed to link together all the deformations (linear and angular) the beam can be subjected to. The latter will be similar to (B3), although their expressions are more complicated. In this case, we apply once again the plane stress hypothesis, and (B4) becomes

$$\begin{bmatrix} e_{xx} \\ e_{yy} \\ e_{zz} \end{bmatrix} = b \begin{bmatrix} u_{xA} \\ u_{yA} \\ u_{zA} \\ u_{\theta_x A} \\ u_{\theta_y A} \\ u_{\theta_z A} \\ u_{xB} \\ u_{yB} \\ u_{zB} \\ u_{\theta_x B} \\ u_{\theta_y B} \\ u_{\theta_z B} \end{bmatrix} \tag{B35}$$

where b is not a vector anymore, but is a 3×12 matrix [169], written as follows:

$$b = \begin{bmatrix} b_{11} & b_{12} & b_{13} & b_{14} & b_{15} & b_{16} & b_{17} & b_{18} & b_{19} & b_{1,10} & b_{1,11} & b_{1,12} \\ b_{21} & b_{22} & b_{23} & b_{24} & b_{25} & b_{26} & b_{27} & b_{28} & b_{29} & b_{2,10} & b_{2,11} & b_{2,12} \\ b_{31} & b_{32} & b_{33} & b_{34} & b_{35} & b_{36} & b_{37} & b_{38} & b_{39} & b_{3,10} & b_{3,11} & b_{3,12} \end{bmatrix} \tag{B36}$$

By substitution of matrix b of (B36) and its transposed matrix b^T into (B9), the stiffness matrix, whose dimensions are 12×12, becomes

$$k = [k_1 | k_2] \tag{B37}$$

where

$$k_1 = \begin{bmatrix} k_{11} & 0 & 0 & 0 & 0 & 0 \\ 0 & k_{22} & 0 & 0 & 0 & k_{26} \\ 0 & 0 & k_{33} & 0 & k_{35} & 0 \\ 0 & 0 & 0 & k_{44} & 0 & 0 \\ 0 & 0 & k_{53} & 0 & k_{55} & 0 \\ 0 & k_{62} & 0 & 0 & 0 & k_{66} \\ k_{71} & 0 & 0 & 0 & 0 & 0 \\ 0 & k_{82} & 0 & 0 & 0 & k_{86} \\ 0 & 0 & k_{93} & 0 & k_{95} & 0 \\ 0 & 0 & 0 & k_{10,4} & 0 & 0 \\ 0 & 0 & k_{11,3} & 0 & k_{11,5} & 0 \\ 0 & k_{12,2} & 0 & 0 & 0 & k_{12,6} \end{bmatrix} \qquad (B38)$$

$$k_2 = \begin{bmatrix} k_{17} & 0 & 0 & 0 & 0 & 0 \\ 0 & k_{28} & 0 & 0 & 0 & k_{2,12} \\ 0 & 0 & k_{39} & 0 & k_{3,11} & 0 \\ 0 & 0 & 0 & k_{4,10} & 0 & 0 \\ 0 & 0 & k_{59} & 0 & k_{5,11} & 0 \\ 0 & k_{68} & 0 & 0 & 0 & k_{6,12} \\ k_{77} & 0 & 0 & 0 & 0 & 0 \\ 0 & k_{88} & 0 & 0 & 0 & k_{8,12} \\ 0 & 0 & k_{99} & 0 & k_{9,11} & 0 \\ 0 & 0 & 0 & k_{10,10} & 0 & 0 \\ 0 & 0 & k_{11,9} & 0 & k_{11,11} & 0 \\ 0 & k_{12,8} & 0 & 0 & 0 & k_{12,12} \end{bmatrix} \qquad (B39)$$

The elements of (B37), also accounting for the shear effect [169], are expressed as follows:

$$k_{11} = \frac{AE}{L} \qquad (B40)$$

$$k_{17} = -\frac{AE}{L} \qquad (B41)$$

$$k_{22} = \frac{12 E I_z}{L^3 (1 + \Phi_y)} \qquad (B42)$$

$$k_{26} = \frac{6 E I_z}{L^2 (1 + \Phi_y)} \qquad (B43)$$

$$k_{28} = -\frac{12 E I_z}{L^3 (1 + \Phi_y)} \qquad (B44)$$

$$k_{2,12} = \frac{6 E I_z}{L^2 (1 + \Phi_y)} \qquad (B45)$$

$$k_{33} = \frac{12 E I_y}{L^3 (1 + \Phi_z)} \qquad (B46)$$

$$k_{35} = -\frac{6EI_y}{L^2(1+\Phi_z)} \tag{B47}$$

$$k_{39} = -\frac{12EI_y}{L^3(1+\Phi_z)} \tag{B48}$$

$$k_{3,11} = -\frac{6EI_y}{L^2(1+\Phi_z)} \tag{B49}$$

$$k_{44} = \frac{GJ_x}{L} \tag{B50}$$

$$k_{4,10} = -\frac{GJ_x}{L} \tag{B51}$$

$$k_{55} = \frac{(4+\Phi_z)EI_y}{L(1+\Phi_z)} \tag{B52}$$

$$k_{59} = \frac{6EI_y}{L^2(1+\Phi_z)} \tag{B53}$$

$$k_{5,11} = \frac{(2-\Phi_z)EI_y}{L(1+\Phi_z)} \tag{B54}$$

$$k_{66} = \frac{(4+\Phi_y)EI_z}{L(1+\Phi_y)} \tag{B55}$$

$$k_{68} = -\frac{6EI_z}{L^2(1+\Phi_y)} \tag{B56}$$

$$k_{6,12} = \frac{(2-\Phi_y)EI_z}{L(1+\Phi_y)} \tag{B57}$$

$$k_{77} = \frac{AE}{L} \tag{B58}$$

$$k_{88} = \frac{12EI_z}{L^3(1+\Phi_y)} \tag{B59}$$

$$k_{8,12} = -\frac{6EI_z}{L^2(1+\Phi_y)} \tag{B60}$$

$$k_{99} = \frac{12EI_y}{L^3(1+\Phi_z)} \tag{B61}$$

$$k_{9,11} = \frac{6EI_y}{L^2(1+\Phi_z)} \tag{B62}$$

$$k_{10,10} = \frac{GJ_x}{L} \tag{B63}$$

$$k_{11,11} = \frac{(4+\Phi_z)EI_y}{L(1+\Phi_z)} \tag{B64}$$

$$k_{12,12} = \frac{(4+\Phi_y)EI_z}{L(1+\Phi_y)} \tag{B65}$$

In (B40)–(B65), G is the shear modulus, expressed as

$$G = \frac{E}{2(1+\nu)} \quad (B66)$$

where ν is the Poisson's ratio, while Φ_y and Φ_z are

$$\Phi_y = \frac{12EI_z}{GWTL^2} \quad (B67)$$

$$\Phi_z = \frac{12EI_y}{GWTL^2} \quad (B68)$$

Furthermore, I_y and I_x are the moments of inertia referred to the y and z axes, respectively, and J_x is the second moment of inertia with respect to the x axis [169]. They can be expressed in the forms

$$I_y = \frac{1}{4}\int_0^W\int_0^T z^2\,dz\,dy = \frac{W^3T}{12} \quad (B69)$$

$$I_z = \frac{1}{4}\int_0^T\int_0^W y^2\,dy\,dz = \frac{WT^3}{12} \quad (B70)$$

$$J_x = \frac{2(WT)^3}{7(W^2+T^2)} \quad (B71)$$

Now, by use of the transposed matrices expressed in (B33) and (B34) and (B37), (B13) becomes

$$\begin{bmatrix} F_{xA} \\ F_{yA} \\ F_{zA} \\ T_{\theta xA} \\ T_{\theta yA} \\ T_{\theta zA} \\ F_{xB} \\ F_{yB} \\ F_{zB} \\ T_{\theta xB} \\ T_{\theta yB} \\ T_{\theta zB} \end{bmatrix} = k \begin{bmatrix} u_{xA} \\ u_{yA} \\ u_{zA} \\ u_{\theta xA} \\ u_{\theta yA} \\ u_{\theta zA} \\ u_{xB} \\ u_{yB} \\ u_{zB} \\ u_{\theta xB} \\ u_{\theta yB} \\ u_{\theta zB} \end{bmatrix} \quad (B72)$$

Finally, it is possible to rewrite (B72) as the corresponding set of equations, which are as follows:

$$F_{xA} = k_{11}u_{xA} + k_{17}u_{xB} \quad (B73)$$

$$F_{yA} = k_{22}u_{yA} + k_{26}u_{\theta zA} + k_{28}u_{yB} + k_{2,12}u_{\theta zB} \quad (B74)$$

$$F_{zA} = k_{33} u_{zA} + k_{35} u_{\theta_y A} + k_{39} u_{zB} + k_{3,11} u_{\theta_y B} \tag{B75}$$

$$T_{\theta_x A} = k_{44} u_{\theta_x A} + k_{4,10} u_{\theta_x B} \tag{B76}$$

$$T_{\theta_y A} = k_{35} u_{zA} + k_{55} u_{\theta_y A} + k_{59} u_{zB} + k_{5,11} u_{\theta_y B} \tag{B77}$$

$$T_{\theta_z A} = k_{26} u_{yA} + k_{66} u_{\theta_z A} + k_{68} u_{yB} + k_{6,12} u_{\theta_z B} \tag{B78}$$

$$F_{xB} = +k_{17} u_{xA} + k_{77} u_{xB} \tag{B79}$$

$$F_{yB} = +k_{28} u_{yA} + k_{68} u_{\theta_z A} + k_{88} u_{yB} + k_{8,12} u_{\theta_z B} \tag{B80}$$

$$F_{zB} = +k_{39} u_{zA} + k_{59} u_{\theta_y A} + k_{99} u_{zB} + k_{9,12} u_{\theta_y B} \tag{B81}$$

$$T_{\theta_x B} = +k_{4,10} u_{\theta_x A} + k_{10,10} u_{\theta_x B} \tag{B82}$$

$$T_{\theta_y B} = k_{3,11} u_{zA} + k_{5,11} u_{\theta_y A} + k_{9,11} u_{zB} + k_{11,11} u_{\theta_y B} \tag{B83}$$

$$T_{\theta_z B} = k_{2,12} u_{yA} + k_{6,12} u_{\theta_z A} + k_{8,12} u_{yB} + k_{12,12} u_{\theta_z B} \tag{B84}$$

B.2.2
The Mass Matrix for 12 DOFs

We now proceed to extend the mass matrix m, already introduced for the case with two DOFs (see (B29)), to the case of 12 DOFs. Subsequently, the introduction of a punctual description of the beam depending on all the stimuli (forces and torques) applied to the ends (A and B) is necessary. Such information is provided by a set of relationships called Hermitian shape functions (HSFs) [169]. Each of them describes the deformation of the beam at a generic point, for instance, along the length in the x direction, depending on the deformation applied to one end. Every HSF is defined by assuming the superposition principle is satisfied. In other words, an HSF determines the effect of a particular DOF related to a certain beam end, assuming all the other stimuli are equal to zero and the other end is constrained. For instance, if we apply a deformation u_{zA} to node A directed along the z axis, the configuration of beam is that shown in Figure B.4.

Introducing now the suitable HSF $\psi_{zA}(x)$ describing the punctual displacement along the z axis at a generic point on the x axis, and supposing u_{zA} is equal to 1, we obtain

$$\psi_{zA}(x) = 1 - 3\left(\frac{x}{L}\right)^2 + 2\left(\frac{x}{L}\right)^3 \tag{B85}$$

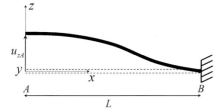

Figure B.4 Configuration of the Euler beam which undergoes a linear deformation u_{zA} applied at node A, with end B anchored.

Appendix B Flexible Straight Beam (Complete Model)

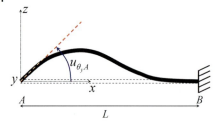

Figure B.5 Configuration of the Euler beam which undergoes an angular deformation $u_{\theta_y A}$ applied around the y axis at node A, with end B anchored.

Reintroducing the nondimensional variable ξ, already defined in (B20), we find that (B85) becomes

$$\psi_{zA}(\xi) = 1 - 3\xi^2 + 2\xi^3 \tag{B86}$$

Let us now examine an angular DOF. In this regard, in Figure B.5 a rotation $u_{\theta_y A}$ around the y axis is applied to beam end A.

Such a stimulus is supposed to be equal to 1. It should be noted that the rotation is applied in the negative y direction. This was done in order not to capsize the schematic in Figure B.5. The local displacement along z at a generic point x induced by $u_{\theta_y A}$ is then expressed as

$$\psi_{\theta_y A}(\xi) = (-\xi + 2\xi^2 - \xi^3)L \tag{B87}$$

By using the same approach for all 12 DOFs, we can extend the vector $\boldsymbol{a}_m$ in (B21) to a 3 × 12 matrix form, which is transposed $\boldsymbol{a}_m^T$ as [169]

$$\boldsymbol{a}_m^T = \begin{bmatrix} 1-\xi & 0 & 0 \\ 6(\xi - \xi^2)\eta & 1 - 3\xi^2 + 2\xi^3 & 0 \\ 6(\xi - \xi^2)\zeta & 0 & 1 - 3\xi^2 + 2\xi^3 \\ 0 & -(1-\xi)L\zeta & -(1-\xi)L\eta \\ (1 - 4\xi + 3\xi^2)L\zeta & 0 & (-\xi + 2\xi^2 - \xi^3)L \\ (-1 + 4\xi - 3\xi^2)L\eta & (\xi - 2\xi^2 + \xi^3)L & 0 \\ \xi & 0 & 0 \\ 6(-\xi + \xi^2)\eta & 3\xi^2 - 2\xi^3 & 0 \\ 6(-\xi + \xi^2)\zeta & 0 & 3\xi^2 - 2\xi^3 \\ 0 & -L\xi\zeta & -L\xi\eta \\ (-2\xi + 3\xi^2)L\zeta & 0 & (\xi^2 - \xi^3)L \\ (2\xi - 3\xi^2)L\eta & (-\xi^2 + \xi^3) & 0 \end{bmatrix} \tag{B88}$$

In (B88) another two nondimensional variables similar to those in (B20) have been introduced:

$$\eta = \frac{y}{L} \tag{B89}$$

B.2 Mechanical Model of the Flexible Beam with 12 DOFs

$$\zeta = \frac{z}{L} \tag{B90}$$

It is now possible to derive the mass matrix m. Its expression (B24) can be now rearranged as follows:

$$m = \rho A L \int_0^1 \int_0^1 \int_0^1 a^T a \, d\xi \, d\eta \, d\zeta \tag{B91}$$

The resolution of the volume integral (B91) leads to the mass matrix expressed as:

$$m = [m_1 | m_2] \tag{B92}$$

where

$$m_1 = \begin{bmatrix} m_{11} & 0 & 0 & 0 & 0 & 0 \\ 0 & m_{22} & 0 & 0 & 0 & m_{26} \\ 0 & 0 & m_{33} & 0 & m_{35} & 0 \\ 0 & 0 & 0 & m_{44} & 0 & 0 \\ 0 & 0 & m_{53} & 0 & m_{55} & 0 \\ 0 & m_{62} & 0 & 0 & 0 & m_{66} \\ m_{71} & 0 & 0 & 0 & 0 & 0 \\ 0 & m_{82} & 0 & 0 & 0 & m_{86} \\ 0 & 0 & m_{93} & 0 & m_{95} & 0 \\ 0 & 0 & 0 & m_{10,4} & 0 & 0 \\ 0 & 0 & m_{11,3} & 0 & m_{11,5} & 0 \\ 0 & m_{12,2} & 0 & 0 & 0 & m_{12,6} \end{bmatrix} \tag{B93}$$

$$m_2 = \begin{bmatrix} m_{17} & 0 & 0 & 0 & 0 & 0 \\ 0 & m_{28} & 0 & 0 & 0 & m_{2,12} \\ 0 & 0 & m_{39} & 0 & m_{3,11} & 0 \\ 0 & 0 & 0 & m_{4,10} & 0 & 0 \\ 0 & 0 & m_{59} & 0 & m_{5,11} & 0 \\ 0 & m_{68} & 0 & 0 & 0 & m_{6,12} \\ m_{77} & 0 & 0 & 0 & 0 & 0 \\ 0 & m_{88} & 0 & 0 & 0 & m_{8,12} \\ 0 & 0 & m_{99} & 0 & m_{9,11} & 0 \\ 0 & 0 & 0 & m_{10,10} & 0 & 0 \\ 0 & 0 & m_{11,9} & 0 & m_{11,11} & 0 \\ 0 & m_{12,8} & 0 & 0 & 0 & m_{12,12} \end{bmatrix} \tag{B94}$$

The elements of (B92) are expressed as follows:

$$m_{11} = \rho A L \left(\frac{1}{3}\right) \tag{B95}$$

$$m_{17} = \rho A L \left(\frac{1}{6}\right) \tag{B96}$$

Appendix B Flexible Straight Beam (Complete Model)

$$m_{22} = \rho A L \left(\frac{13}{35} + \frac{6I_z}{5AL^2} \right) \tag{B97}$$

$$m_{26} = \rho A L \left(\frac{11L}{210} + \frac{I_z}{10AL} \right) \tag{B98}$$

$$m_{28} = \rho A L \left(\frac{9}{70} - \frac{6I_z}{5AL^2} \right) \tag{B99}$$

$$m_{2,12} = \rho A L \left(-\frac{13L}{420} + \frac{I_z}{10AL} \right) \tag{B100}$$

$$m_{33} = \rho A L \left(\frac{13}{35} + \frac{6I_y}{5AL^2} \right) \tag{B101}$$

$$m_{35} = \rho A L \left(-\frac{11L}{210} - \frac{I_y}{10AL} \right) \tag{B102}$$

$$m_{39} = \rho A L \left(\frac{9}{70} - \frac{6I_y}{5AL^2} \right) \tag{B103}$$

$$m_{3,11} = \rho A L \left(\frac{13L}{420} - \frac{I_y}{10AL} \right) \tag{B104}$$

$$m_{44} = \rho A L \left(\frac{J_x}{3A} \right) \tag{B105}$$

$$m_{4,10} = \rho A L \left(\frac{J_x}{6A} \right) \tag{B106}$$

$$m_{55} = \rho A L \left(\frac{L^2}{105} + \frac{2I_y}{15A} \right) \tag{B107}$$

$$m_{59} = \rho A L \left(-\frac{13L}{420} + \frac{I_y}{10AL} \right) \tag{B108}$$

$$m_{5,11} = \rho A L \left(-\frac{L^2}{140} - \frac{I_y}{30A} \right) \tag{B109}$$

$$m_{66} = \rho A L \left(\frac{L^2}{105} + \frac{2I_z}{15A} \right) \tag{B110}$$

$$m_{68} = \rho A L \left(\frac{13L}{420} - \frac{I_z}{10AL} \right) \tag{B111}$$

$$m_{6,12} = \rho A L \left(-\frac{L^2}{140} - \frac{I_z}{30A} \right) \tag{B112}$$

$$m_{77} = \rho A L \left(\frac{1}{3} \right) \tag{B113}$$

$$m_{88} = \rho A L \left(\frac{13}{35} + \frac{6I_z}{5AL^2} \right) \tag{B114}$$

$$m_{8,12} = \rho A L \left(-\frac{11L}{210} - \frac{I_z}{10AL} \right) \tag{B115}$$

$$m_{99} = \rho A L \left(\frac{13}{35} + \frac{6 I_y}{5 A L^2} \right) \tag{B116}$$

$$m_{9,11} = \rho A L \left(\frac{11 L}{210} + \frac{I_y}{10 A L} \right) \tag{B117}$$

$$m_{10,10} = \rho A L \left(\frac{J_x}{3A} \right) \tag{B118}$$

$$m_{11,11} = \rho A L \left(\frac{L^2}{105} + \frac{2 I_y}{15 A} \right) \tag{B119}$$

$$m_{12,12} = \rho A L \left(\frac{L^2}{105} + \frac{2 I_z}{15 A} \right) \tag{B120}$$

Equation (B17) for the case with 12 DOFs becomes

$$\begin{bmatrix} F_{xA} \\ F_{yA} \\ F_{zA} \\ T_{\theta_x A} \\ T_{\theta_y A} \\ T_{\theta_z A} \\ F_{xB} \\ F_{yB} \\ F_{zB} \\ T_{\theta_x B} \\ T_{\theta_y B} \\ T_{\theta_z B} \end{bmatrix} = m \begin{bmatrix} \ddot{u}_{xA} \\ \ddot{u}_{yA} \\ \ddot{u}_{zA} \\ \ddot{u}_{\theta_x A} \\ \ddot{u}_{\theta_y A} \\ \ddot{u}_{\theta_z A} \\ \ddot{u}_{xB} \\ \ddot{u}_{yB} \\ \ddot{u}_{zB} \\ \ddot{u}_{\theta_x B} \\ \ddot{u}_{\theta_y B} \\ \ddot{u}_{\theta_z B} \end{bmatrix} \tag{B121}$$

Finally, the set of equations originating from the mass matrix and representing relationship (B30) are as follows:

$$F_{xA} = m_{11} \ddot{u}_{xA} + m_{17} \ddot{u}_{xB} \tag{B122}$$

$$F_{yA} = m_{22} \ddot{u}_{yA} + m_{26} \ddot{u}_{\theta_z A} + m_{28} \ddot{u}_{yB} + m_{2,12} \ddot{u}_{\theta_z B} \tag{B123}$$

$$F_{zA} = m_{33} \ddot{u}_{zA} + m_{35} \ddot{u}_{\theta_y A} + m_{39} \ddot{u}_{zB} + m_{3,11} \ddot{u}_{\theta_y B} \tag{B124}$$

$$T_{\theta_x A} = m_{44} \ddot{u}_{\theta_x A} + m_{4,10} \ddot{u}_{\theta_x B} \tag{B125}$$

$$T_{\theta_y A} = m_{35} \ddot{u}_{zA} + m_{55} \ddot{u}_{\theta_y A} + m_{59} \ddot{u}_{zB} + m_{5,11} \ddot{u}_{\theta_y B} \tag{B126}$$

$$T_{\theta_z A} = m_{26} \ddot{u}_{yA} + m_{66} \ddot{u}_{\theta_z A} + m_{68} \ddot{u}_{yB} + m_{6,12} \ddot{u}_{\theta_z B} \tag{B127}$$

$$F_{xB} = m_{17} \ddot{u}_{xA} + m_{77} \ddot{u}_{xB} \tag{B128}$$

$$F_{yB} = m_{28} \ddot{u}_{yA} + m_{68} \ddot{u}_{\theta_z A} + m_{88} \ddot{u}_{yB} + m_{8,12} \ddot{u}_{\theta_z B} \tag{B129}$$

$$F_{zB} = m_{39} \ddot{u}_{zA} + m_{59} \ddot{u}_{\theta_y A} + m_{99} \ddot{u}_{zB} + m_{11,9} \ddot{u}_{\theta_y B} \tag{B130}$$

$$M_{\theta_x B} = m_{4,10} \ddot{u}_{\theta_x A} + m_{10,10} \ddot{u}_{\theta_x B} \tag{B131}$$

$$M_{\theta_y B} = m_{3,11} \ddot{u}_{zA} + m_{5,11} \ddot{u}_{\theta_y A} + m_{9,11} \ddot{u}_{zB} + m_{11,12} \ddot{u}_{\theta_y B} \tag{B132}$$

$$M_{\theta_z B} = m_{2,12} \ddot{u}_{yA} + m_{6,12} \ddot{u}_{\theta_z A} + m_{8,12} \ddot{u}_{yB} + m_{12,12} \ddot{u}_{\theta_z B} \tag{B133}$$

B.3
Complete Mechanical Model of the Euler Beam with 12 DOFs

After all the relationships related to the elastic and inertial behavior of the Euler beam have been defined, the model must be linked to the global reference system in order to keep the consistency mentioned in Appendix A. In the most general case, the beam can assume a position as shown in Figure B.6, where all the orientation angles around the global frame axes (θ_X, θ_Y, and θ_Z) have nonzero values.

Moreover, since we are dealing with a deformable body, the rotations of one end around the global frame axes XYZ can be different from those of the other end. Let $u_{\theta_X A}$, $u_{\theta_Y A}$, and $u_{\theta_Z A}$ be the triplet of angles referring to end A, and $u_{\theta_X B}$, $u_{\theta_Y B}$, and $u_{\theta_Z B}$ be the triplet related to end B. If we now introduce three initial orientation angles for the beam around the global system, and we address them as $\theta_{X_{INIT}}$, $\theta_{Y_{INIT}}$, and $\theta_{Z_{INIT}}$, the three instantaneous angles of end A around the global axes can be defined as

$$\theta_{XA} = \theta_{X_{INIT}} + u_{\theta_X A} \tag{B134}$$

$$\theta_{YA} = \theta_{Y_{INIT}} + u_{\theta_Y A} \tag{B135}$$

$$\theta_{ZA} = \theta_{Z_{INIT}} + u_{\theta_Z A} \tag{B136}$$

and the other three instantaneous angles referring to end B can be defined as

$$\theta_{XB} = \theta_{X_{INIT}} + u_{\theta_X B} \tag{B137}$$

$$\theta_{YB} = \theta_{Y_{INIT}} + u_{\theta_Y B} \tag{B138}$$

$$\theta_{ZB} = \theta_{Z_{INIT}} + u_{\theta_Z B} \tag{B139}$$

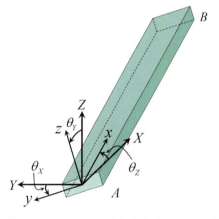

Figure B.6 Schematic of the Euler beam with orientation angles θ_X, θ_Y, and θ_Z. Such rotations define the relationships between the global reference system (XYZ) and the local one (xyz).

B.3 Complete Mechanical Model of the Euler Beam with 12 DOFs

Now we have to define the relationships between the global reference system and the two local systems linked to beam ends A and B. The approach consists in transforming the linear and angular deformation, applied at the two ends in the global frame into the corresponding deformation in the local system at ends A and B. The notation necessary to perform such a transformation is as follows:

$$\begin{bmatrix} u_{xA} \\ u_{yA} \\ u_{zA} \\ u_{\theta_x A} \\ u_{\theta_y A} \\ u_{\theta_z A} \\ u_{xB} \\ u_{yB} \\ u_{zB} \\ u_{\theta_x B} \\ u_{\theta_y B} \\ u_{\theta_z B} \end{bmatrix} = \begin{bmatrix} \Omega_{LA}^T & 0 & 0 & 0 \\ 0 & \Omega_{LA}^T & 0 & 0 \\ 0 & 0 & \Omega_{LB}^T & 0 \\ 0 & 0 & 0 & \Omega_{LB}^T \end{bmatrix} \begin{bmatrix} u_{XA} \\ u_{YA} \\ u_{ZA} \\ u_{\theta_X A} \\ u_{\theta_Y A} \\ u_{\theta_Z A} \\ u_{XB} \\ u_{YB} \\ u_{ZB} \\ u_{\theta_X B} \\ u_{\theta_Y B} \\ u_{\theta_Z B} \end{bmatrix} \quad (B140)$$

where Ω_{LA}^T and Ω_{LB}^T are calculated as a function of θ_{XA}, θ_{YA}, and θ_{ZA} at node A, and as a function of θ_{XB}, θ_{YB}, and θ_{ZB} at node B. Finally, 0 is a 3×3 matrix with all elements zero. For instance, from (B140), we find

$$u_{xA} = \cos\theta_{YA}\cos\theta_{ZA} u_{XA} + \cos\theta_{YA}\sin\theta_{ZA} u_{YA} - \sin\theta_{YA} u_{ZA} \quad (B141)$$

When all the deformations have been transformed into the local system, the corresponding forces and torques are determined with (B73)–(B84). Finally, these magnitudes are transferred back into the global reference system as follows:

$$\begin{bmatrix} F_{XA} \\ F_{YA} \\ F_{ZA} \\ T_{\theta_X A} \\ T_{\theta_Y A} \\ T_{\theta_Z A} \\ F_{XB} \\ F_{YB} \\ F_{ZB} \\ T_{\theta_X B} \\ T_{\theta_Y B} \\ T_{\theta_Z B} \end{bmatrix} = \begin{bmatrix} \Omega_{LA} & 0 & 0 & 0 \\ 0 & \Omega_{LA} & 0 & 0 \\ 0 & 0 & \Omega_{LB} & 0 \\ 0 & 0 & 0 & \Omega_{LB} \end{bmatrix} \begin{bmatrix} F_{xA} \\ F_{yA} \\ F_{zA} \\ T_{\theta_x A} \\ T_{\theta_y A} \\ T_{\theta_z A} \\ F_{xB} \\ F_{yB} \\ F_{zB} \\ T_{\theta_x B} \\ T_{\theta_y B} \\ T_{\theta_z B} \end{bmatrix} \quad (B142)$$

Concerning the inertial effects, the approach to be followed is the same. First, all the linear and angular accelerations are transferred into the local frame:

$$\begin{bmatrix} \ddot{u}_{xA} \\ \ddot{u}_{yA} \\ \ddot{u}_{zA} \\ \ddot{u}_{\theta_x A} \\ \ddot{u}_{\theta_y A} \\ \ddot{u}_{\theta_z A} \\ \ddot{u}_{xB} \\ \ddot{u}_{yB} \\ \ddot{u}_{zB} \\ \ddot{u}_{\theta_x B} \\ \ddot{u}_{\theta_y B} \\ \ddot{u}_{\theta_z B} \end{bmatrix} = \begin{bmatrix} \Omega_{LA}^T & 0 & 0 & 0 \\ 0 & \Omega_{LA}^T & 0 & 0 \\ 0 & 0 & \Omega_{LB}^T & 0 \\ 0 & 0 & 0 & \Omega_{LB}^T \end{bmatrix} \begin{bmatrix} \ddot{u}_{XA} \\ \ddot{u}_{YA} \\ \ddot{u}_{ZA} \\ \ddot{u}_{\theta_X A} \\ \ddot{u}_{\theta_Y A} \\ \ddot{u}_{\theta_Z A} \\ \ddot{u}_{XB} \\ \ddot{u}_{YB} \\ \ddot{u}_{ZB} \\ \ddot{u}_{\theta_X B} \\ \ddot{u}_{\theta_Y B} \\ \ddot{u}_{\theta_Z B} \end{bmatrix} \qquad \text{(B143)}$$

Then, (B122)–(B133) are calculated, and are finally transformed into the global frame by means of (B142).

The mechanical model discussed thus far is linear. Indeed, it is reasonable to assume the elastic behavior of the beam is linear as long as we are dealing with small deformations. This is what typically happens in the functional situations in which we are interested. For instance, in a microelectromechanical system (MEMS) switch based on a suspended rigid plate with Euler beams joined at the corners, a typical plate-to-substrate gap is in the range of a few microns. If the beams are relatively long, it is sensible to assume a vertical deformation of a few microns (due to the pull-in) as this falls within the linear elastic regime. However, a model which accounts for the nonlinearities due to large deformations is also implemented. This is the one proposed by Jing [170], but it is not described here.

B.4
Electrostatic Model of the Euler Beam with 12 DOFs

Before introducing the electrostatic model of the Euler beam, we have to mention that a model describing the contact with the substrate is also implemented for this component. The contact force is applied to beam end A or/and beam end B, depending on which node(s) is (are) involved in the collision. Such a model is based on the principle already discussed in Appendix A, and for this reason, it is not described here.

Moving the focus now onto the electrostatic model for the Euler beam, we need to know the punctual distance between a generic point along the beam length and the substrate. First of all, we refer to the displacements of the beam centerline, lying on its lower face, as shown in Figure B.7. Such a line identifies the vertical displacement along L, as defined by the HSF $\psi_{zA}(\xi)$.

We are interested in defining the local vertical displacement due to all the stimuli applied to ends A and B which give a contribution along the z axis. Looking at

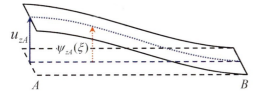

Figure B.7 The vertical displacement of the beam centerline (dotted line) is determined by means of the Hermitian shape functions.

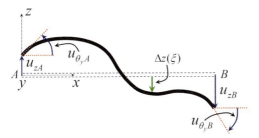

Figure B.8 Schematic of a beam where all the DOFs giving a contribution along the z axis are stimulated.

the right column of (B88) (corresponding to the z direction), and assuming the vertical displacement contribution given by the torsion around the x axis to be negligible, we find the local z displacement is due to the linear deformation along z, and the torsion around the y axis at ends A and B. The corresponding HSFs are, respectively,

$$\psi_{zA}(\xi) = 1 - 3\xi^2 + 2\xi^3 \tag{B144}$$

$$\psi_{\theta_y A}(\xi) = (-\xi + 2\xi^2 - \xi^3)L \tag{B145}$$

$$\psi_{zB}(\xi) = 3\xi^2 - 2\xi^3 \tag{B146}$$

$$\psi_{\theta_y B}(\xi) = (\xi^2 - \xi^3)L \tag{B147}$$

Let us now assume the beam is stimulated in all the above-mentioned DOFs, yielding contributions to the deformation along the local z axis, as shown in Figure B.8.

The local z displacement $\Delta z(\xi)$ with respect to the initial beam position (dashed line in Figure B.8) is given by the sum of (B144)–(B147), multiplied by the corresponding perturbation magnitude:

$$\Delta z(\xi) = \psi_{zA}(\xi) u_{xA} + \psi_{\theta_y A}(\xi) u_{\theta_x A} + \psi_{zB}(\xi) u_{xB} + \psi_{\theta_y B}(\xi) u_{\theta_x B} \tag{B148}$$

Equation (B148) describes the punctual z deformation of the beam centerline in Figure B.7 in the local system xyz. We now have to derive an expression which defines the vertical distance of each beam centerline point with respect to the substrate in the global reference system as a function of the initial orientation angles

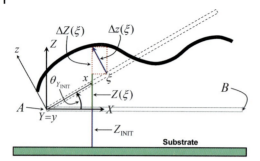

Figure B.9 Total displacement of a generic point ξ along the beam length, expressed in the global Z direction.

$\theta_{X_{INIT}}$, $\theta_{Y_{INIT}}$, and $\theta_{Z_{INIT}}$ (see Figure B.6). The contributions to the vertical distance between a generic point ξ and the substrate due to an initial orientation angle $\theta_{Y_{INIT}}$ are depicted in Figure B.9.

There are three such contributions. The first is the projection of $\Delta z(\xi)$ along the global Z axis, called $\Delta Z(\xi)$. The second is the projection of the local x coordinate of the point ξ along the Z axis. The third is the initial gap between the nondisplaced and nondeformed beam with respect to the substrate. If we now define $Z_{AIR}(\xi)$ as the total vertical abscissa of point ξ, we obtain

$$Z_{AIR}(\xi) = \Delta Z(\xi) + Z(\xi) + Z_{INIT} \tag{B149}$$

where

$$\Delta Z(\xi) = \cos\theta_{X_{INIT}} \cos\theta_{Y_{INIT}} \Delta z(\xi) \tag{B150}$$

and

$$Z(\xi) = -\sin\theta_{Y_{INIT}} \xi \tag{B151}$$

Still using the HSF of (B88), and also accounting for the initial rotation angle around the X axis $\theta_{X_{INIT}}$, we can express the distribution of the total torsion around the X axis as a function of ξ as follows:

$$\theta_X(\xi) = (1-\xi)u_{\theta_X A} + \xi u_{\theta_X B} + \theta_{X_{INIT}} \tag{B152}$$

Such a distribution of torsion is supposed to be similar to the one of the x deformations reported in (B88). For the calculation of the capacitance and electrostatic force, considering the beam area partitioned into a suitable number of rectangular subelements, the subelements are assumed to be rigid and with a certain orientation in space, as shown in Figure B.10.

Once the local x abscissa has been determined (see Figure B.11), the orientation angle around the Y axis for each subelement can be derived with simple trigonometric considerations. Figure B.12 shows the cross section of one element.

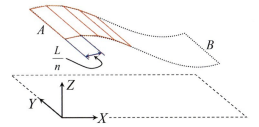

Figure B.10 The beam length is divided into n elements of length L/n. Each element is assumed to be rigid.

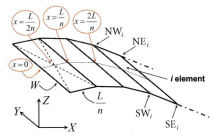

Figure B.11 Local abscissa value along the x-axis centerline for the discretized beam elements.

First of all, we need to express the local centerline abscissa of each element as a function of ξ. If the generic length of each element

$$x = \frac{L}{n} \tag{B153}$$

where n is the number of elements, is substituted into (B20), the latter becomes

$$\xi = \frac{1}{n} \tag{B154}$$

After such notation has been introduced, let us consider a generic element i. Its ξ abscissa, referred to the center point along the x direction, is

$$\xi = \frac{3i}{2n} \tag{B155}$$

while the abscissa of the edge of the ith element, shared with the adjacent $(i+1)$th element, is

$$\xi = \frac{i+1}{n} \tag{B156}$$

The distances of the center point and edge of the ith element along the Z axis are defined by (B149), and are

$$Z_{\text{AIR}}\left[\frac{1}{n}\left(i+\frac{1}{2}\right)\right] \quad \text{and} \quad Z_{\text{AIR}}\left(\frac{i+1}{n}\right),$$

respectively, in Figure B.12.

Appendix B Flexible Straight Beam (Complete Model)

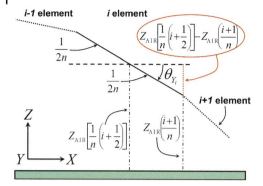

Figure B.12 Cross section of one subelement of the beam. The rotation angle around the Y axis is determined by means of simple calculations.

By using the basic triangle relationships, and considering that half the length of each element is equal to $1/2n$, we can determine the angle θ_{Y_i} as

$$\theta_{Y_i} = \arcsin\left(2n\left\{Z_{\text{AIR}}\left[\frac{1}{n}\left(i+\frac{1}{2}\right)\right] - Z_{\text{AIR}}\left(\frac{i+1}{n}\right)\right\}\right) \tag{B157}$$

Finally, the angle around the X axis is calculated with (B152) and, choosing it as the one at the center point of each rectangle, referring to the ith element, we obtain

$$\theta_{X_i} = \theta_X\left[\frac{1}{n}\left(i+\frac{1}{2}\right)\right] \tag{B158}$$

As already shown in the calculation of electrostatic effects for the rigid plate, the rotation angle around the Z axis is not involved in any formula for the capacitance and the attraction force. This means that the angles θ_{X_i} and θ_{Y_i} are sufficient to determine the electrostatic effects for each element along the beam. It is possible to apply once again the formulas developed for the capacitance and electrostatic force for a rigid plate. Each subelement of the beam is assumed to be a small rigid plate. Still considering the ith element, we find the local coordinates of the four corners (see Figure B.11) are

$$\text{NE}_i(\xi, y) = \left(\frac{i+1}{n}, \frac{W}{2}\right) \tag{B159}$$

$$\text{SE}_i(\xi, y) = \left(\frac{i+1}{n}, -\frac{W}{2}\right) \tag{B160}$$

$$\text{SW}_i(\xi, y) = \left(\frac{i}{n}, -\frac{W}{2}\right) \tag{B161}$$

$$\text{NW}_i(\xi, y) = \left(\frac{i}{n}, \frac{W}{2}\right) \tag{B162}$$

The capacitance of each element is calculated with (A149). In this case, in order to take into consideration the overall contributions of the capacitance, we have to include a sum over the n subelements as follows:

$$C_{AIR} = L \frac{\varepsilon_{AIR}}{\sigma_{Arc}} \sum_{i=0}^{n-1} \int_{\frac{i}{n}}^{\frac{i+1}{n}} \int_{-\frac{W}{2}}^{\frac{W}{2}} \frac{d\xi\, dy}{Z_L(\xi, y, \theta_{X_i}, \theta_{Y_i})} \quad \text{(B163)}$$

where σ_{Arc} is the coefficient taken as the curvature of the electric field lines, as already shown for the rigid plate. Moreover, the local Z_L is defined as

$$Z_L = Z_{AIR} \left[\frac{1}{n}\left(i + \frac{1}{2}\right) \right] - \sin\theta_{X_i} L\xi + \sin\theta_{X_i} \cos\theta_{Y_i} y \quad \text{(B164)}$$

where Z_{AIR} is as in (B149) and is calculated at the middle point of each element ($\xi = 1/n(i + 1/2)$), while the other two terms represent the local Z displacement of each point on the surface of the subelement (see Figure A.18). Dealing now with the electrostatic force, we can rewrite (A150) in the case of the flexible beam divided into subelements. It is necessary to consider that the electrostatic force is a distributed effect, even though in the end it has to be punctually applied to beam nodes A and B. Since the beam is a flexible structure, the electrostatic force cannot be directly transferred to the two ends, but it should be somehow weighted by means of suitable functions. The approach proposed by Jing [170] uses once again the HSFs in order to weight the electrostatic force to be applied at one beam end depending on the distance of the element (where it is calculated) from the node of application. This means that the electrostatic force must be multiplied by the HSFs which determine the punctual beam displacement along the Z axis, and that the force must be calculated using different functions for the two ends A and B, as follows:

$$F_{zA_{EL}} = \frac{L}{2} \frac{\varepsilon_{AIR}}{\sigma_{Arc}^2} \sum_{i=0}^{n-1} \left\{ \psi_{zA}\left[\frac{1}{n}\left(i + \frac{1}{2}\right)\right] \int_{\frac{i}{n}}^{\frac{i+1}{n}} \int_{-\frac{W}{2}}^{\frac{W}{2}} \frac{V_{AIR}^2\, d\xi\, dy}{Z_L(\xi, y, \theta_{X_i}, \theta_{Y_i})^2} \right\}$$

(B165)

$$F_{zB_{EL}} = \frac{L}{2} \frac{\varepsilon_{AIR}}{\sigma_{Arc}^2} \sum_{i=0}^{n-1} \left\{ \psi_{zB}\left[\frac{1}{n}\left(i + \frac{1}{2}\right)\right] \int_{\frac{i}{n}}^{\frac{i+1}{n}} \int_{-\frac{W}{2}}^{\frac{W}{2}} \frac{V_{AIR}^2\, d\xi\, dy}{Z_L(\xi, y, \theta_{X_i}, \theta_{Y_i})^2} \right\}$$

(B166)

where $\psi_{zA}[1/n(i + 1/2)]$ and $\psi_{zB}[1/n(i + 1/2)]$ are (B144) and (B146), respectively, calculated at the middle point of each element along ξ. Equations (B165) and (B166) must then be added to the respective forces applied to the two nodes along the local z axis in (B31). The contribution of the electrostatic force will be automatically translated back into the global frame, together with the elastic and inertial effects, by means of (B142). With the same approach it is possible to determine the contribution to the torque around the x and y axes.

Hence, (A151) and (A152) can be rewritten, using (B145) and (B147), as follows:

$$T_{\theta x A_{EL}} = \frac{L \, \varepsilon_{AIR}}{2 \, \sigma_{Arc}^2} \sum_{i=0}^{n-1} \left\{ \left[1 - \frac{1}{n}\left(i + \frac{1}{2}\right)\right] \int_{\frac{i}{n}}^{\frac{i+1}{n}} \int_{-\frac{W}{2}}^{\frac{W}{2}} \frac{V_{AIR}^2 \, y \, d\xi \, dy}{Z_L(\xi, y, \theta_{X_i}, \theta_{Y_i})^2} \right\} \tag{B167}$$

$$T_{\theta x B_{EL}} = \frac{L \, \varepsilon_{AIR}}{2 \, \sigma_{Arc}^2} \sum_{i=0}^{n-1} \left\{ \left[\frac{1}{n}\left(i + \frac{1}{2}\right)\right] \int_{\frac{i}{n}}^{\frac{i+1}{n}} \int_{-\frac{W}{2}}^{\frac{W}{2}} \frac{V_{AIR}^2 \, y \, d\xi \, dy}{Z_L(\xi, y, \theta_{X_i}, \theta_{Y_i})^2} \right\} \tag{B168}$$

$$T_{\theta y A_{EL}} = \frac{L \, \varepsilon_{AIR}}{2 \, \sigma_{Arc}^2} \sum_{i=0}^{n-1} \left\{ \psi_{\theta_y A} \left[\frac{1}{n}\left(i + \frac{1}{2}\right)\right] \int_{\frac{i}{n}}^{\frac{i+1}{n}} \int_{-\frac{W}{2}}^{\frac{W}{2}} \frac{V_{AIR}^2 \, \xi \, d\xi \, dy}{Z_L(\xi, y, \theta_{X_i}, \theta_{Y_i})^2} \right\} \tag{B169}$$

$$T_{\theta y B_{EL}} = \frac{L \, \varepsilon_{AIR}}{2 \, \sigma_{Arc}^2} \sum_{i=0}^{n-1} \left\{ \psi_{\theta_y B} \left[\frac{1}{n}\left(i + \frac{1}{2}\right)\right] \int_{\frac{i}{n}}^{\frac{i+1}{n}} \int_{-\frac{W}{2}}^{\frac{W}{2}} \frac{V_{AIR}^2 \, \xi \, d\xi \, dy}{Z_L(\xi, y, \theta_{X_i}, \theta_{Y_i})^2} \right\} \tag{B170}$$

In (B167) and (B168) a torque distribution around the x axis, similar to the one for the deformation along x at the two ends reported in (B88), is chosen. Also (B167)–(B170) are transformed back into the global system by means of the same procedure previously discussed for the electrostatic force.

B.4.1
Fringing Effect Model

In the electrostatic model of the beam the contribution given by the fringing field is also accounted for. The proposed model is the same as that presented in detail when describing the rigid plate electromechanical transducer (see Appendix A). All the rigid subelements into which the beam is divided are affected by the fringing effect close to the two minor edges, that is, the ones along the x direction. Furthermore, for the first and last elements (which represent the two ends A and B) one of the two major edges (along the y axis) contributes to the fringing field as well. In the next formulas, σ_{ArcF} is defined by (A168), while the area in which the fringing is confined, which was indicated by ζ in Appendix A, is indicated here by λ so

B.4 Electrostatic Model of the Euler Beam with 12 DOFs

that it is not confused with ζ in (B90). It is now possible to rewrite (B163) for the capacitance also accounting for the fringing field effect as follows

$$C_{AIR} = L \frac{\varepsilon_{AIR}}{\sigma_{Arc}} \sum_{i=0}^{n-1} \int_{\frac{i}{n}}^{\frac{i+1}{n}} \int_{-\frac{W}{2}}^{\frac{W}{2}} \frac{d\xi\, dy}{Z_L(\xi, y, \theta_{X_i}, \theta_{Y_i})}$$

$$+ \lambda \frac{\varepsilon_{AIR}}{\sigma_{ArcF}} \int_{-\frac{W}{2}}^{\frac{W}{2}} \frac{dy}{Z_L(0, y, \theta_{X_i}, \theta_{Y_i})}$$

$$+ \lambda \frac{\varepsilon_{AIR}}{\sigma_{ArcF}} \int_{-\frac{W}{2}}^{\frac{W}{2}} \frac{dy}{Z_L(1, y, \theta_{X_i}, \theta_{Y_i})}$$

$$+ \lambda L \frac{\varepsilon_{AIR}}{\sigma_{ArcF}} \sum_{i=0}^{n-1} \int_{\frac{i}{n}}^{\frac{i+1}{n}} \frac{d\xi}{Z_L(\xi, -\frac{W}{2}, \theta_{X_i}, \theta_{Y_i})}$$

$$+ \lambda L \frac{\varepsilon_{AIR}}{\sigma_{ArcF}} \sum_{i=0}^{n-1} \int_{\frac{i}{n}}^{\frac{i+1}{n}} \frac{d\xi}{Z_L(\xi, \frac{W}{2}, \theta_{X_i}, \theta_{Y_i})} \quad (B171)$$

In the previous equation the second and third terms represent the fringing on the boundary major edges of beam ends A and B (along the y axis), respectively. On the other hand, the fourth and last terms are the fringing capacitance on the lower and upper minor edges (along the x axis) of all the elements, respectively. Following the same approach, (B165) and (B166) become

$$F_{zA_{EL}} = \frac{L\,\varepsilon_{AIR}}{2\,\sigma_{Arc}^2} \sum_{i=0}^{n-1} \left\{ \psi_{zA}\left[\frac{1}{n}\left(i+\frac{1}{2}\right)\right] \int_{\frac{i}{n}}^{\frac{i+1}{n}} \int_{-\frac{W}{2}}^{\frac{W}{2}} \frac{V_{AIR}^2\, d\xi\, dy}{Z_L(\xi, y, \theta_{X_i}, \theta_{Y_i})^2} \right\}$$

$$+ \lambda \frac{\varepsilon_{AIR}}{\sigma_{ArcF}^2} \int_{-\frac{W}{2}}^{\frac{W}{2}} \frac{V_{AIR}^2\, dy}{Z_L(0, y, \theta_{X_i}, \theta_{Y_i})^2}$$

$$+ \lambda \frac{L\,\varepsilon_{AIR}}{2\,\sigma_{ArcF}^2} \sum_{i=0}^{n-1} \left\{ \psi_{zA}\left[\frac{1}{n}\left(i+\frac{1}{2}\right)\right] \int_{\frac{i}{n}}^{\frac{i+1}{n}} \frac{V_{AIR}^2\, d\xi}{Z_L(\xi, -\frac{W}{2}, \theta_{X_i}, \theta_{Y_i})^2} \right\}$$

$$+ \lambda \frac{L\,\varepsilon_{AIR}}{2\,\sigma_{ArcF}^2} \sum_{i=0}^{n-1} \left\{ \psi_{zA}[\frac{1}{n}\left[\left(i+\frac{1}{2}\right)\right] \int_{\frac{i}{n}}^{\frac{i+1}{n}} \frac{V_{AIR}^2\, d\xi}{Z_L(\xi, \frac{W}{2}, \theta_{X_i}, \theta_{Y_i})^2} \right\}$$

$$(B172)$$

$$F_{z\,B_{EL}} = \frac{L}{2} \frac{\varepsilon_{AIR}}{\sigma_{Arc}^2} \sum_{i=0}^{n-1} \left\{ \psi_{zB} \left[\frac{1}{n}\left(i+\frac{1}{2}\right) \right] \int_{\frac{i}{n}}^{\frac{i+1}{n}} \int_{-\frac{W}{2}}^{\frac{W}{2}} \frac{V_{AIR}^2 \, d\xi \, dy}{Z_L(\xi, y, \theta_{X_i}, \theta_{Y_i})^2} \right\}$$

$$+ \lambda \frac{\varepsilon_{AIR}}{\sigma_{ArcF}^2} \int_{-\frac{W}{2}}^{\frac{W}{2}} \frac{V_{AIR}^2 \, dy}{Z_L(1, y, \theta_{X_i}, \theta_{Y_i})^2}$$

$$+ \lambda \frac{L}{2} \frac{\varepsilon_{AIR}}{\sigma_{ArcF}^2} \sum_{i=0}^{n-1} \left\{ \psi_{zB} \left[\frac{1}{n}\left(i+\frac{1}{2}\right) \right] \int_{\frac{i}{n}}^{\frac{i+1}{n}} \frac{V_{AIR}^2 \, d\xi}{Z_L(\xi, -\frac{W}{2}, \theta_{X_i}, \theta_{Y_i})^2} \right\}$$

$$+ \lambda \frac{L}{2} \frac{\varepsilon_{AIR}}{\sigma_{ArcF}^2} \sum_{i=0}^{n-1} \left\{ \psi_{zB} \left[\frac{1}{n}\left(i+\frac{1}{2}\right) \right] \int_{\frac{i}{n}}^{\frac{i+1}{n}} \frac{V_{AIR}^2 \, d\xi}{Z_L(\xi, \frac{W}{2}, \theta_{X_i}, \theta_{Y_i})^2} \right\}$$

(B173)

In (B172), the second term is the fringing field contribution given by the major edge (along the y axis) at node A, while the last two terms account for the fringing along x at the lower ($y = -W/2$) and higher ($y = W/2$) beam edges, respectively. In (B173) the second term describes the fringing force on the major edge at node B, and the last two terms have the same meaning as in (B172).

To conclude the discussion of the electrostatic model of the flexible beam, an additional feature has to be introduced. Thus far, we have considered the substrate electrode deployed beneath the whole length of the suspended beam. However, there are different possibilities for placing the electrode. For instance, it is possible to define a certain region along the beam where there is no electrode below it. Differently, the electrode can be placed in an area along the beam without including one or both of ends A and B. The previous formulas for the capacitance and electrostatic force suitably modified to account for this additional feature are not shown here for the sake of brevity. All the different possible configurations for the electrode placement are shown in Figure B.13. The lengths of the two electrodes, including the left end (A) and the right end (B), are EL_{lft} and EL_{rgt}, respectively. Moreover, a Boolean parameter is also defined. It is labeled as AB_{in}, and when it is true (i.e., equal to 1), the electrodes are considered to be the two segments EL_{lft} and EL_{rgt}, including beam ends A and B (see Figure B.13). Differently, when AB_{in} is false (i.e., equal to 0), the electrode does not include a beam end.

B.5
Viscous Damping Model

The achievement of a reasonably detailed model for describing the behavior of the flexible beam, also accounting for the dynamic regime, requires the inclusion of the viscous damping effect. This results in a force counteracting the motion of a

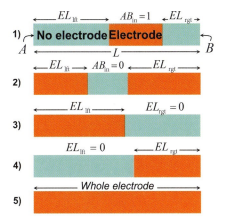

Figure B.13 Schematic of all the possible configurations (accounted for by the model discussed) for placing the electrode(s) beneath the flexible beam.

body immersed within a fluid with a certain viscosity (e.g., air) due to the friction of the particles of the media when touching the body.

To make easier the introduction and explanation of basis of the viscous damping theory, we now refer for a short while to a rigid body. The study of the viscous damping effect is based on the Reynolds equation. In the case of a body characterized by certain dimensions on the xy plane which is moving along the z direction, we have

$$\frac{\partial^2 P}{\partial x^2} + \frac{\partial^2 P}{\partial y^2} = \frac{12\mu}{z^3}\frac{\partial z}{\partial t} \tag{B174}$$

where P is the pressure produced by the fluid (of viscosity μ) on the surface of the body when its distance from the substrate is z and it is moving toward it with velocity $\partial z/\partial t$. Still referring to a rigid body which is parallel to the substrate, we cannot easily integrate (B174) in a closed form. Nevertheless, in the case of a flexible beam it can be simplified. Indeed, since $L \gg W$ (see Figure B.1), the pressure variation along the beam length (x axis) can be assumed to be negligible with respect to the variation of P along y [224]. This allows us to rewrite (B174) as

$$\frac{\partial^2 P}{\partial y^2} = \frac{12\mu}{z^3}\frac{\partial z}{\partial t} \tag{B175}$$

which after integration becomes

$$P(y) = -\frac{6\mu}{z^3}(W^2 - y^2)V_z \tag{B176}$$

where

$$V_z = \frac{\partial z}{\partial t} \tag{B177}$$

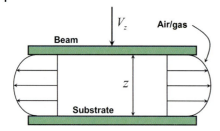

Figure B.14 In the squeeze-film damping description, the fluid is squeezed as the beam moves toward the surface of the substrate.

The force generated by the pressure applied on the surface of the beam is derived by integration over the area:

$$F_z = L \int_{-\frac{W}{2}}^{\frac{W}{2}} P(y) dy \qquad (B178)$$

Substituting (B176) in (B178), and developing the integral, we obtain

$$F_z = -\frac{\mu L W^3}{z^3} V_z \qquad (B179)$$

This type of damping is called squeeze-film damping as the fluid, assumed to be incompressible, is squeezed out of the beam when it moves toward the substrate, as shown in Figure B.14 [223].

Now that the squeeze-film damping has been expressed in a closed form, we have to cope with the fact that the Euler beam is a flexible structure. Starting from the theory discussed above, we propose an extension of the model accounting for the flexibility of the beam. By still using the HSFs as shown in the electrostatic model, we can define a local speed for each subelement.

Equation (B148) can be rewritten by replacing the vertical displacements and angular rotations with the corresponding linear and angular velocities at ends A and B:

$$V_z(\xi) = \psi_{zA}(\xi)\dot{u}_{xA} + \psi_{\theta_y A}(\xi)\dot{u}_{\theta_x A} + \psi_{zB}(\xi)\dot{u}_{xB} + \psi_{\theta_y B}(\xi)\dot{u}_{\theta_x B} \qquad (B180)$$

This expression yields the instantaneous velocity of each point along the beam length depending on the dynamics of both ends. Under appropriate assumptions, the squeeze-film damping theory can be applied to the model of the beam divided into subelements. First of all, we assume a homogeneous vertical velocity over the surface of a rectangular subelement, as well as small orientation angles. Because of these hypotheses, we use both the vertical distance of a certain element from the substrate and its speed calculated at the center point, determined by (B164) and (B180), respectively, calculated for $\xi = 1/n(i + 1/2)$.

Assuming each element is quasi-parallel to the substrate because of the small θ_{X_i} and θ_{Y_i} angles, and also assuming the velocity difference among adjacent subelements is small enough that there is no need for singularly applying boundary conditions to each element, we can consider (B179) to be still valid. This allows us to write (B179) in a form similar to that of (B165) and (B166) for the two ends A and B as follows:

$$F_{zA_{VD}} = -\mu \sum_{i=0}^{n-1} \psi_{zA} \left[\frac{1}{n} \left(i + \frac{1}{2} \right) \right]$$
$$\times \frac{V_z \left[\frac{1}{n} \left(i + \frac{1}{2} \right) \right] \frac{LW^3}{n}}{\left\{ Z_{AIR} \left[\frac{1}{n} \left(i + \frac{1}{2} \right) \right] - \sin\theta_{X_i} L \left[\frac{1}{n} \left(i + \frac{1}{2} \right) \right] \right\}^3} \quad \text{(B181)}$$

$$F_{zB_{VD}} = -\mu \sum_{i=0}^{n-1} \psi_{zB} \left[\frac{1}{n} \left(i + \frac{1}{2} \right) \right]$$
$$\times \frac{V_z \left[\frac{1}{n} \left(i + \frac{1}{2} \right) \right] \frac{LW^3}{n}}{\left\{ Z_{AIR} \left[\frac{1}{n} \left(i + \frac{1}{2} \right) \right] - \sin\theta_{X_i} L \left[\frac{1}{n} \left(i + \frac{1}{2} \right) \right] \right\}^3} \quad \text{(B182)}$$

The negative sign of (B181) and (B182) is due to the fact that the viscous damping force acts in the direction opposite to the movement of the beam. The corresponding contributions in the global frame are then derived with the usual transformation via the rotation matrix, expressed by (B142). With a similar approach, the torques around the y axis at the two ends produced by the squeeze-film damping are determined, but are not shown here.

Concerning the lateral movements of the Euler beam (xy plane), the proposed model is the same as that described Appendix A. This is based on the assumption of a Couette-type laminar flow distribution of the fluid underneath the beam. Moreover, the beam deformations are supposed to be not too large in order to assume the Couette hypothesis is still valid. Given these considerations, the viscous damping forces for the in-plane movements are expressed by (A212) and (A213).

Finally, the gas rarefaction effect for very small gaps (comparable with the mean free path of fluid molecules) is accounted for by replacing μ with (A214) in all the previous formulas.

B.6
Conclusions

The topic of this appendix was the detailed description of the MEMS deformable straight beam implemented in the Verilog-A programming language based library introduced in Chapter 5. The level of detail reported in this appendix was very deep, and for this reason, the information presented is intended for those readers who wish to know all the specific features of the model. The mechanical, electrostatic, and dynamic modeled features were presented, completing the detailed explanation of the implemented model based on six DOFs for each end of the beam.

References

1 Jaeger, R.C. (1988) *Microelectronic Fabrication*, Addison-Wesley.
2 Hobstetter, J.H. (1960) Mechanical properties of semiconductors. *Properties of Crystalline Solids*, ASTM Special Technical Publication 283, ASTM Publisher p. 40.
3 Beams, J.W., Freazeale, J.B., and Bart, W.L. (1955) Mechanical strength of thin films of metals. *Physical Review Letters, APS*, **100** (6), 1657–1661.
4 Neugebauer, C.A. (1960) Tensile properties of thin, evaporated gold films. *Journal of Applied Physics*, **31** (6), 1096–1101.
5 Blakely, J.M. (1964) Mechanical properties of vacuum-deposited gold. *Journal of Applied Physics*, **35** (6), 1756–1759.
6 Bassous, E. (1978) Fabrication of novel three-dimensional microstructures by the anisotropic etching of (100) and (110) Silicon. *IEEE Transactions on Electron Devices*, **25** (10), 1178–1185.
7 Wen, H.K., Hynecek, J., and Boettcher, S.F. (1979) Development of a miniature pressure transducer for biomedical applications. *IEEE Transactions on Electron Devices*, **26** (12), 1896–1905.
8 Roylance, L.M. and Angell, B.J. (1979) A batch-fabricated silicon accelerometer. *IEEE Transactions on Electron Devices*, **26** (12), 1911–1917.
9 Chen, P., Muller, R., Shiosaki, T., and White, R. (1979) WP-B6 silicon cantilever beam accelerometer utilizing a PI-FET capacitive transducer. *IEEE Transactions on Electron Devices*, **26** (11), 1857–1857.
10 Petersen, K.E. (1979) Micromechanical membrane switches on silicon. *IBM Journal of Research and Development*, **23** (4), 376–385.
11 Petersen, K.E. (1977) Micromechanical light modulator array fabricated on silicon. *Applied Physics Letters*.
12 Petersen, K.E. (1982) Silicon as a mechanical material. *Proceedings of the IEEE*, **70** (5), 420–457.
13 Bustillo, J., Howe, R., and Muller, R. (1998) Surface micromachining for microelectromechanical systems. *Proceedings of the IEEE*, **86** (8), 1552–1574.
14 Howe, R.T. and Muller, R.S. (1983) Polycrystalline silicon micromechanical beams. *Journal of the Electrochemical Society*, **130** (6), 1420–1423.
15 Howe, R. and Muller, R. (1986) Resonant-microbridge vapor sensor. *IEEE Transactions on Electron Devices*, **33** (4), 499–506.
16 Fan, L.S., Tai, Y.C., and Muller, R.S. (1987) Pin joints, gears, springs, cranks, and other novel micromechanical structures, http://www.osti.gov/energycitations/product.biblio.jsp?osti_id=5974687, pp. 853–856.
17 Fan, L.S., Tai, Y.C., and Muller, R.S. (1989) IC-processed electrostatic micromotors. *Sensors and Actuators*, **20** (1/2), 41–47.
18 Besser, L. and Gilmore, R. (2003) Practical RF circuit design for modern wireless systems, in *Passive Circuits and Systems*, vol. 1, Artech House.
19 Haykin, S. (2001) *Communication Systems*, John Wiley & Sons, Inc., New York.

20. Martin, N. and Downing, B. (1986) Effect of varactor Q-factor on tuning sensitivity of microwave oscillators, including reverse tuning. *Electronics Letters*, **22** (6), 306–307, doi:10.1049/el:19860209.

21. Grajal, J., Krozer, V., Gonzalez, E., Maldonado, F., and Gismero, J. (2000) Modeling and design aspects of millimeter-wave and submillimeter-wave Schottky diode varactor frequency multipliers. *IEEE Transactions on Microwave Theory and Techniques*, **48** (4), 700–711, doi:10.1109/22.841962.

22. Peroulis, D. and Katehi, L. (2003) *Electrostatically-tunable analog RF MEMS varactors with measured capacitance range of 300%*. Microwave Symposium Digest, vol. 3, IEEE MTT-S International, pp. 1793–1796, doi:10.1109/MWSYM.2003.1210488.

23. Rebeiz, G. (2003) *RF MEMS: Theory, Design, and Technology*, John Wiley & Sons, Inc., Hoboken.

24. Fouladi, S., Bakri-Kassem, M., and Mansour, R. (2007) *An integrated tunable band-pass filter using MEMS parallel-plate variable capacitors implemented with 0.35 μm CMOS technology*. Microwave Symposium, IEEE/MTT-S International, pp. 505–508, doi:10.1109/MWSYM.2007.380518.

25. Frigo, S.P., Levine, Z.H., and Zaluzec, N.J. (2002) Submicron imaging of buried integrated circuit structures using scanning confocal electron microscopy. *Applied Physics Letters*, **81** (11), 2112–2114, doi:10.1063/1.1506010.

26. Yang, F. (2002) *Membrane Modeling of Pull-in Instability in MEMS Sensors and Actuators. Sensors*, Proceedings of IEEE, IEEE.

27. Elshurafa, A. and El-Masry, E. (2006) Finite-element modeling of low-stress suspension structures and applications in RF MEMS parallel-plate variable capacitors. *IEEE Transactions on Microwave Theory and Techniques*, **54** (5), 2211–2219, doi:10.1109/TMTT.2006.872787.

28. Lakshminarayanan, B. and Weller, T. (2003) *Tunable Bandpass Filter using Distributed MEMS Transmission Lines*. Microwave Symposium Digest, vol. 3, IEEE MTT-S International, IEEE, pp. 1789–1792, doi:10.1109/MWSYM.2003.1210487.

29. Sahu, A. and Sarkar, B. (2009) *A Novel Low Actuation Voltage RF MEMS Shunt Capacitive Switch*. Applied Electromagnetics Conference (AEMC), IEEE, pp. 1–3, doi:10.1109/AEMC.2009.5430612.

30. Novak, E., Der-Shen, W., Unruh, P., and Schurig, M. (2003) *MEMS Metrology using a Strobed Interferometric System*. Proceedings of the XVII IMEKO World Congress, pp. 178–182.

31. Park, J., Kim, G., Chung, K., and Bu, J. (2000) *Electroplated RF MEMS Capacitive Switches*. 13th Annual International Conference on Micro Electro Mechanical Systems, MEMS 2000, IEEE, pp. 639–644.

32. Goldsmith, C.L., Malczewski, A., Yao, Z.J., Chen, S., Ehmke, J., and Hinzel, D.H. (1999) RF MEMs variable capacitors for tunable filters. *International Journal of RF and Microwave Computer-Aided Engineering*, **9** (4), 362–374.

33. Liang, Y., Domier, C.W., and Luhmann Jr., N.C.L. (2008) RF MEMS extended tuning range varactor and varactor based true time delay line design. *PIERS Online*, **4** (4), 433–436, doi:10.2529/PIERS070730142758.

34. Mahameed, M.A.E.T.R. and Rebeiz, G.M. (2010) A zipper RF MEMS tunable capacitor with interdigitated RF and actuation electrodes. *Journal of Micromechanics and Microengineering*, **20** (3), 1–6.

35. Farinelli, P., Solazzi, F., Calaza, C., Margesin, B., and Sorrentino, R. (2008) *A Wide Tuning Range MEMS Varactor Based on a Toggle Push–Pull Mechanism*. 38th European Microwave Conference, EuMC 2008, EUMA, pp. 1501–1504, doi:10.1109/EUMC.2008.4751752.

36. Fang, D.M., Yuan, Q., Li, X.H., Zhang, H.X., Zhou, Y., and Zhao, X.L. (2010) *High Performance MEMS Spiral Inductors*. 5th IEEE International Conference on Nano/Micro Engineered and Molecular Systems (NEMS), IEEE, pp. 1033–1035, doi:10.1109/NEMS.2010.5592582.

37. Blondy, P., Palego, C., Houssini, M., Pothier, A., and Crunteanu, A. (2007) *RF-MEMS Reconfigurable Filters on*

Low Loss Substrates for Flexible Front Ends. Asia-Pacific Microwave Conference, APMC 2007, IEEE, pp. 1–3, doi:10.1109/APMC.2007.4554997.

38 van Beek, J., van Delden, M., van Dijken, A., van Eerd, P., van Grootel, M., Jansman, A., Kemmeren, A., Rijks, T., Steeneken, P., den Toonder, J., Ulenaers, M., den Dekker, A., Lok, P., Pulsford, N., van Straten, F., van Teeffelen, L., de Coster, J., and Puers, R. (2003) *High-Q Integrated RF Passives and Micromechanical Capacitors on Silicon*. Proceedings of the Bipolar/BiCMOS Circuits and Technology Meeting, IEEE, pp. 147–150, doi:10.1109/BIPOL.2003.1274955.

39 Mizuochi, Y., Amakawa, S., Ishihara, N., and Masu, K. (2009) *Study of Air-Suspended RF MEMS Inductor Configurations for Realizing Large Inductance Variations*. Argentine School of Micro-Nanoelectronics, Technology and Applications, EAMTA 2009, IEEE, pp. 50–55, doi:10.1109/EAMTA.2009.5288900.

40 Park, E.C., Yoon, J.B., Hong, S., and Yoon, E. (2002) *A 2.6 GHz Low Phase-Noise VCO Monolithically Integrated with High Q MEMS Inductors*. Proceedings of the 28th European Solid-State Circuits Conference, ESSCIRC 2002, pp. 147–150, doi:10.1109/ESSCIR.2002.186730.

41 Li, X., Gu, L., and Wu, Z. (2008) *(INVITED) High-performance RF Passives using Post-CMOS MEMS Techniques for RF SoC*. IEEE Radio Frequency Integrated Circuits Symposium, RFIC 2008, IEEE.

42 Pham, N.P., Sarro, P., Ng, K., and Burghartz, J. (2001) IC-compatible two-level bulk micromachining process module for RF silicon technology. *IEEE Transactions on Electron Devices*, **48** (8), 1756–1764, doi:10.1109/16.936704.

43 Choi, D.H., Lee, H.S., and Yoon, J.B. (2009) *Linearly Variable Inductor with RF MEMS Switches to Enlarge a Continuous Tuning Range*. Solid-State Sensors, International Actuators and Microsystems Conference, TRANSDUCERS 2009, IEEE, pp. 573–576, doi:10.1109/SENSOR.2009.5285389.

44 Zhou, S., Sun, X.Q., and Carr, W. (1997) *A Micro Variable Inductor Chip using MEMS Relays*. International Conference on Solid State Sensors and Actuators, TRANSDUCERS 1997 Chicago, IEEE, pp. 1137–1140, doi:10.1109/SENSOR.1997.635404.

45 Lubecke, V., Barber, B., Chan, E., Lopez, D., and Gammel, P. (2000) Self-assembling MEMS variable and fixed RF inductors. Asia-Pacific Microwave Conference, Sydney, Australia.

46 Okada, K., Sugawara, H., Ito, H., Itoi, K., Sato, M., Abe, H., Ito, T., and Masu, K. (2006) On-chip high-Q variable inductor using wafer-level chip-scale package technology. *IEEE Transactions on Electron Devices*, **53** (9), 2401–2406, doi:10.1109/TED.2006.880815.

47 Gmati, I., Boussetta, H., Kallala, M., and Besbes, K. (2008) *Wide-Range RF MEMS Variable Inductor using Micro Pump Actuator*, 2nd International Conference on Signals, Circuits and Systems, SCS 2008, IEEE, pp. 1–4, doi:10.1109/ICSCS.2008.4746937.

48 Caekenberghe, K.V. and Sarabandi, K. (2008) A self-aligned fabrication process for capacitive fixed-fixed beam RF MEMS components. *Journal of Microelectromechanical Systems*, **17** (3), 747–754, doi:10.1109/JMEMS.2008.924259.

49 Thakur, S., SumithraDevi, K., and Ranjitha, I. (2009) *Performance of Low Loss RF MEMS Fixed–Fixed Capacitive Switch Characterization*. Applied Electromagnetics Conference (AEMC), IEEE, pp. 1–4, doi:10.1109/AEMC.2009.5430610.

50 Iannacci, J., Giacomozzi, F., Colpo, S., Margesin, B., and Bartek, M. (2009) *A General Purpose Reconfigurable MEMS-Based Attenuator for Radio Frequency and Microwave Applications*. IEEE EUROCON 2009, IEEE, pp. 1197–1205, doi:10.1109/EURCON.2009.5167788.

51 Shen, H., Gong, S., and Barker, N. (2008) *DC-contact RF MEMS switches using thin-film cantilevers*. European Microwave Integrated Circuit Conference, EuMIC 2008, EUMA, pp. 382–385, doi:10.1109/EMICC.2008.4772309.

52 Shalaby, M., Wang, Z., Chow, L.W., Jensen, B., Volakis, J., Kurabayashi, K., and Saitou, K. (2009) Robust design of RF-MEMS cantilever switches using contact physics model-

ing. *IEEE Transactions on Industrial Electronics*, **56** (4), 1012–1021, doi:10.1109/TIE.2008.2006832.

53 Patel, C. and Rebeiz, G. (2010) *An RF-MEMS switch with mN contact forces*. International Microwave Symposium Digest (MTT), 2010 IEEE MTT-S, pp. 1242–1245, doi:10.1109/MWSYM.2010.5517237.

54 Mahameed, R. and Rebeiz, G. (2010) A high-power temperature-stable electrostatic RF MEMS capacitive switch based on a thermal buckle-beam design. *Journal of Microelectromechanical Systems*, **19** (4), 816–826, doi:10.1109/JMEMS.2010.2049475.

55 Martinez, J., Blondy, A., Pothier, A., Bouyge, D., Crunteanu, A., and Chatras, M. (2007) *Surface and Bulk Micromachined RF MEMS Capacitive Series Switch for Watt-Range Hot Switching Operation*. European Microwave Conference, EUMA, pp. 1237–1240, doi:10.1109/EUMC.2007.4405424.

56 Mollah, M. and Karmakar, N. (2001) RF-MEMS switches: paradigms of microwave switching. Asia-Pacific Microwave Conference, APMC 2001, Taipei, Taiwan.

57 Lahiri, S., Saha, H., and Kundu, A. (2009) *RF MEMS Switch: An Overview At-a-glance*. 4th International Conference on Computers and Devices for Communication, CODEC 2009, pp. 1–5.

58 Hwang, J. (2007) *Reliability of Electrostatically Actuated RF MEMS Switches*. IEEE International Workshop on Radio-Frequency Integration Technology, RFIT 007, IEEE, pp. 168–171, doi:10.1109/RFIT.2007.4443942.

59 Pozar, D.M. (2004) *Microwave Engineering*, Wiley Academic.

60 Reines, I., Pillans, B., and Rebeiz, G. (2011) Thin-film aluminum RF MEMS switched capacitors with stress tolerance and temperature stability. *Journal of Microelectromechanical Systems*, **20** (1), 193–203, doi:10.1109/JMEMS.2010.2090505.

61 Yamazaki, H., Ikehashi, T., Saito, T., Ogawa, E., Masunaga, T., Ohguro, T., Sugizaki, Y., and Shibata, H. (2010) *A High Power-Handling RF MEMS Tunable Capacitor using Quadruple Series Capacitor Structure*. International Microwave Symposium Digest (MTT), IEEE MTT-S, IEEE, pp. 1138–1141, doi:10.1109/MWSYM.2010.5518185.

62 Lee, Y.S., Jang, Y.H., Kim, J.M., and Kim, Y.K. (2010) *A 50–110 GHz Ohmic Contact RF MEMS Silicon Switch with High Isolation*. 23rd International Conference on Micro Electro Mechanical Systems (MEMS), 2010 IEEE, IEEE, pp. 759–762, doi:10.1109/MEMSYS.2010.5442295.

63 Paredes, F., Gonzalez, G., Bonache, J., and Martin, F. (2010) Dual-band impedance-matching networks based on split-ring resonators for applications in RF identification (RFID). *IEEE Transactions on Microwave Theory and Techniques*, **58** (5), 1159–1166, doi:10.1109/TMTT.2010.2045449.

64 Iannacci, J., Masotti, D., Kuenzig, T., and Niessner, M. (2011) *A Reconfigurable Impedance Matching Network Entirely Manufactured in RF-MEMS Technology*. Proceeding of the SPIE Microtechnologies Conference, vol. 8066, SPIE, p. 12, doi:10.1117/12.886186.

65 Domingue, F., Fouladi, S., and Mansour, R.R. (2010) *A Reconfigurable Impedance Matching Network using Dual-Beam MEMS Switches for an Extended Operating Frequency Range*. International Microwave Symposium Digest (MTT), 2010 IEEE MTT-S, IEEE, p. 1, doi:10.1109/MWSYM.2010.5515472.

66 Unlu, M., Topalli, K., Atasoy, H., Temocin, E., Istanbulluoglu, I., Bayraktar, O., Demir, S., Civi, O., Koc, S., and Akin, T. (2006) *A Reconfigurable RF MEMS Triple Stub Impedance Matching Network*. 36th European Microwave Conference, EUMA, pp. 1370–1373, doi:10.1109/EUMC.2006.281272.

67 Fouladi, S. and Mansour, R.R. (2009) *Reconfigurable Amplifier with Tunable Impedance Matching Networks Based on CMOS-MEMS Capacitors in 0.18-µm CMOS Technology*. 2nd Microsystems and Nanoelectronics Research Conference, MNRC 2009, IEEE, pp. 33–36.

68 Ocera, A., Gatti, R., Mezzanotte, P., Farinelli, P., and Sorrentino, R. (2005) *A MEMS Programmable Power Divider/Combiner for Reconfigurable An-*

tenna Systems. European Microwave Conference, vol. 1, EUMA, p. 4, doi:10.1109/EUMC.2005.1608933.

69. Ocera, A., Farinelli, P., Cherubini, F., Mezzanotte, P., Sorrentino, R., Margesin, B., and Giacomozzi, F. (2007) *A MEMS-reconfigurable power divider on high resistivity silicon substrate*. International Microwave Symposium, IEEE/MTT-S, pp. 501–504, doi:10.1109/MWSYM.2007.380517.

70. Caekenberghe, K.V. (2009) RF MEMS on the radar. *IEEE Microwave Magazine*, **10** (6), 99–116, doi:10.1109/MMM.2009.933596.

71. Iannacci, J., Faes, A., Mastri, F., Masotti, D., and Rizzoli, V. (2010) *A MEMS-Based Wide-Band Multi-State Power Attenuator for Radio Frequency and Microwave Applications*. Proceedings of TechConnect World, NSTI Nanotech 2010, vol. 2, NSTI, pp. 328–331.

72. Ko, Y., Park, J., and Bu, J. (2003) *Integrated 3-bit RF MEMS Phase Shifter with Constant Phase Shift for Active Phased Array Antennas in Satellite Broadcasting Systems*. 12th International Conference on TRANSDUCERS, Solid-State Sensors, Actuators and Microsystems, vol. 2, IEEE, pp. 1788–1791, doi:10.1109/SENSOR.2003.1217133.

73. Pillans, B., Eshelman, S., Malczewski, A., Ehmke, J., and Goldsmith, C. (1999) Ka-band RF MEMS phase shifters. *IEEE Microwave and Guided Wave Letters*, **9** (12), 520–522, doi:10.1109/75.819418.

74. Buck, T. and Kasper, E. (2010) *RF MEMS Phase Shifters for 24 and 77 GHz on High Resistivity Silicon*. Topical Meeting on Silicon Monolithic Integrated Circuits in RF Systems (SiRF), IEEE, pp. 224–227, doi:10.1109/SMIC.2010.5422951.

75. Puyal, V., Dubuc, D., Grenier, K., Bordas, C., Vendier, O., and Cazaux, J.L. (2007) Which architecture to choose for robust RF-MEMS phase shifters? International Semiconductor Conference, CAS 2007, Sinaia, Romania.

76. Kim, M., Hacker, J., Mihailovich, R., and DeNatale, J. (2001) A DC-to-40 GHz four-bit RF MEMS true-time delay network. *IEEE Microwave and Wireless Components Letters*, **11** (2), 56–58, doi:10.1109/7260.914301.

77. Hacker, J., Mihailovich, R., Kim, M., and DeNatale, J. (2003) A Ka-band 3-bit RF MEMS true-time-delay network. *IEEE Transactions on Microwave Theory and Techniques*, **51** (1), 305–308, doi:10.1109/TMTT.2002.806508.

78. Farinelli, P., Giacomozzi, F., Mannocchi, G., Marcelli, R., Margesin, B., Mezzanotte, P., Nardo, S.D., Russer, P., Sorrentino, R., Vitulli, F., and Vietzorreck, L. (2004) *RF-MEMS SPDT Switch on Silicon Substrate for Space Applications*. Topical Meeting on Silicon Monolithic Integrated Circuits in RF Systems, Digest of Papers, pp. 151–154, doi:10.1109/SMIC.2004.1398190.

79. Yamane, D., Yamashita, K., Seita, H., Fujita, H., Toshiyoshi, H., and Kawasaki, S. (2009) *A Dual-SPDT RF-MEMS Switch on a Small-Sized LTCC Phase Shifter for Ku-Band Operation*. Asia Pacific Microwave Conference, APMC 2009, IEEE, pp. 555–558, doi:10.1109/APMC.2009.5384169.

80. Casini, F., Farinelli, P., Mannocchi, G., DiNardo, S., Margesin, B., Angelis, G.D., Marcelli, R., Vendier, O., and Vietzorreck, L. (2010) *High Performance RF-MEMS SP4T Switches in CPW Technology for Space Applications*, European Microwave Conference (EuMC), EUMA, pp. 89–92.

81. Stehle, A., Georgiev, G., Ziegler, V., Schoenlinner, B., Prechtel, U., Schmid, U., and Seidel, H. (2009) *Broadband Single-Pole Multithrow RF-MEMS Switches for Ka-Band*. German Microwave Conference, pp. 1–4, doi:10.1109/GEMIC.2009.4815911.

82. Yamane, D., Seita, H., Sun, W., Kawasaki, S., Fujita, H., and Toshiyoshi, H. (2009) *A 12-GHz DPDT RF-MEMS Switch with Layer-Wise Waveguide/Actuator Design Technique*. IEEE 22nd International Conference on Micro Electro Mechanical Systems, MEMS 2009, IEEE, pp. 888–891, doi:10.1109/MEMSYS.2009.4805526.

83. McErlean, E., Hong, J.S., Tan, S., Wang, L., Cui, Z., Greed, R., and Voyce, D.

(2005) *2 × 2 RF MEMS Switch Matrix*. IEE Proceedings Microwaves, Antennas and Propagation, pp. 449–454, doi:10.1049/ip-map:20050104.

84. Fomani, A. and Mansour, R. (2009) *Miniature RF MEMS Switch Matrices*. International Microwave Symposium Digest, IEEE MTT-S, IEEE, pp. 1221–1224, doi:10.1109/MWSYM.2009.5165923.

85. Chan, K.Y., Daneshmand, M., Mansour, R., and Ramer, R. (2008) *Monolithic Crossbar MEMS Switch Matrix*. International Microwave Symposium Digest, IEEE MTT-S, IEEE, pp. 129–132, doi:10.1109/MWSYM.2008.4633120.

86. Hughes, T.J.R. (2000) *The Finite Element Method: Linear Static and Dynamic Finite Element Analysis*, Dover Publications.

87. Ricart, J., Pons, J., Gorreta, S., and Dominguez, M. (2009) *Digital Resonance for MEMS*. Spanish Conference on Electron Devices, CDE 2009, IEEE, pp. 312–315, doi:10.1109/SCED.2009.4800494.

88. Malik, A., Shoaib, M., Naseem, S., and Riaz, S. (2008) *Modeling and Designing of RF MEMS Switch Using ANSYS*. 4th International Conference on Emerging Technologies, ICET 2008, IEEE, pp. 44–49, doi:10.1109/ICET.2008.4777472.

89. Peyrou, D., Pons, P., Granier, H., Leray, D., Ferrand, A., Yacine, K., Saadaoui, M., Nicolas, A., Tao, J., and Plana, R. (2006) *Multiphysics Softwares Benchmark on Ansys/Comsol Applied for RF MEMS Switches Packaging Simulations*. 7th International Conference on Thermal, Mechanical and Multiphysics Simulation and Experiments in Micro-Electronics and Micro-Systems, EuroSime 2006, IEEE, pp. 1–8, doi:10.1109/ESIME.2006.1644011.

90. Zhao, X., Wang, L., Tan, Y., Sun, G., and Lu, G. (2007) *A Novel Design Methodology for MEMS Device*. 7th IEEE Conference on Nanotechnology, IEEE-NANO 2007, IEEE, pp. 39–44, doi:10.1109/NANO.2007.4601136.

91. McCorquodale, M., Gebara, F., Kraver, K., Marsman, E., Senger, R., and Brown, R. (2003) A top-down microsystems design methodology and associated challenges. Design, Automation and Test in Europe Conference and Exhibition, Munich, Germany.

92. Senturia, S.D. (2000) *Microsystem Design*, Springer Publications.

93. Tin, L.D., Iannacci, J., Gaddi, R., Gnudi, A., Rudnyi, E., Greiner, A., and Korvink, J. (2007) *Non Linear Compact Modeling of RF-MEMS Switches by Means of Model Order Reduction*. International Solid-State Sensors, Actuators and Microsystems Conference, TRANSDUCERS 2007, IEEE, pp. 635–638, doi:10.1109/SENSOR.2007.4300210.

94. Rudnyi, E.B., Lienemann, J., Greiner, A., and Korvink, J.G. (2004) *mor4ansys: Generating Compact Models Directly from ANSYS Models*. Technical Proceedings of the 2004 Nanotechnology Conference and Trade Show, Nanotech 2004, vol. 2, NSTI, pp. 279–282.

95. Jungwirth, M., Hofinger, D., and Weinzierl, H. (2010) *A Comparison of Model Order Reduction Methods used in Different FE Software Tools*. 11th International Conference on Thermal, Mechanical Multi-Physics Simulation, and Experiments in Microelectronics and Microsystems (EuroSimE), IEEE, pp. 1–5, doi:10.1109/ESIME.2010.5464612.

96. Fouladi, S., Hajhosseini, A., Mesgar, H., and Kahrizi, M. (2005) *Behavioral Modeling of Microelectromechanical Filters*. International Conference Physics and Control, Proceedings 2005, IEEE, pp. 419–424, doi:10.1109/PHYCON.2005.1514019.

97. Fedder, G. and Jing, Q. (1999) A hierarchical circuit-level design methodology for microelectromechanical systems. *IEEE Transactions on Circuits and Systems II: Analog and Digital Signal Processing*, **46** (10), 1309–1315, doi:10.1109/82.799682.

98. Zhou, L., Herve, Y., Chapuis, Y.A., and Fujita, H. (2006) *VHDL-AMS Modeling for Simulation of MEMS Array-Based Smart Surface Applied in Microfluid Air-Flow Environment*. 5th IEEE Conference on Sensors, IEEE, pp. 807–810, doi:10.1109/ICSENS.2007.355590.

99. Mukherjee, T. and Fedder, G.K. (1999) Hierarchical mixed-domain circuit

simulation, synthesis and extraction methodology for MEMS. *The Journal of VLSI Signal Processing*, **21** (3), 233–249, doi:10.1023/A:1008122921631.

100 Cadence Design System Inc. (2000) Affirma spectre circuit simulator user guide. Product version 4.4.6.

101 Cadence Design System Inc. (1999) Affirma Verilog-A language reference. Product version 4.4.5.

102 Gaddi, R. (2003) *The Hierarchical HDL-Based Design of an Integrated MEMS-CMOS Oscillator*. Symposium on Design, Test, Integration and Packaging of MEMS/MOEMS 2003, pp. 176–180, doi:10.1109/DTIP.2003.1287031.

103 Coventor Inc. (2011) ARCHITECT Data Sheet, www.coventor.com/pdfs/ARCHITECT.pdf, accessed 27 April 2013.

104 IntelliSense Inc. (2011) Application Note. Tang Resonator, http://intellisense.com/upload/201212160211242420.pdf, accessed 27 April 2013.

105 Imhoff, A. (1999) Packaging technologies for RFICs: current status and future trends. IEEE Radio Frequency Integrated Circuits (RFIC) Symposium.

106 Gilleo, K. (2005) *MEMS/MOEM Packaging*, McGraw-Hill Professional.

107 Varadan, V.K., Vinoy, K.J., and Jose, K.A. (2002) *RF MEMS and their Applications*, Wiley Publisher.

108 Park, Y.K., Park, H.W., Lee, D.J., Park, J.H., Song, I.S., Kim, C.W., Song, C.M., Lee, Y.H., Kim, C.J., and Ju, B.K. (2002) A novel low-loss wafer-level packaging of the RF-MEMS devices, 15th IEEE International Conference on Micro Electro Mechanical Systems, Las Vegas, Nevada, USA.

109 Tilmans, H., Ziad, H., Jansen, H., Monaco, O.D., Jourdain, A., Raedt, W.D., Rottenberg, X., Backer, E.D., Decaussernaeker, A., and Baert, K. (2001) Wafer-level packaged RF-MEMS switches fabricated in a CMOS fab. International Electron Devices Meeting, IEDM Technical Digest, 2001 International Electron Devices Meeting, Washington, DC, December 2–5, 2001.

110 Margomenos, A., Peroulis, D., Herrick, K.J., and Katehi, L.P.B. (2001) *Silicon Micromachined Packages for RF MEMS Switches*. 31st European Microwave Conference, EUMA, pp. 1–4, doi:10.1109/EUMA.2001.339130.

111 Niklaus, F., Enoksson, P., Griss, P., Kalvesten, E., and Stemme, G. (2001) Low-temperature wafer-level transfer bonding. *Journal of Microelectromechanical Systems*, **10** (4), 525–531, doi:10.1109/84.967375.

112 Jaeger, R. and Blalock, T. (2010) *Microelectronic Circuit Design*, McGraw-Hill Science/Engineering/Math.

113 Lau, J.H., Lee, C.K., Premachandran, C.S., and Aibin, Y. (2010) *Advanced MEMS Packaging*, McGraw-Hill Professional.

114 Zimin, Y. and Ueda, T. (2010) *Low-temperature anodic bonding of silicon and crystal quartz wafers for MEMS application*. IEEE Sensors, IEEE, pp. 269–272, doi:10.1109/ICSENS.2010.5689857.

115 Na, K., Kim, I.H., Lee, E., Kim, H.C., Lee, Y.H., and Chun, K. (2006) *Wafer Level Package using Polymer Bonding of Thick SU-8 Photoresist*. The 2006 International Conference on MEMS, NANO and Smart Systems, IEEE, pp. 31–34, doi:10.1109/ICMENS.2006.348211.

116 Zine-El-Abidine, I. and Okoniewski, M. (2009) A low-temperature SU-8 based wafer-level hermetic packaging for MEMS devices. *IEEE Transactions on Advanced Packaging*, **32** (2), 448–452, doi:10.1109/TADVP.2008.2006757.

117 Wilkerson, P., Kranz, M., Przekwas, A., and Hudson, T. (2001) Flip-chip hermetic packaging of RF MEMS. Microelectromechanical Systems Conference, 24–26 August 2001, Berkeley, California, USA.

118 Shimooka, Y., Inoue, M., Endo, M., Obata, S., Kojima, A., Miyagi, T., Sugizaki, Y., Mori, I., and Shibata, H. (2008) *Robust Hermetic Wafer Level Thin-Film Encapsulation Technology for Stacked MEMS/IC Package*. 58th Electronic Components and Technology Conference, ECTC 2008, IEEE, pp. 824–828, doi:10.1109/ECTC.2008.4550071.

119 Barriere, F., Crunteanu, A., Bessaudou, A., Pothier, A., Cosset, F., Mardivirin, D., and Blondy, P. (2010) Zero Lev-

el Metal Thin Film Package for RF MEMS. Topical Meeting on Silicon Monolithic Integrated Circuits in RF Systems (SiRF), IEEE, pp. 148–151, doi:10.1109/SMIC.2010.5422957.
120 Kear, F.W. (1992) *Hybrid Assemblies and Multichip Modules*, CRC Press.
121 Tian, J., Iannacci, J., Sosin, S., Gaddi, R., and Bartek, M. (2006) *RF-MEMS Wafer-Level Packaging using Through-Wafer via Technology*. 8th Electronics Packaging Technology Conference, EPTC '06, IEEE, pp. 441–447, doi:10.1109/EPTC.2006.342755.
122 Nguyen, C.T.C. (1994) *Micromechanical Signal Processors*, Ph.D. Thesis dissertation, University of California, Berkeley.
123 Jin, Y., Wang, Z., Zhao, L., Lim, P.C., Wei, J., and Wong, C.K. (2004) Zr/V/Fe thick film for vacuum packaging of MEMS. *Journal of Micromechanics and Microengineering*, **14** (5), 687–692, doi:10.1088/0960-1317/14/5/005.
124 Chen, S., Sun, D., and Lin, L. (2007) *Nanogetters for MEMS Hermetic Packaging*. 7th IEEE Conference on Nanotechnology, IEEE-NANO 2007, IEEE, pp. 921–924, doi:10.1109/NANO.2007.4601334.
125 Benvenuti, C. (1998) *Non-Evaporable Getters: from Pumping Strips to thin Film Coatings*. Proceedings of EPAC 1998, pp. 200–204.
126 Sparks, D., Massoud-Ansari, S., and Najafi, N. (2003) *Reliable Vacuum Packaging using NanoGetters and Glass Frit Bonding*. Proceedings of Reliability, Testing, and Characterization of MEMS/MOEMS III, SPIE, pp. 70–78, doi:10.1117/12.530414.
127 Longoni, G., Conte, A., Moraja, M., and Fourrier, A. (2006) *Stable and Reliable Q-Factor in Resonant MEMS with Getter Film*. Proceedings 44th Annual IEEE International Reliability Physics Symposium, IEEE, pp. 416–420, doi:10.1109/RELPHY.2006.251254.
128 Nokia Networks Oy. (2002) GSM Architecture, Training Document, available at: www.roggeweck.net/uploads/media/ Student_-_GSM_Architecture.pdf, accessed 27 April 2013.
129 Rizvi, S.A.P. (2011) *Reconfigurable RF Front-End for Multi-standard Communication*, LAP Lambert Academic Publishing AG & Co KG.
130 J. Laskar, B.M. and Sudipto, C. (2004) *Modern Receiver Front-Ends: Systems, Circuits, and Integration*, 1st edn, Wiley-Interscience.
131 Nguyen, C.C. (2001) *Transceiver front-end architectures using vibrating micromechanical signal processors*. Topical Meeting on Silicon Monolithic Integrated Circuits in RF Systems, Digest of Papers 2001, IEEE, pp. 23–32, doi:10.1109/SMIC.2001.942335.
132 Rangra, K., Marcelli, R., Soncini, G., Giacomozzi, F., Margesin, B., Lorenzelli, L., Mulloni, M., and Collini, C. (2004) *Micromachined Low Actuation Voltage RF MEMS Capacitive Switches, Technology and Characterization*. International Semiconductor Conference, CAS 2004 Proceedings, vol. 1, IEEE, pp. 165–168, doi:10.1109/SMICND.2004.1402830.
133 Ocera, A., Farinelli, P., Cherubini, F., Mezzanotte, P., Sorrentino, R., Margesin, B., and Giacomozzi, F. (2007) *A MEMS-Reconfigurable Power Divider on High Resistivity Silicon Substrate*. International Microwave Symposium, IEEE/MTT-S, IEEE, pp. 501–504, doi:10.1109/MWSYM.2007.380517.
134 Kolesar, E., Htun, T., Least, B., and Tippey, J. (2008) *Design and Performance Comparison of Single- and Double-Hot Arm Polysilicon Surface Micromachined Electrothermal Actuators and Arrays Applied to Realize a Microengine*. 8th IEEE Conference on Nanotechnology, NANO '08, IEEE, pp. 444–447, doi:10.1109/NANO.2008.136.
135 SoftMEMS (2007) MEMS Pro v6.0 Data Sheet, SoftMEMS. www.softmems.com/ MPv60_DS_letter2007.pdf, accessed 27 April 2013.
136 Gaddi, R., Bellei, M., Gnudi, A., Margesin, B., and Giacomozzi, F. (2004) *Interdigitated Low-Loss Ohmic RF-MEMS Switches*. Technical Proceedings of the 2004 NSTI Nanotechnology Conference and Trade Show, vol. 2, NSTI, pp. 327–330.
137 Muldavin, J., Bozler, C., Rabe, S., and Keast, C. (2004) *Large Tuning Range Analog and Multi-Bit*

MEMS Varactors. International Microwave Symposium Digest, IEEE MTT-S, vol. 3, IEEE, pp. 1919–1922, doi:10.1109/MWSYM.2004.1338984.

138 Iannacci, J., Gaddi, R., and Gnudi, A. (2007) *Non-Linear Electromechanical RF Model of a MEMS Varactor Based on VerilogA© and Lumped-Element Parasitic Network.* European Microwave Conference, EUMA, pp. 1342–1345, doi:10.1109/EUMC.2007.4405451.

139 Iannacci, J., Gaddi, R., and Gnudi, A. (2010) Experimental validation of mixed electromechanical and electromagnetic modeling of RF-MEMS devices within a standard IC simulation environment. *Journal of Microelectromechanical Systems,* **19** (3), 526–537, doi:10.1109/JMEMS.2010.2048417.

140 Hartzell, A.L., da Silva, M.G., and Shea, H.R. (2010) *MEMS Reliability (MEMS Reference Shelf),* Springer.

141 Papaioannou, G. and Papapolymerou, J. (2007) *Dielectric Charging Mechanisms in RF-MEMS Capacitive Switches.* European Microwave Conference, EUMA, pp. 1157–1160, doi:10.1109/EUMC.2007.4405404.

142 Kim, B. and Kasim, R. (2007) *Advanced MEMS Development for High Power Sensor Application.* IEEE Sensors, IEEE, pp. 811–814, doi:10.1109/ICSENS.2007.4388524.

143 Melle, S., Bordas, C., Dubuc, D., Grenier, K., Vendier, O., Muraro, J., Cazaux, J., and Plana, R. (2007) *Investigation of Stiction Effect in Electrostatic Actuated RF MEMS Devices.* Topical Meeting on Silicon Monolithic Integrated Circuits in RF Systems, IEEE, pp. 173–176, doi:10.1109/SMIC.2007.322787.

144 Repchankova, A. and Iannacci, J. (2009) *Heat-Based Recovery Mechanism to Counteract Stiction of RF-MEMS Switches.* Symposium on Design, Test, Integration Packaging of MEMS/MOEMS, MEMS/MOEMS '09, EDA Publishing, pp. 176–181.

145 Iannacci, J., Repchankova, A., Faes, A., Tazzoli, A., Meneghesso, G., and Niessner, M. (2010) Experimental investigation on the exploitation of an active mechanism to restore the operability of malfunctioning RF-MEMS switches. *Procedia Engineering,* **5**, 734–737, doi:10.1016/j.proeng.2010.09.213.

146 Iannacci, J., Repchankova, A., Faes, A., Tazzoli, A., Meneghesso, G., and Betta, G.F.D. (2010) Enhancement of RF-MEMS switch reliability through an active anti-stiction heat-based mechanism. *Microelectronics Reliability,* **50** (9/11), 1599–1603, doi:10.1016/j.microrel.2010.07.108.

147 Larcher, L., Brama, R., Ganzerli, M., Iannacci, J., Margesin, B., Bedani, M., and Gnudi, A. (2009) *A MEMS Reconfigurable Quad-Band Class-E Power Amplifier for GSM Standard.* IEEE 22nd International Conference on Micro Electro Mechanical Systems, MEMS 2009, pp. 864–867, doi:10.1109/MEMSYS.2009.4805520.

148 Larcher, L., Brama, R., Ganzerli, M., Iannacci, J., Bedani, M., and Gnudi, A. (2009) *A MEMS Reconfigurable Quad-Band Class-E Power Amplifier for GSM Standard.* Design, Automation Test in Europe Conference Exhibition, DATE '09, EDAA, pp. 364–368.

149 Iannacci, J., Gnudi, A., Margesin, B., Giacomozzi, F., Larcher, L., and Brama, R. (2009) *Reconfigurable RF MEMS Based Impedance Matching Network for a CMOS Power Amplifier, in New Developments in Micro Electro Mechanical Systems for Radio Frequency and Millimeter Wave Applications,* series in Micro and Nanoengineering, Editura Academiei Romane, Bucharest.

150 Tanner EDA (2011) Tanner EDA Software Tools – L-Edit Layout for Analog IC & Mixed Signal Design, available at: www.tannereda.com/l-edit-pro, accessed 27 April 2013.

151 Calma Company (1987) GDSII™ Stream Format Manual, available at: http://bitsavers.informatik.uni-stuttgart.de/pdf/calma/GDS_II_Stream_Format_Manual_6.0_Feb87.pdf, accessed 27 April 2013.

152 Bourke, P. (2011) SAT Save File Format, available at: http://paulbourke.net/dataformats/sat/sat.pdf, accessed 27 April 2013.

153 ANSYS Inc. (2011) ANSYS, available at: www.ansys.com/, accessed 27 April 2013.

154 Yin, J., Xie, Y., and Chen, P. (2009) *Modal Analysis Comparison of Beam and Shell Models for Composite Blades*. Asia-Pacific Power and Energy Engineering Conference, APPEEC 2009, IEEE, pp. 1–4, doi:10.1109/APPEEC.2009.4918484.

155 Masresha, M. (2007) Bending-torsion decoupling in eigenmode of subsonic and supersonic composite monocoque aircraft wing, available at: http://etd.aau.edu.et/dspace/bitstream/123456789/1540/1/Muluken%20Masresha.pdf, accessed 27 April 2013.
A thesis submitted to the School of Graduate Studies of Addis Ababa University in partial fulfillment of the requirements for the Degree of Masters of Science in Mechanical Engineering (Applied Mechanics Stream).

156 Tazzoli, A. and Meneghesso, G. (2011) Acceleration of microwelding on ohmic RF-MEMS switches. *Journal of Microelectromechanical Systems*, **20** (3), 552–554, doi:10.1109/JMEMS.2011.2140360.

157 ANSYS Inc. (2013) ANSYS HFSS, available at: www.ansys.com/Products/Simulation+Technology/Electromagnetics/High-performance+Electronic+Design/ANSYS+HFSS, accessed 27 April 2013.

158 Kaleeba, P., Tennant, A., and Ide, J. (2003) *Modelling a Planar Phase Switched Structure (PSS) in Ansoft HFSS (High Frequency Structure Simulator)*. Twelfth International Conference on Antennas and Propagation, ICAP 2003, Conf. Publ. No. 491, vol. 1, IEE, Institute of Electrical Engineering, pp. 257–261.

159 Clarke, R., Quraishi, J., and Ridler, N. (2007) *A Bilateral Comparison of On-Wafer S-Parameter Measurements at Millimeter Wavelengths*. 69th ARFTG Conference, IEEE, pp. 1–7.

160 Rochester Institute of Technology (2009) Project 1: Rectangular Waveguide (HFSS), available at: www.rit.edu/kgcoe/eta/docs/Project-1-HFSS-tutorial-Rectangular%20WG.pdf, accessed 27 April 2013.

161 Hailongk, W., Guangbao, S., and Youbao, L. (2006) *Lumped Behavioral Modeling for Suspended MEMS*. Proceedings of the 8th International Conference on Solid-State and Integrated Circuit Technology, IEEE, pp. 679–681.

162 Sciavicco, L. and Siciliano, B. (2000) *Modelling and Control of Robot Manipulators*, Technology & Industrial Arts, Springer.

163 Iannacci, J. and Gaddi, R. (2010) *Mixed-Domain Simulation and Wafer-Level Packaging of RF-MEMS Devices*, Lambert Academic Publishing.

164 Iannacci, J., Gaddi, R., and Gnudi, A. (2004) *Nodal Modelling of Uneven Electrostatic Transduction in MEMS*. Proceedings of the 9th Italian Conference Sensors and Microsystems, AISEM 2004, World Scientific, pp. 385–389.

165 Griffin, W., Richardson, H., and Yamanami, S. (1966) *A Study of Fluid Squeeze-Film Damping*. ASME Transaction of Basic Engineering, ASME, pp. 451–456.

166 Bao, M., Yang, H., and Sun, Y. (2003) A modified reynolds equation for squeeze-film air damping of hole plate. *IOP Journal of Micromechanics and Microengineering*, **13** (6), 795–800.

167 Cho, Y.H., Pisano, A., and Howe, R. (1994) Viscous damping model for laterally oscillating microstructures. *IEEE/ASME Journal of Microelectromechanical Systems*, **3** (2), 81–87.

168 Veijola, T., Kuisma, H., and Lahdenpera, J. (1997) Model for gas film damping in a silicon accelerometer. *Proceedings of the International Conference on Solid-State Sensors Actuators*, **2**, 1097–1100.

169 Przemieniecki, J.S. (1968) *Theory of Matrix Structural Analysis*, McGraw-Hill.

170 Jing, Q. (2003) Modeling and simulation for design of suspended MEMS. Ph.D. Dissertation in Electrical and Computer Engineering, Carnegie Mellon University Pittsburgh.

171 Schauwecker, B., Strohm, K.M., Simon, W., Mehner, J., and Luy, J.F. (2002) Toggle-Switch – A new type of RF MEMS switch for power applications. *Technical Digest of IEEE MTT-S International Microwave Symposium*, **1**, 219–222.

172 Rangra, K.J., Margesin, B., Giacomozzi, F., Lorenzelli, L., and Collini, C. (2004) *Toggle Switch – A New Type of RF MEMS Switch for Telecommunication Applications*. Proceedings of the 18th European conference on Solid-State Transducers Eurosensors XVIII 2004, Elsevier, pp. 505–514.

173 Farinelli, P., Solazzi, F., Calaza, C., Margesin, B., and Sorrentino, R. (2008) A Wide Tuning Range MEMS Varactor Based on a Toggle Push-Pull Mechanism. European Microwave Integrated Circuit Conference, EuMIC 2008, EUMA, pp. 474–477.

174 Gaddi, R., Gnudi, A., Franchi, E., Guermandi, D., Tortori, P., Margesin, B., and Giacomozzi, F. (2005) *Reconfigurable MEMS-enabled LC-tank for Multi-Band CMOS Oscillator*. Proceedings of the IEEE MTT-S 2005 International Microwave Symposium, IMS2005, IEEE, pp. 1353–1356.

175 Tiebout, M. (2005) *Low Power Vco Design in CMOS*, Springer Technology & Industrial Arts.

176 AMS (2005) 0.35 µm HBT BiCMOS RF SPICE Models. Austriamicrosystems (AMS), Release date 3 November 2005, www.austriamicrosystems.com, accessed 1 March 2007.

177 Gaddi, R., Bellei, M., Gnudi, A., Margesin, B., and Giacomozzi, F. (2004) *Interdigitated Low-Loss Ohmic RF-MEMS Switches*. Technical Proceedings of the 2004 NSTI Nanotechnology Conference and Trade Show, vol. 2, NSTI, pp. 327–330.

178 Kristiansen, H. and Liu, J. (1997) *Overview of Conductive Adhesive Interconnection Technologies for LCD's*. 1st IEEE International Symposium on Polymeric Electronics Packaging, IEEE, pp. 223–232, doi:10.1109/PEP.1997.656494.

179 Liu, J., Lai, Z., Kristiansen, H., and Khoo, C. (1998) *Overview of Conductive Adhesive Joining Technology in Electronics Packaging Applications*. Proceedings of 3rd International Conference on Adhesive Joining and Coating Technology in Electronics Manufacturing, IEEE, pp. 1–18, doi:10.1109/ADHES.1998.741996.

180 Laermer, F. and Urban, A. (2005) *Milestones in Deep Reactive Ion Etching*. 13th International Conference on Solid-State Sensors, Actuators and Microsystems, Digest of Technical Papers. TRANSDUCERS '05, vol. 2, IEEE, pp. 1118–1121, doi:10.1109/SENSOR.2005.1497272.

181 Allan, G. and Sutherland, K. (1962) A preliminary study of the lapping process. *Production Engineer*, **41** (4), 195–202, doi:10.1049/tpe:19620027.

182 Guzzo, P., Daniel, J., and de Mello, J. (1999) Effect of crystal orientation on lapping and polishing processes of natural quartz. IEEE transactions on ultrasonics, ferroelectrics, and frequency control, **47** (5), 1217–1227, IEEE 2000.

183 Li, X., Yi, J., Seet, H., Yin, J., Thongmee, S., and Ding, J. (2007) Effect of Sputtered Seed Layer on Electrodeposited NiFe/Cu Composite Wires. *IEEE Transactions on Magnetics*, **43** (6), 2983–2985, doi:10.1109/TMAG.2007.893801.

184 Hsu, H.J., Huang, J.T., Chao, P.S., Wu, C.S., Shih, S.H., and Yang, S.Y. (2006) *Singular Uniformity after Reflow of Varied-Shaped Flip Chip Solder Bump on Single Substrate*. International Conference on Electronic Materials and Packaging, EMAP 2006, IEEE, pp. 1–7, doi:10.1109/EMAP.2006.4430608.

185 Baum, M., Jia, C., Haubold, M., Wiemer, M., Schneider, A., Rank, H., Trautmann, A., and Gessner, T. (2010) *Eutectic Wafer Bonding for 3-D Integration*. 3rd Electronic System-Integration Technology Conference, ESTC, pp. 1–6, doi:10.1109/ESTC.2010.5642870.

186 Iannacci, J., Bartek, M., Tian, J., Gaddi, R., and Gnudi, A. (2008) Electromagnetic optimization of an RF-MEMS wafer-level package. *Sensors and Actuators A: Physical*, **142** (1), 434–441, doi:10.1016/j.sna.2007.08.018.

187 Cao, L., Lai, Z., and Liu, J. (2005) Interfacial adhesion of anisotropic conductive adhesives on polyimide substrate. *Journal of Electronic Packaging*, **127** (1), 43–46, doi:10.1115/1.1846066.

188 Aschenbrenner, R., Ostmann, A., Motulla, G., Zakel, E., and Reichl, H. (1997)

Flip chip attachment using anisotropic conductive adhesives and electroless nickel bumps. *IEEE Transactions on Components, Packaging, and Manufacturing Technology, Part C*, **20** (2), 95–100, doi:10.1109/3476.622879.

189 Lu, S.T. and Chen, W.H. (2010) Reliability and flexibility of Ultra-Thin Chip-on-Flex (UTCOF) interconnects with Anisotropic Conductive Adhesive (ACA) joints. *IEEE Transactions on Advanced Packaging*, **33** (3), 702–712, doi:10.1109/TADVP.2010.2052806.

190 Xie, B., Shi, X., and Ding, H. (2008) Investigation of mechanical and electrical characteristics for cracked conductive particle in Anisotropic Conductive Adhesive (ACA) assembly. *IEEE Transactions on Components and Packaging Technologies*, **31** (2), 361–369, doi:10.1109/TCAPT.2008.916802.

191 Zine-El-Abidine, I. and Okoniewski, M. (2009) A low-temperature SU-8 based wafer-level hermetic packaging for MEMS devices. *IEEE Transactions on Advanced Packaging*, **32** (2), 448–452, doi:10.1109/TADVP.2008.2006757.

192 Seok, S., Rolland, N., and Rolland, P.A. (2006) *A New BCB Film Zero-Level Packaging for RF Devices*. 36th European Microwave Conference, EUMA, pp. 1118–1121, doi:10.1109/EUMC.2006.281152.

193 Huang, A., Chou, C.K., and Chen, C. (2006) Hermetic packaging using eutectic SnPb solder and Cr/Ni/Cu metallurgy layer. *IEEE Transactions on Advanced Packaging*, **29** (4), 760–765.

194 Tian, J., Sosin, S., Iannacci, J., Gaddi, R., and Bartek, M. (2008) RF-MEMS wafer-level packaging using through-wafer interconnect. *Sensors and Actuators A: Physical*, **142** (1), 442–451, doi:10.1016/j.sna.2007.09.004.

195 Emerson & Cuming Microwave Products (2002) CE 3103 WLV Electrically Conductive Adhesive Datasheet, www.eccosorb.com/, accessed 27 April 2013.

196 Imhoff, A. (1999) Packaging technologies for RFICs: current status and future trends. IEEE Radio Frequency Integrated Circuits (RFIC) Symposium.

197 Iannacci, J. and Bartek, M. (2006) *RF Behaviour of Anisotropic Conductive Adhesive in Wafer Level Packaging for RF MEMS*. Technical Digest The 17th Micromechanics Europe Workshop, MME 2006, pp. 93–96.

198 Honeycombe, R.W.K. (1984) *Plastic Deformation of Metals*, Edward Arnold.

199 Zhen, P., Xiaobin, Y., Hwang, J., Forehand, D., and Goldsmith, C. (2007) *Dielectric Charging of RF MEMS Capacitive Switches under Bipolar Control–Voltage Waveforms*. Proceedings of the IEEE/MTT-S International Microwave Symposium, EUMA, pp. 1817–1820.

200 Agilent Technologies (2001) Agilent 8719C Microwave Network Analyzer Data Sheet, www.mrtestequipment.com/getfile.php?s=Agilent+8719C+Microwave+Network+Analyzer+Data+Sheet.pdf, accessed 27 April 2013.

201 Lord, A. (2006) Advanced RF Calibration Techniques. Cascade Microtech, http://ekv.epfl.ch/files/content/sites/ekv/files/mos-ak/wroclaw/MOS-AK_AL.pdf, accessed 27 April 2013.

202 Dambrine, G., Cappy, A., Heliodore, F., and Playez, E. (1988) A new method for determining the FET small-signal equivalent circuit. *IEEE Transanction on Microwave Theory and Technology*, **36** (7), 1151–1159, doi:10.1109/22.3650.

203 Agilent Technologies (2013) Advanced Design System (ADS) by Agilent. www.home.agilent.com/en/pc-1475688/agilent-eesof-eda-design-software?nid=-34360.0&cc=US&lc=eng, accessed 27 April 2013.

204 Agilent Technologies (2009) Agilent 4155C Semiconductor Parameter Analyzer, Agilent 4156C Precision Semiconductor Parameter Analyzer, Data Sheet, http://cp.literature.agilent.com/litweb/pdf/5988-9238EN.pdf, accessed 27 April 2013.

205 Agilent Technologies (2008) Agilent 4285A Precision LCR Meter, Data Sheet, http://cp.literatur.agilent.com/litweb/pdf/5952-1431.pdf, accessed 27 April 2013.

206 Caekenberghe, K.A.P.A.V. (2007) *RF MEMS Technology for Millimeter-Wave*

Radar Sensors, A dissertation submitted in partial fulfillment of the requirements for the degree of Doctor of Philosophy (Electrical Engineering) in The University of Michigan 2007, http://deepblue.lib.umich.edu/bitstream/handle/2027.42/61348/vcaeken.pdf;jsessionid=849C7CA05273EC8F948264E458999E85?sequence=1, accessed 27 April 2013.

207. Agilent Technologies (2011) Agilent 423B, 8470B, 8472B, 8473B/C Low Barrier Schottky Diode Detectors, Data Sheet, http://cp.literature.agilent.com/litweb/pdf/5952-8299.pdf, accessed 27 April 2013.

208. Larcher, L., Brama, R., Ganzerli, M., Iannacci, J., Margesin, B., Bedani, M., and Gnudi, A. (2009) *A MEMS Reconfigurable Quad-Band Class-E Power Amplifier for GSM Standard*. IEEE 22nd International Conference on Micro Electro Mechanical Systems, MEMS 2009, IEEE, pp. 864–867, doi:10.1109/MEMSYS.2009.4805520.

209. GEIA/IBIS Open Forum (2007) Touchstone 2.0 File Format Specification Rev 1.0, available at: www.vhdl.org/ibis/interconnect_wip/touchstone_spec2_draft3.pdf, , accessed 27 April 2013.

210. Remke, R.L. and Burdick, G.A. (1974) *Spiral Inductors for Hybrid and Microwave Applications*. Proceeding of the 24th Electron Components Conference, Washington, pp. 152–161.

211. Sokal, N. and Sokal, A. (1975) Class E-A new class of high-efficiency tuned single-ended switching power amplifiers. *IEEE Journal of Solid-State Circuits*, **10** (3), 168–176, doi:10.1109/JSSC.1975.1050582.

212. Sokal, N.O. (2001) Class-E RF power amplifiers. *QEX/Communications Quarterly*, pp. 9–20.

213. Larcher, L., Sanzogni, D., Brama, R., Mazzanti, A., and Svelto, F. (2006) *Oxide Breakdown After RF Stress: Experimental Analysis and Effects on Power Amplifier Operation*. Proceedings of the 44th Annual IEEE International Reliability Physics Symposium, 2006, IEEE, pp. 283–288, doi:10.1109/RELPHY.2006.251229.

214. Mazzanti, A., Larcher, L., Brama, R., and Svelto, F. (2006) Analysis of reliability and power efficiency in cascode class-E PAs. *IEEE Journal of Solid-State Circuits*, **41** (5), 1222–1229, doi:10.1109/JSSC.2006.872734.

215. Laskar, J., Pinel, S., Dawn, D., Sarkar, S., Sen, P., Perunama, B., and Yeh, D. (2007) *FR-4 and CMOS: Enabling Technologies for Consumer Volume Millimeterwave Applications*. IEEE International Electron Devices Meeting, IEDM 2007, pp. 981–984.

216. Bahmanyar, P., Mafinezhad, K., and Bahmanyar, M. (2010) *Switching Performance Analysis in RF MEMS Capacitive Shunt Switches by Geometric Parameters Trade-Offs*. IEEE Asia Pacific Conference on Circuits and Systems, APCCAS, IEEE, pp. 831–834.

217. Halliday, D., Resnick, R., and Walker, J. (2004) *Fundamentals of Physics*, John Wiley and Sons; 7th edition.

218. (2006) *MEMS and Microsystems System-Level Design*, Reference of CoventorWare ARCHITECT, Coventor Inc.

219. Loper, O.E. and Tedsen, E. (1999) *Direct Current Fundamentals*, 6th edn., Delmar Cengage Learning.

220. Leus, V. and Elata, D. (2004) *Fringing Field Effect in Electrostatic Actuators*, TECHNION – Israel Institute of Technology, Faculty of Mechanical Engineering.

221. Hu, Y.C. and Wei, C.S. (2007) An analytical model considering the fringing fields for calculating the pull-in voltage of micro curled cantilever beams. *Journal of Micromechanics and Microengineering*, **17** (1), 61–67, doi:110.1088/0960-1317/17/1/008.

222. Iannacci, J., Tin, L.D., Gaddi, R., Gnudi, A., and Rangra, K.J. (2005) *Compact Modeling of a MEMS Toggle-Switch Based on Modified Nodal Analysis*. Proceedings of Symposium on Design, Test, Integration and Packaging of MEMS/MOEMS, DTIP 2005, TIMA Labs, pp. 411–416.

223. Rouse, H. (2011) *Elementary Mechanics of Fluids*, Dover Publications.

224. Fedder, G. (1994) *Simulation of Microelectromechanical Systems*, Ph.D. Thesis dissertation, University of California at Berkeley.

Index

A

acceleration, 158–159, 161, 242, 244, 252
accelerometer, 3
actuation mechanism, 145
ADC/DAC block, 35
ADS `TLines-Microstrip` library, 221
Agilent Advanced Design System (ADS) framework, 205, 221
Agilent 4156C semiconductor parameter analyzer, 206
air gap, 5, 9f, 89, 94, 134, 136, 138, 146, 198, 202, 232
alloy, 172, 174
anchored *see* anchor point
anchoring area, 120
anchoring point, 5–6, 13, 88, 197
angular deformation, 298
Anisotropic Conductive Adhesive (ACA), 176
 – influence on performance, 191–194
anisotropic etching, 3
ANSYS HFSS (high-frequency structure simulator), 121–130
 – air block, defined in, 123
 – Boolean operation, 123
 – boundary condition, 126
 – calibration line in, 125
 – Dev B1 3D model, 122
 – `Discrete` simulation, 127
 – `Driven Modal` option, 124
 – `Eigenmode` option, 124
 – electromagnetic simulation of capacitive switch, 229–233
 – electromagnetic simulation of power attenuators, 234–238
 – `Fast Discrete` simulation, 127
 – feature, 121–122
 – frequency, defining, 126
 – `Frequency Setup` mask, 127
 – integration line, 125
 – `Interpolating` simulation, 127
 – mask for selection, definition, and modification of materials, 123
 – `Maximum Delta S` field, 127
 – `Maximum Number of Passes` field, 127
 – of a CPW surrounded by uncompressed ACA, 193
 – on-wafer characterization of RF-MEMS devices, 124
 – postprocessing operation, 128
 – `Radiation` condition, 126
 – S11 characteristic, 128–130
 – S21 characteristic, 128–130
 – sequence of commands in, 129
 – `Solution Frequency`, 126
 – `Solution Setup` window, 126
 – S-parameter simulation in, 121
 – user-defined electromagnetic excitations in, 124
 – variable capacitor microdevice in, 122
 – variable capacitor under test after an air block, 123
 – `Wave Port` excitation, 124–126
ANSYS Parametric Design Language (APDL), 86
APDL *see* ANSYS Parametric Design Language
arbitrary waveform generator, 216
arching, 200
ARCHITECT, 31
architecture, 35
arc length, 266–268, 273
attenuation, 23–26, 234, 237–238
attenuator *see* RF power attenuator
automatic gain control, 36

B

benzocyclobutene protective ring, 31
bias, 5–6

Practical Guide to RF-MEMS, First Edition. Jacopo Iannacci.
© 2013 WILEY-VCH Verlag GmbH & Co. KGaA. Published 2013 by WILEY-VCH Verlag GmbH & Co. KGaA.

biasing electrode, 214
biasing voltage, 151
block diagram, 35–36
Bosch Deep Reactive Ion Etching (DRIE) process, 170, 179
bowing, 200, 211
BRIDGE rectangle, 65–66
buffer amplifier, 36
bump *see* solder bump

C

Cadence IC development framework, 31
Cadence simulation environment, 134, 197
cantilever-type RF-MEMS switch, 13–14
 – DC biasing of, 24
 – ohmic switch, 21
capacitance, 137, 138, 141, 259
 – due to fringing effect, 274–277
 – of Euler beam with 12 DOFs, 306, 308
 – of flexible beam, 306, 308–309, 312
 – of parallel plate capacitor, 136
 – of parallel plates under four DOFs, 260–261
 – SE–NE vertical face, 279–280
 – with air as dielectric material, 260–261
capacitive coupling, 191
capacitor, 4–10
 – C_{max}/C_{min} tuning range of, 205
 – lumped, 4
 – metal–insulator–metal, 20
 – Q factor, 4
 – suspended plate of, 8
 – tunability range, 4
 – variable, based on MEMS technology, 4–10
cell phone, 33, 35
charge conservation, 261
clamped–clamped (or fixed–fixed) RF-MEMS switch, 12–13, 20
closed switch, 19
close-up, 53–54, 96, 107, 114, 119–120, 122, 125, 235–236
CMOS, 180
CMOS circuitry, 32
CMOS class E PA, 224–227
CMOS PA *see* GSM CMOS power amplifier
collision condition at rigid plate, 256
color scale, 148, 196
coming-in surface, 266
coming-out surface, 266
commercial tool, 85–131 *see also* ANSYS Multiphysics
 – Advanced Design System, 205, 220
 – ANSYS Multiphysics, 28, 86–100, 105
  – analysis of device Dev B2 in, 101–104
  – boundary condition, commands for, 114
  – boundary condition, defining, 88, 95–96
  – central window, 86
  – clamped–clamped suspended movable gold membrane, simulation of, 105
  – deformation of gold membrane after simulation, display of, 120
  – 3D schematic of deformed structure, 105–106
  – 3D schematic of structure of ohmic switch, 107
  – dynamic property of a mechanical structure under vibrational excitation, 101
  – electrical and thermal properties, defining, 87
  – electromechanical coupling, generation of elements in, 93–95
  – execution of simulation, 99
  – generation of mesh, 92–93
  – geometrical feature, simulation of, 87–92
  – Graphical User Interface (GUI) of, 86–87
  – mechanical property, defining, 87
  – meshed 3D structure, 114
  – meshing the 3D structure, commands for, 113
  – modal analysis of RF-MEMS suspended structures, 101–104
  – of ohmic switch with microheaters, 104–121
  – of thermal conduction and convection through the silicon, 114–115
  – postprocessing and visualization of results, 99–100
  – 25 resonant modes predicted by, 104
  – simulation type, defining, 97–99
  – steps used in, 88
  – temperature distribution after simulation, 119
  – voltage at polysilicon microheater terminals, 114
 – Cadence Virtuoso, 146–147, 149, 153–157
  – toggle switch, schematic of, 146–147
  – two RF-MEMS varactors, schematic of, 150
  – Voltage-Controlled Oscillator (VCO), schematic of, 149

– COMSOL Multiphysics, 28, 139, 141, 142–143
– CoventorWare, 28, 146, 146–148
– discipline *see* Verilog-A
– HFSS, 121–130, 125, 130, 192, 232–233, 235–236
– L-Edit, 65–68, 78
– MEMS Pro, 44, 77
– Spectre, 31, 141, 142–143, 147–148, 151, 197, 199, 201, 212, 214
– Verilog-A, 31, 139, 151–165, 249, 253, 256, 264

compact modeling, 28–29
– of varactor, 196–202

comparison, 210–211

complex network
– delay line, 26
– impedance-matching network, 20–23, 55, 218–229, 223, 226. *see also* reconfigurable impedance-matching networks
  – electromagnetic design and optimization of, 219–224
– phase shifter, 26
– power attenuator, 23–26
  – electromagnetic simulation of, 234–238
  – with different resistive loads, 52–54
– reconfigurable impedance-matching networks, 20–23
– reconfigurable RF power attenuators, 23–26
– RF power attenuator, 23–26, 52, 234–238
  – 3D schematic of, 52
  – features of, 24
  – general-purpose of, 23
  – high-resistivity polysilicon of, 53
  – low-resistivity polysilicon of, 53–54
  – microphotograph of, 24
  – multimetal layer scheme of, 54
  – radar application of, 23
  – realizations of, 23
  – with different resistive loads, 52–54
– single pole, double throw, 27
– single pole, multiple throw, 27
– switching matrix, 26–27

complex network, in RF-MEMS technology
– reconfigurable impedance-matching network, 20–23
– reconfigurable phase shifters and delay lines, 26
– reconfigurable switching matrix, 26–27
– RF power attenuator and power divider, 23–26

conductive particle *see* ECA
contact finger, 14, 21
contact force *see* contact model
contact modeling of rigid plates with four and six DOFs, 255–258
contact realized in RF-MEMS switches
– contactless, 15
– metal-to-metal contact, 14–15
controlling bias, 18
Couette-type damping theory, 281
CPU, 34
cross section, 5, 145, 173, 231
curved electric field lines model, 266–268
C–V characteristic, 9
C–V characteristics of RF-MEMS variable capacitor, 9, 206–207
C–V measured characteristic, 206, 210
C–V measurement *see* C–V characteristic

D

decoupling resistor *see* RF-DC ground decoupling resistors
degree of freedom *see* DOF
design
– Boolean operation, 70–71, 74, 89, 123
– cell, 62, 67, 75
– 2D layout, 77–82
– 3D model *see* three-dimensional schematic
– 3D schematic *see* three-dimensional schematic
– design rule, 61–63
– dimple, 69
– GDS, 59
– hierarchy, 67, 74
– instance, 68
– layout, 68, 75, 220
– lithography, 58–59, 92
– mask, 59, 69
– minimum distance, 63
– minimum feature, 61–62
– minimum size, 63
– misalignment, 58
– optimum, 58, 188–189, 191, 220–221, 224
– overlap, 61, 63
– physical property, 60
– reconfigurability, 4, 11–12, 33
– reconfigurable *see* reconfigurability,
– reliability, 1, 50, 53, 224
– RF ground frame, 76
– roughness, 181, 199
– SAT format, 89, 122
– selectivity, 4

– simulation, 28–31, 88, 97–100,118,128, 142, 144–148, 187–191
  – AC simulation, 142
  – boundary condition, 88, 95–96, 114–116, 126
    – ANSYS HFSS, 126
    – ANSYS Multiphysics, 88, 95–96, 114
  – constraint, 115
  – eigenmode, 101
  – electromagnetic simulation, 234–238
  – electromechanical simulation, 86–88, 196–202
  – finite element method, 28, 143, 187
  – geometry, 88
  – mechanical constraint, 96
  – mixed-domain, 30–31
  – modal analysis, simulation, 101–104
  – optimization, 189
  – optimization software tool, 205
  – simulation time, 147
  – thermoelectromechanical simulation, 104–121
  – transition, 73–74, 207
  – parameterization, 86, 89, 210, 236
  – RF simulation, 121–130
  – TRANS126 element, 95
  – transient analysis, 217
  – transient simulation, 151
  – validation, 144–148
  – visualization of results, 88, 99–100
  – wave port, 125
    – excitation in ANSYS HFSS, 124–126
– smoothing, 200
  – S-parameter simulation, 121–130, 186–187, 205, 207, 209
– temporary fictitious layer, 66
– three-dimensional *see* three-dimensional schematic
– three-dimensional schematic, 44, 51, 77–83, 89, 122, 236
  – in ANSYS HFSS, 121–122, 124, 126, 235–236
  – in ANSYS Multiphysics, 107
  – in SAT format, 234
  – input and output ports of, 124
  – transition, 73–74, 207
design of RF-MEMS device, 57, 60–77, 219–224
  – BRIDGE rectangle, 65–66
  – CONHO strip, 69–71
  – design phase, 57–58
  – design philosophy, 61
  – design rule, 58–60

  – 3D model, 77–83
  – features and distances of FLOMET, 62–63, 65, 72
  – features and distances of POLY, CONHO, and VIA, 62, 64–65
  – GDS number, 65–66
  – in/out fixed contact, 69
  – minimum distance between BRIDGE/CPW and SPACER, 63, 74–75
  – opening and spacing in suspended gold membrane, 63
  – polysilicon actuation electrode finger, generation of, 67
  – pull-in voltage, 57
  – RF–DC decoupling resistor, 75–76
  – series ohmic switch, 60–77
  – set of specifications, 57–58
  – SPACER overlap around VIA, 63, 65
  – TIN geometry, 72–73
Device Under Test (DUT), 197, 199–200
dielectric constant, 60
displacement, 4–5, 7, 10, 31, 70, 95, 99, 134, 136, 139–140, 142, 146–148, 152–154, 157–158, 161, 199–201, 205, 216–217, 242, 246–248, 250, 252–254, 259, 262, 283, 290–291, 297–298, 304–306, 309, 314
3D measured profile *see* 3D static profile
DOF, 136, 143, 157, 160, 189, 242–249
  – 12 DOFs, 157
  – six Degrees Of Freedom (DOF), 137, 153–154, 249
    – extension of, 249–253
Double Pole, Double Throw (DPDT) commutation devices, 27
3D plot, 188
drain efficiency, 228–229
DRIE *see* deep reactive ion etching
`Driven Modal` option of ANSYS HFSS, 124
dummy wafer, 175
DUT *see* Device Under Test
duty cycle, 201, 218
dynamic behavior *see* dynamic response
dynamic response, 201

E

eddy current, 190
effective dielectric constant, 262
effective viscosity, 285
`Eigenmode` option ANSYS HFSS, 124
elasticity theory, 28
Electrically Conductive Adhesives (ECA), 175–177
electric field lines model, 266–268

electrode, 50, 52, 69, 73–74, 215
electrodeposition of metal, 2
   – of copper, 171
   – of first gold layer, 48
electrode, 53–54
electromagnetic excitation, 124
electromechanical coupling, 93–95
electrostatically controlled, 5
electrostatic attraction force, 5, 259, 263–264, 267, 279–280 *see* electrostatic force
   – due to fringing field, 274–277
   – due to the fringing field, 276–277
   – for the oxide layer, 262
   – of Euler beam with 12 DOFs, 306, 308
   – of flexible beam, 306, 308–309, 312
   – of parallel plate capacitor, 136
electrostatic force, 137, 138, 163
electrostatic model of Euler beam with 12 DOFs, 304–310
encapsulation of RF-MEMS device, 177–181
   – etching of narrow through-wafer peripheral grooves, 179
   – package-to-device wafer electrical interfacing, 178
   – protective ring, developing, 177–178
   – vacuum, 178
   – wafer-to-wafer alignment, issue of, 178–179
   – with CMOS chip, 180–181
entrapment *see* charge accumulation
equivalent electrical circuit, 29–30
etching *see* etched
Euler beam with 12 DOFs, 302–304
   – electrostatic model of, 304–310
evaporated *see* evaporation
evaporation, 2
exploitation of RF-MEMS device, 33–38
exploiting complex RF-MEMS *see* exploitation of RF-MEMS devices

F

fabrication, 3, 169–173
   – bulk micromachining, 10–11
   – buried, 47
   – chemical vapor deposition/physical vapor deposition, 2
   – copper bump, 172
   – deep reactive ion etching, 170
   – electrodeposited *see* electroplated
   – electrodeposition *see* electroplated
   – electroplated, 43, 82
   – electroplated gold layer, 48
   – epitaxial growth, 2
   – etched, 170, 269
   – etched opening, 47
   – evaporation, 2
   – FBK RF-MEMS technology, 42–44 *see* Fondazione Bruno Kessler (FBK) RF-MEMS
   – gold layer, 43, 47, 49, 82
   – grinding, 170–171
   – implantation, 51, 59
   – multimetal, 42, 46, 50, 54, 80
   – opening, 9, 46, 64, 80–81
   – oxide layer, 79, 261
   – photoresist, 43
   – photoresist film, 171–172
   – physical vapor deposition, 2
   – polysilicon, 42, 45, 53, 95
   – sacrificial layer, 43, 81, 138, 231, 269
   – seed layer, 171–172
   – silicon oxide, 42, 45, 80
   – silicon substrate, 42
   – silicon wafer, 170
   – sputtered, 2, 42, 171
   – surface micromachining, 9–11, 41–42, 59, 268, *see also* fabrication
   – titanium oxide, 205
   – vertical vias, 170, 172
   – via diameter, 183, 188–190, 205
fabrication process *see* fabrication
fabrication step *see* fabrication
Finite Element Method (FEM), 28
flexible straight beam, multiphysical model of
   – Euler beam with 12 DOFs, 302–304
   – fringing effect, 310–312
   – mass matrix, 290–292, 297–301
   – stiffness matrix, 288–290, 293–297
   – viscous damping model, 312–315
   – with 12 DOFs, 292–301
   – with two DOFs, 287–292
   – Z displacement, 309
flexible suspension *see* flexible beam
folded beam, 50
folded suspension *see* folded beam
Fondazione Bruno Kessler (FBK) RF-MEMS, 41, 61, 64, 77, 203, 214
   – design rule, 58–61
   – introduction, 42
   – micromachining process, 42–43
   – ohmic switch, 50–52
   – RF-MEMS series ohmic switch, 44–49
   – RF power attenuators, 52–54
   – three-dimensional schematic of, 44
   – varactor, 49–50
force acting on body, 242

– inertia effect, 243
– Z-axis torque of center of mass, 245–247
frequency range, 128
fringing effect, 272–281
– and capacitance, 274–277
– and electrostatic attraction force, 274–277
– of flexible beam, 310–312
– on vertical faces of plate, 278–281
fringing field *see* fringing effect model

G

GDS Number, 61
General Purpose Interface Bus, 198
global reference system (*XYZ*), 244–250, 264, 283
gold inductor, 10, 12, 20
good agreement, 208
graphical user interface, 87
ground–signal–ground, 196
GSG *see* ground–signal–ground
GSM, 55
GSM (Global System for Mobile Communications), 33
GUI *see* Graphical User Interface

H

HBT BiCMOS S35 technology, 149
hermeticity, 33
hierarchical structural analysis, 30
higher-order switching matrix, 27
high-performance, 1, 10, 12, 23–23f, 37
high-Q-factor inductor, 11
history of MEMS, 2–3
hole, 138, 269–270
HP 8719C Vector Network Analyzer (VNA), 202
human kind, progress of, 1
hybrid CMOS–MEMS functional block, 22
hybrid packaging of RF-MEMS device, 32–33, 180–181
hybrid RF-MEMS/CMOS PA circuit, 227–229
hybrid RF-MEMS/CMOS VCO, 134, 149–151
– capacitance change, 150
– oscillation frequency, 150–151
– tuning characteristic, 150

I

IF *see* intermediate frequency
IF filter, 37
implantation/diffusion technique, 2
inductance
– of RF-MEMS metal coil, 11
– of spiral air-core coil planar inductor, 221–222

inductor, 10–12
– gold-based MEMS, 10, 12
– high-Q-factor, 11
– metal spiral, 10
– modeled using `MSIND` (microstrip round spiral inductor) element, 221
– Q factor, 10
– RF-MEMS-based, 10–11
– lumped component, 222
– suspended MEMS, 10
inner electrode *see* toggle switch
input/output RF ohmic pad, 105
input/output transfer function of an RF-MEMS shunt switch, 14
instantaneous position, 247
instantaneous position of NE node, 252
instantiation *see* instance
Integrated Circuit (IC), 2, 20
– fabrication of, 2–3
– in RF identification (RFID) system, 20
integration, 20, 180
– wire-bonded, 227
integration line, 125
interdigitated, 45, 214
intermediate frequency, 36
intrinsic MEMS impedance, 203
ion implantation, 2
isolation, 211, 233
Isotropic Conductive Adhesive (ICA), 176

K

Ka-band, phase shifter for, 26
Kirchhoff flow law, 30
Kirchhoff voltage law, 30

L

layer of RF-MEMS structures, program for
– FBK_BRIDGE, 82
– FBK_BRIDGE_MASK, 82
– FBK_CONHO, 80
– FBK_CPW, 82
– FBK_CPW_MASK, 82
– FBK_FIELD_OXIDE, 79
– FBK_FLOMET, 81
– FBK_FLOMET_MASK, 81
– FBK_HIGH_POLY_MASK, 79
– FBK_LOW_POLY_MASK, 79
– FBK_LTO, 80
– FBK_POLY_HIGH_RES, 79–80
– FBK_POLY_LOW_RES, 79
– FBK_SILICON_SUB, 78–79
– FBK_SPACER, 81
– FBK_SPACER_MASK, 81

– FBK_TEOS, 80
  – FBK_TIN, 80
  – FBK_TIN_MASK, 80
  – FBK_VIA, 81
layout editor of L-Edit, 65–87
LCR meter, 206
L-Edit software tool, 60, 67
linear deformation, 297
losses and mismatch, 185, 186, 190
lumped component, 3–20
  – capacitive switch, 15, 49–55, 229–233
  – capacitive shunt switch, 18, 102
  – MEMS varactor, 196–202
  – ohmic shunt switch, 17
  – ohmic switch, 15, 50–52, 107, 236
    – cantilever-type, 21
    – 3D schematic of, 51
    – gold membrane of, 51
    – in/out pad, 51
    – polysilicon layer, 51–52
    – RF power attenuator in cantilever-type, 23
    – self-release mechanism, 50
    – series see series ohmic switch
    – transmission parameter (S21) of, 16
    – TTD network in, 26
    – with microheater, simulation using ANSYS Multiphysics, 104–121
    – with silicon wafer, 106–107
  – resistor, 25
  – series ohmic switch, 13, 44–49, 64, 213–218
  – spiral suspended inductor, 12
  – toggle switch, 144–148, 147
    – boundary condition, 145
    – Cadence Virtuoso schematic of, 146–147
    – on a central suspended rigid plate, 144
    – CoventorWare schematic of, 148
    – dimensions of toggle structure, 146
    – pull-in characteristic, 146–148
    – rotation of controlling plates, 145
    – schematic view of, 144–145
    – static DC simulation, 146
    – as a varactor, 145
    – vertical displacement, 146
  – varactor, 4, 49–55, 50–51, 145, 202–213
    – design, 49–50
    – Ground–Signal–Ground (GSG) input/output Coplanar Waveguide (CPW)-like access lines, 196
    – multimetal fixed capacitor electrode, 50
  – pull-in/pull-out characteristic, 198–199
  – with a lumped element network, RF modeling of, 202–213
  – with compact model, electromechanical simulation, 196–202
  – with movable gold electrode, 50
  – zipper RF-MEMS, 9
– variable capacitor, 4–10, 122, 207, 209, 230, 232
  – as a capacitive switch for RF signal, 6–7
  – by suspended MEMS structure, 8–9
  – Coplanar Waveguide (CPW) structure, 8
  – C–V characteristic, 9, 206–207
  – displacement of capacitor plate, 4, 10
  – implementation of, 4
  – in shunt-to-ground configuration, 6
  – increasing capacitance, strategy to, 5
  – pull-in/pull-out quasi-static characteristic, 215–216
  – surface micromachining of, 9
  – topology, 10
  – working principle of, 4–5

M

material property, 86, 90, 106, 122–123, 127, 196
mathematical model, 133
Maxwell equations for electromagnetic problem, 28
mean free path of gas molecules, 285
measurement, 7
  – 3D static profile, 197
  – electromagnetic characterization, 185–191
  – experimental result, 227–229
  – microphotograph, 25, 27, 179, 181–182, 227, 234–235
  – scanning electron microscopy, 173, 269
mechanics
  – center of mass, 134–136, 241–248, 251–253, 255–257, 260, 263–265, 280–281, 284
  – damping matrix, 140
  – force, 154, 244–250, 264
  – inertia, 201
  – mass matrix, 140
    – of six DOFs, 297–301
    – of two DOFs, 290–292
  – Poisson's ratio, 87, 90, 108, 296
  – stiffness matrix, 140

- of six DOFs, 293–297
- of two DOFs, 288–290
- stress–strain curve, 140
- torque, 154
- von Mises stress, 142
- Young's modulus, 60, 87, 90, 108

memory, 34

MEMS compact model library, 134
- flexible beam, 140–144
- hybrid RF-MEMS/CMOS VCO, 149–151
- library feature, 134
- mechanical constraints and stimuli, 134
- suspended rigid plate electromechanical transducer, 134–139
- toggle switch, 144–148

M_2 gate-source capacitance, 226

microelectromechanical systems (MEMS) technology
- definition, 2
- fabrication process of, 2–3
- pull-in/pull-out characteristic of a MEMS switch, 5–8
- sensor and actuator, 3

microheater, 51–52, 54, 104–121

micromachined RF-MEMS device
- FBK RF-MEMS process, 42–43
- variable capacitor, 9

microrelay (closed switch), 14, 16

microswitches (ohmic and capacitive), 12–20
- cantilever-type, 13–14
- capacitance realized in on/off state, 15
- clamped–clamped, 12–13
- contacts in, 14–15
- implementation of, 15
- performance and characteristic, 15
- reflection parameter (S11), 15
- RF behavior, 15
- series, 14
- shunt, 14
- transmission parameter (S21), 16

mixer, 37

model
- anchor point, 152–154
- compact model, 134–148, 196–202, 208, 213–218
- contact model, 255–258
- curved electric field line, 266
- electrostatic model of rigid plate
  - arc length of point, 267–268
  - curved electric field lines model, 266–268
  - for two parallel plates, 259
  - four DOFs, 258–262

- four DOFs condition, 258–262
- instantaneous gap between plates, 259–260
- of plate with hole, 268–272
- six DOFs, 262–266
- six DOFs condition, 262–266
- torque balance equation, 265
- with fringing effect, 272–281
- with hole, 268–272
- Euler beam, 297–298
- electrostatic model, 304–310
- with 12 DOFs, 302–304
- flexible beam, 140–144, 157–165
- COMSOLMultiphysics schematic of, 141–143
- DOF, 140
- electrostatic model for, 140–141
- error of predicting frequency, 144
- expression of static and dynamic behavior of, 140
- model for elastic behavior of, 143
- resonant frequency of the mode, 143–144
- Spectre schematic for, 141–143
- total capacitance and electrostatic force, 141
- total electrostatic attraction force, 141
- Verilog-A code for, 157–165
- flexible beam with 12 DOFs, 292–301
- mass matrix, 297–301
- stiffness matrix, 293–297
- flexible beam with two DOFs, 287–292
- mass matrix, 290–292
- stiffness matrix, 288–290
- flexible straight beam, 157
- force source, 154–157
- fringing effect model
  - capacitance, 311
  - electrostatic model of rigid plate with, 272–277
  - on vertical faces of plate, 278–281
- mechanical model of rigid plate, 242–249
- rigid plate, 135
- rigid plate vertex, 243
- rigid plate electromechanical, 134–139
- after a rotation around Z axis, 244
- placement of nodes along edges, 253–255
- with eight nodes, 254
- with forces and torques applied to four vertexes, 246
- with four DOFs, 247–248, 255–258
- with six DOFs, 249, 255–258

Index

- viscous damping, 201, 312–315
  - effect of gas rarefaction, 285
  - for lateral movement, 284–285
  - of flexible beam, 312–315
  - of fluid underneath the plate, 284–285
  - of model with six DOFs, 283
  - of rigid plate with four DOFs, 284
  - squeeze-film damping model, 281–284
  - without holes, squeeze-film damping model, 284
- modeling, 28–31
  - bottom-up approach, 28–29
  - compact modeling, 28–29
  - electromagnetic modeling, 218–229
  - electromechanical modeling, 213–218
  - equivalent electrical circuit, 29–30
  - Hermitian shape function, 141
  - Kirchhoff law, 134
  - lumped element network, 202–213
  - lumped element scheme, 203, 220
  - model order reduction, 29
  - model parameter, 155–156
  - node, 152–153, 156, 254
  - RF modeling, 202–213
  - subnetwork, 204
  - through value, 31
  - top-down approach, 28–30
- modeling and simulation of RF-MEMS device
  - compact modeling, 28–29
  - equivalent electrical circuit, 29–30
  - Finite Element Method (FEM), 28
  - mixed-domain electromechanical simulation environment, 30–31
  - model order reduction method, 29
- moment of inertia of the plate, 245
- MOSFET capacitance, 225
- MOSFET power loss ($P_{L,MOS}$), 226
- multistate phase shifter, 26

N
nonideality, 204

O
- off state, 14
- off-state
  - capacitance of capacitive switch, 19
- ohmic contact, 14, 47
- ohmic microrelay, 105
- ohmic shunt microrelay, 19
- on/off state of switch, 14
  - capacitance realized, 15
- on state, 14
- open switch, 19

- optical interferometry, 196
- optical profiling system, 230
- oscillation frequency, 150
- oscilloscope, 216
- outer electrode *see* toggle switch
- output network quality factor, 224
- output power, 228–229
- output voltage, 151
- overpass, 180
- oxide permittivity, 198

P
- package, 190 *see also* packaging
- packaging
  - ACA, 176, 191–194
  - alignment, 173, 182
  - alignment mark, 179
  - benzocyclobutene, 177
  - bump, 178
  - capped CPW, 183–184, 188
  - cavity, 180
  - chip-to-wafer bonding, 182
  - ECA, 168
  - eutectic bonding, 174
  - first-level packaging, 32
  - *getter*, 33
  - harmful factor, 31
  - hermetic sealing, 178
  - hybrid packaging, 180–181
  - ICA, 182f
  - packaged MEMS/RF-MEMS, 169
  - packaged test structure, 181–185, 185, 187–191
  - package electrical interconnect, 178
  - peripheral ring, 177
  - reflow, 174–175
  - ring, 177
  - sealing ring, 178, 193
  - solder bump, 174–175
  - standard surface-mount technology, 169
  - SU-8, 177
  - through-wafer interconnect, 168
  - ultrasonic cleaning, 179
  - uncompressed ACA, 192–194
  - vacuum packaging solution, 33, 178, 180
  - wafer-to-wafer alignment, 178–179
  - wafer-to-wafer bonding, 32, 169, 173–177
  - WLP *see* Wafer-Level Packaging (WLP) approach to packaging
  - zero-level packaging, 32
- packaging of RF-MEMS devices, 31–33
  - based on a glass substrate, 31
  - benzocyclobutene protective ring, 31

- concerning issue, 31
- electromagnetic characterization of package, 185–191
- encapsulation of, 177–181
- fabrication run of packaged test structure, 181–185
- first-level, 32
- hermeticity and vacuum sealing, 33
- hybrid, 32–33
- hybrid solution for, 180–181
- influence of uncompressed ACA on performance, 191–194
- packaged MEMS/RF-MEMS, 169
- packaged test structure, 181–187, 187–191
- package electrical interconnect, 178
- peripheral ring, 177
- silicon capping wafer, 31–32
- thin film capping, 32
- Wafer-Level Packaging (WLP) approach, 167–168
- zero-level, 32

pad, 71–72, 182
paradox, 2
parameterized *see* parameterization
parasitic, 191, 203–204, 207
peripheral driver, 35
permittivity of air, 136
phase-locked loop, 37
phase shifter, 26
polymer *see* ECA
polysilicon heater, 119
polysilicon microheater, 51, 106, 114
polysilicon resistor, 23–24
power amplifier, 36
power detector, 216
power divider, 23
power envelope, 217
pressure sensor, 3
program for designing layer *see* layers of RF-MEMS structures, program for
progress of human kind, 1
pull-in phenomenon, 5–8, 18, 98, 100, 147–148, 199–200, 216, 218, 231
pull-in/pull-out characteristic of a MEMS switch, 5–8, 51
pull-in voltage (V_{PI}), 5–6, 18, 57, 201
pull-out phenomenon, 7, 98, 100, 198–200, 216, 218, 231
pull-out voltage (V_{PO}), 198, 200
pulse, 218

Q

Q factor *see* quality factor

quality factor, 33, 220, 224–225, 229
- capacitor, 4
- inductor, 10

R

radiation, 126
Radio Frequency (RF)-MEMS devices
- capacitor, 4–10
- category, 4
- complex network, 20–27
- inductor, 10–12
- microswitch (ohmic and capacitive), 12–20

reconfigurable impedance-matching networks with CMOS power amplifier
- electromechanical modeling of, 218–229
- example of, 22
- feature of, 20–21
- GSM impedance matching network, 55
- integration and exploitation of, 22
- LC ladder in, 55
- planar inductor of, 55
- reactive stages, 21
- Smith chart, 22–23
- two-state capacitor of, 55

reference frame of origin, 241–242, 245–246
reference system, 135, 137, 161–162, 164, 242–243, 250, 266
reflection parameter (S11), 16, 18, 206–213
- in actuated state of the capacitive shunt switch, 19
- for a CPW, 193
- of RF-MEMS capacitive shunt switch, 18
- of RF-MEMS series switch, 15–17

relative displacement of node, 247
- of NE node, 253

reliability
- charge accumulation, 9, 65, 200
- microwelding, 51, 105

residual air gap, 198–199, 202, 207, 210–211, 215, 230, 232
resistive load, 53
resonant mode, 104–106, 143–144
restoring force, 30, 52, 106
rest position, 5
RF and DC contact, 46
RF and microwave application, 34
RF choke inductor, 226
RF component and system, 33
- active element, 36–37
- analog to digital converter, 35
- choke inductor, 203
- Coplanar Waveguide (CPW), 24, 180–181, 193, 196

- CPW *see* Coplanar Waveguide (CPW)
- digital to analog converter, 35
- filter, 34, 37–38, 225
- GSM CMOS power amplifier, 218–229
- hybrid MEMS/CMOS PA, 228
- hybrid RF-MEMS/CMOS PA, 227–229
- hybrid RF-MEMS/CMOS VCO, 149–151
- LC section, 225
- LC tank, 34, 149
- low-noise amplifier, 36
- metal–insulator–metal, 220
- passive element, 37
- phase-locked loop, 37
- radio receiver, 34
- reconfigurable class E PA, 224–227
- RF–DC decoupling resistor, 54, 64, 75
- RF–DC ground decoupling resistor, 45
- RF transceiver, 29, 35–37, 85, 169
- superheterodyne RF receiver, 38
- telecommunication system, 4
- Voltage-Controlled Oscillator (VCO), 36–37, 134, 149, 151

RF filter, 37
RF identification (RFID) system, 20
2 × 2 RF-MEMS-based switching matrix, 27
RF-MEMS series ohmic switch
- conductive multimetal layer, 46
- Coplanar Waveguide (CPW) frame, 44
- DC bias, 44
- electrodeposition of gold layer, 48–49
- evaporated gold layer, 47
- fabrication step, 45–49
- high-resistivity polysilicon layer, 45–46
- oxide layer, 46–47
- pad of RF port, 44
- pull-in voltage, 44
- sacrificial layer (photoresist), 47–48
- silicon oxide layer, 45

RF power, 15
RF signal, 124, 196–197, 203, 214
- as output, 10, 12, 14, 16, 19
- conditioning of, 11, 20
- from the DC ground, 45, 50
- in MEMS microrelay, 17
- in ohmic switch, 216
- in pulled-in position, 6
- in reconfigurable power attenuator, 234
- in shunt-to-ground capacitive switch, 230
- in telecommunication platform, 26
- in TTD network, 26
- inductor winding and, 10
- low impedance path to, 49
- response of series and shunt switches, 14

- RF transceiver and, 35
- shunt switch and, 14
- SPDT switch and, 27

rigid plate electromechanical transducer, mechanical model
- contact modeling with four and six DOFs, 255–258
- electric field lines model, 266–268
- electrostatic model of, 258–262
- extension of mechanical model, 249–253
- extension of six DOFs, 262–266
- flag variable, 255–256
- global reference system (XYZ), 244–250
- instantaneous and initial positions of each node, 252–253
- instantaneous displacement of center of mass along Z, 255–256
- mean free path of gas molecules, 285
- moment of inertia of the plate, 245
- of fringing effect, 272–281
- placement of nodes along edges of a rigid plate, 253–255
- position of each node in global frame, 248
- relative displacement of each node, 247
- rotation matrix element, 250–251
- squeeze-film damping model, 281–284
- torque at the center of mass around X and Y axes, 251–252, 257–258
- vertical face of the plate, fringing effect of, 277–281
- viscous damping for lateral movement, 284–285
- with an offset D_y applied to NE node, 254–255
- with four DOFs, 242–249
- with hole, 268–272
- with six DOFs, 249–253

rotation matrix element, 250–251
routine, 152

s

S11 *see* reflection parameter
S21 *see* transmission parameter
satellite for telecommunication *see* telecommunication system
schematic, 123, 135, 138, 140, 142, 197, 203, 214–215
- ANSYS HFSS of the RF-MEMS variable capacitor, 232
- in ANSYS HFSS of a capped CPW, 188
- for flexible beam model, 157
- of a CPW surrounded by a layer of uncompressed ACA, 193

– of a die on RF-MEMS device wafer surrounded by a protective peripheral ring, 177
– of a flexible beam with 12 DOFs, 140
– of a MEMS toggle switch, 144–148
– of a planarizing photoresist material, 231, 235
– of a plate with holes, 270
– of a plate with six DOFs with holes, 270–271
– of a radio frequency (RF) microelectromechanical system (MEMS) switch, 91
– of a reconfigurable class E PA, 224–225
– of anchor point with six DOFs, 153
– of capacitive switch/varactor with two serpentine-shaped flexible suspensions, 51
– of deformed structure, 105–106
– of Dev B1 imported into HFSS as a file in SAT format, 122
– of gold-to-oxide nonideal surface contact, 199
– of GSM impedance matching network, 55
– of lumped element network, 203
– of MEMS-based reconfigurable RF power attenuator, 235
– of meshed RF-MEMS ohmic switch, 92–94
– of plate surface with a matrix of rectangular holes, 138
– of plate with six degrees of freedom (DOF), 137
– of possible configurations for placing electrodes, 313
– of RF-MEMS-based reconfigurable power attenuator, 52
– of RF-MEMS capacitive shunt switch, 102
– of RF-MEMS device and capping wafer, 168
– of RF-MEMS ohmic switch, 96, 100
– of RF-MEMS ohmic switch with microheater, 51, 107, 111
– of RF-MEMS series ohmic switch, 44
– of RF-MEMS variable capacitor, 123
– of rigid plate, 135
– of stimulus source (force) with six DOFs, 154
– of structure highlighting thermal boundary conditions, 115
– of suspended MEMS inductors, 11
– of variable capacitor, 5

– of Voltage-Controlled Oscillator (VCO), 149–150
– Spectre of cantilever implemented with the compact models, 142–143
– top view of a MEMS toggle switch, 144
second-order effect, 201
SEM *see* Scanning Electron Microscopy
semiconductor parameter analyzer, 206
semiconductor technology, 3
SE–NE vertical face, 278
sensor and actuator, 1, 3
series ohmic switch, 14
– BRIDGE rectangle, 65–66
– CONHO strip, 69–71
– contact, 68–69
– feature and distance of FLOMET, 62–63, 65, 72
– feature and distance of POLY, CONHO, and VIA, 62, 64–65
– GDS number, 65–66
– mask, 64–65, 69
– minimum distance between BRIDGE/CPW and SPACER, 63, 74–75
– opening and spacing in suspended gold membrane, 63
– polysilicon actuation electrode finger, generation of, 67
– S21 parameter (transmission), 16
– SPACER overlap around VIA, 63, 65
– TIN geometry, 72–73
– with compact model, electromechanical modeling of, 213–218
series switch, 14, 20
– S11 parameter, 15, 17
– transmission parameter (S21) of an ohmic switch, 16
– type of contact, 14–15
serpentine, 215
sheet resistance, 60, 76
shunt switch, 14, 20
– capacitive coupling and RF signaling, 17
– frequency plot for, 18
– reflection parameter (S11) of, 18–19
– simulation using ANSYS Multiphysics, 101–104
– transmission parameter (S21), 19
silicon capping wafer, 31–32
simulated result of experiments
– capacitive switch, 229–233
– hybrid RF-MEMS/CMOS PA circuit, 227–229
– impedance-matching network with CMOS power amplifier, 218–229

- power attenuator, 234–238
- RF modeling of varactor with a lumped element network, 202–213
- series ohmic switch with compact model, 213–218
- varactor with compact model, 196–202

simulation tool for RF-MEMS device
- ANSYS HFSS (high-frequency structure simulator), 121–130
- ANSYS Multiphysics, 86–100
- based on compact modeling, 133–134
- FEM approach, 85–86, 106, 133
- modal analysis of RF-MEMS suspended structure, 101–104
- simplified and approximate mathematical model, 133–134

Single Pole, Double Throw (SPDT) switch, 27
Single Pole, Four Throw (SP4T) switch, 27
Smith chart, 23, 185, 222–223
solution, 126–127
S11 parameter (reflection parameter), 15–17, 206–213
- for a CPW, 193
- in actuated state of the capacitive shunt switch, 19
- of RF-MEMS capacitive shunt switch, 18
- of RF-MEMS series switch, 15–17

S21 parameter (transmission), 24–26, 206–213
- for nonactuated state of switch, 19
- of an RF-MEMS capacitive shunt switch, 19
- of an RF-MEMS series ohmic switch, 16
- of a CPW surrounded by uncompressed ACA, 193

sputtering *see* sputtered
squeeze-film damping theory, 139, 281
Standard ACIS Text (SAT) format, 77
structural mechanic, 30
subelement, 141, 144
suspended MEMS inductor, 10, 48
suspended rigid plate electromechanical transducer, simulation of, 134–139
- capacitance and electrostatic force expression, 138–139
- coordinates of each vertex, 134
- degrees of freedom (DOF), 136–137
- displacements and rotations along and around X, Y, and Z axes, 136
- distances S_x and S_y between adjacent holes, 137–138
- effective forces applied to SE node along the X and Y axes, 135
- effective viscosity value, 139

- initial nondisplaced plate position, 134
- on/off states of the reactive force, 136
- rotation matrix of reference system, 134–135
- torque around the Z axis, 135
- transduction between electrical and mechanical domains, 136
- viscous damping contribution, 139

sweep, 98
switch, 3, 37 *see also* cantilever-type RF-MEMS switches; ohmic switches
- based on suspended electrostatic transducer, 30

symbol, 153–154, 157, 197
SYNPLE, 31

T

Tanner Tool, 60
technological advance, impact on human beings, 2
telecommunication platform *see* telecommunication system
temperature distribution, 119
testing *see* measurement
thermal expansion, 120
through quantity *see* through value
time derivative, 158, 259
toggle-switch geometry, 10
torque balance equation, 265
torques at the center of mass, 250, 257
trade-off, 86, 190
transceiver *see* RF transceiver
transient behavior, 218
transistor, 2–3, 226
- fabrication of, 3
- modeling, 203

transmission parameter (S21), 16, 24–26, 189, 206–213
- for nonactuated state of switch, 19
- of a CPW surrounded by uncompressed ACA, 193
- of an RF-MEMS capacitive shunt switch, 16
- of an RF-MEMS series ohmic switch, 16

True-Time Delay (TTD) networks, 26
tunable, 34, 37, 149, 200
tuning characteristic, 150
tuning range, 4, 6, 9, 205
two-port device, 124
two-state shunt capacitor, 26

U

underpass, 12, 186

V

vacuum sealing, 33
vector network analyzer, 202
velocity, 139, 158–159, 161
  – difference among adjacent subelements, 315
  – of center of mass, 282–283
  – of Couette-type flow, 285
Verilog-A code for MEMS models, 135, 139, 249, 256
  – analog code block (routine core), 152
  – anchor point, 152–154
  – discipline/s associated with nodes, 152
  – flexible beam model, 157–165
  – force stimulus source, 154–157
  – in Cadence simulation environment, 134
  – internal node, 152
  – module name, 152
  – node type, 152
  – parameters and their default value, 152
  – sequence and arrangement of the code block, 152
  – variable and constant, 152
vertical deformation, 100
vertical displacement, 148, 217
vertical distance, 263
vertical position, 6
vertical profile, 198
vertical upward deformation, 120
viscous damping
  – Couette-type flow, 139, 140, 285
  – damped oscillation, 201
  – Knudsen number, 139, 139, 285
  – squeeze-film damping model, 140, 281–284, 314
  – without hole, 284
  – viscosity, 285
VNA *see* vector network analyzer

W

Wafer-Level Packaging (WLP) approach to packaging, 167–177
  – Anisotropic Conductive Adhesive (ACA), 176
  – Bosch DRIE process, 170
  – bottom capping side of the vias, 172
  – capping part, 173
  – chromium metallization, 171
  – contact surface of, 174
  – Electrically Conductive Adhesive (ECA), 175–177
  – etching of recesses, 170
  – fabrication process, steps, 169–173
  – Isotropic Conductive Adhesive (ICA), 176
  – seed layer, 170–171
  – solder bump, 174–175
  – top-side copper pad, 171
  – wafer-to-wafer bonding solution, 173–177
wet and dry etching, 2
White Light Interferometry (WLI) profiling system, 21, 24
window of L-Edit, 65–68, 70–71

Z

Z-axis torque of center of mass, 244–245
zipper RF-MEMS varactor, 9